I0488148

NUREG-1811
Supplement 1

Draft Environmental Impact Statement for an Early Site Permit (ESP) at the North Anna ESP Site

Draft Report for Comment

Manuscript Completed: June 2006
Date Published: July 2006

Division of Regulatory Improvement Programs
Office of Nuclear Reactor Regulation
U.S. Nuclear Regulatory Commission
Washington, DC 20555-0001

COMMENTS ON DRAFT REPORT

Any interested party may submit comments on this report for consideration by the NRC staff. Comments may be accompanied by additional relevant information or supporting data. This draft report is being issued by the U.S. Nuclear Regulatory Commission for a 45 day comment period. The comment period begins on the date that the U.S. Environmental Protection Agency publishes a Notice of Filing in the *Federal Register*; such Notices are published every Friday. The Notice will identify the comment period end date. Please specify the report number NUREG-1811, Supplement 1, draft, in your comments, and send them by the comment period end date to the following address:

Chief, Rules Review and Directives Branch
U.S. Nuclear Regulatory Commission
Mail Stop T6-D59
Washington, DC 20555-0001

Electronic comments may be submitted to the NRC by the Internet at
North_Anna_Comments@nrc.gov

For any questions about the material in this report, please contact:

J. Cushing
OWFN 11 F-1
U.S. Nuclear Regulatory Commission
Washington, DC 20555-0001
Phone: 301-415-1424
E-mail: JXC9@nrc.gov

Abstract

The staff of the U.S. Nuclear Regulatory Commission (NRC) has prepared this Supplement to the *Draft Environmental Impact Statement (EIS) for an Early Site Permit (ESP) at the North Anna ESP Site* (SDEIS) because Dominion Nuclear North Anna, LLC (Dominion or applicant) amended its ESP application, as described in Revision 6 to its application for an ESP. In Revision 6 (which was submitted to the NRC on April 13, 2006), Dominion described a new approach for cooling its proposed Unit 3. Under the revised approach, Unit 3 would use a closed-cycle cooling system, rather than the originally proposed once-through cooling system. The newly-proposed system would not use the 1376-ha (3400-ac) waste heat treatment facility for cooling. Dominion also proposed to increase the power level of both proposed Units 3 and 4 from 4300 megawatts-thermal (MW(t)) to 4500 MW(t).

The proposed action requested in Dominion's North Anna ESP application is for the NRC to (1) approve a site within the existing North Anna Power Station (NAPS) boundaries as suitable for the construction and operation of one or more new nuclear power generating facilities and (2) issue an ESP for the proposed site located at NAPS. The proposed action does not include any decision or approval to construct or operate one or more units; these are matters that would be considered only upon the filing of applications for a construction permit and an operating license, or an application for a combined license.

In its application, Dominion proposes a plan for redressing the environmental effects of certain site preparation and construction activities; that is, those activities enumerated by Title 10 of the Code of Federal Regulations (CFR) 50.10(e)(1), which an ESP holder may perform under 10 CFR 52.25. In accordance with the site redress plan, the site would be redressed if the NRC issues the requested ESP (containing the site redress plan), the ESP holder performs these site preparation and preliminary construction activities, the ESP is not referenced in an application for a construction permit or combined license, and no alternative use is found for the site.

This SDEIS includes the NRC staff's analysis that considers and weighs the environmental impacts of constructing and operating a closed-cycle cooling system for Unit 3 and the increase in power for proposed Units 3 and 4 and mitigation measures available for reducing or avoiding adverse impacts. It also includes the staff's preliminary recommendation to the Commission regarding the proposed action.

The staff's preliminary recommendation, in view of the environmental impacts described in the Draft EIS, and the impacts reviewed in this SDEIS in relation to the changes presented in ER Revision 6, is that the ESP for North Anna Units 3 and 4 should be issued. This recommendation is based on (1) the ER submitted by Dominion, as revised; (2) consultation with Federal, State, Tribal and local agencies; (3) the staff's independent review; (4) the assessments summarized in the Draft EIS and this SDEIS, including the potential mitigation measures identified in the ER and in both the Draft EIS and SDEIS. In addition, in making its

1   recommendation, the staff has concluded that alternative sites considered are not obviously
2   superior to the proposed site. Finally, the staff concludes that the site preparation and
3   preliminary construction activities allowed by 10 CFR 50.10(e)(1) would not result in any
4   significant adverse environmental impact that cannot be redressed.

Contents

Figures

Tables

Executive Summary

1
2
3
4   *This Executive Summary was revised to reflect the changes Dominion Nuclear North Anna, LLC*
5   *presented in Revision 6 of the Environmental Report for its early site permit (ESP) application for*
6   *proposed North Anna Units 3 and 4.*
7
8   On September 25, 2003, the U.S. Nuclear Regulatory Commission (NRC) received an
9   application from Dominion Nuclear North Anna, LLC (Dominion) for an early site permit (ESP) for
10   two units located adjacent to the North Anna Power Station (NAPS), Units 1 and 2. The North
11   Anna ESP site is located in Louisa County, Virginia, approximately 10 km (6 mi) northeast of the
12   town of Mineral. On April 13, 2006, Dominion submitted Revision 6 to its application, which
13   included a revised Environmental Report (ER). The staff, in its review of Revision 6 of the
14   application, requested additional information (RAI) from Dominion. Dominion responded to the
15   RAIs and on June 21, 2006 submitted Revision 7 to the application, which included the
16   necessary information from the RAI responses.
17
18   In Revision 6 to the North Anna ESP application, Dominion proposed (1) changing its approach
19   for cooling proposed Unit 3 from the once-through cooling system, as described in previous
20   versions of the ER, to a closed-cycle system and (2) increasing the maximum power output per
21   unit from megawatts-thermal 4300 (MW(t)) to 4500 MW(t) for proposed Units 3 and 4 (hereafter
22   referred to as Units 3 and 4). Under the revised cooling system approach, Unit 3 would use a
23   closed-cycle, combination wet and dry cooling system. The proposed increase in power level
24   corresponds to the revision of the designed maximum power of an economic simplified boiling
25   water reactor (ESBWR), one of the reactor designs included in the plant parameter envelope
26   (PPE) and evaluated in the Draft Environmental Impact Statement (EIS), which was issued in
27   December 2004.
28
29   The NRC staff determined that the changes to the proposed action were substantial; therefore,
30   the staff decided to prepare a Supplement to its Draft EIS (referred to as the SDEIS) pursuant to
31   10 CFR 51.72. On May 16, 2006, following receipt of Dominion's ER Revision 6, the staff
32   published a Notice of Intent to prepare a Supplement to the Draft EIS for the North Anna ESP
33   application in the *Federal Register* (71 FR 28392). The scope of this SDEIS is limited to the
34   environmental impacts associated with the change in the cooling system for Unit 3 and the
35   increase in the maximum power level for both units. The evaluation presented in this SDEIS
36   replaces the evaluation of the impacts associated with the originally proposed once-through
37   cooling for Unit 3 and modifies the analysis of impacts related to the power level increase.
38   These revised evaluations, along with public comments received on the analysis presented in
39   this SDEIS, will be incorporated into the Final EIS together with comments and responses
40   received concerning the original Draft EIS and the staff's consideration of such comments.

1   An ESP is a Commission approval of a site or sites for one or more nuclear power facilities.
2   Issuance of an ESP is an action separate from the issuance of a construction permit (CP) or a
3   combined construction permit and operating license (combined license or COL) for such a
4   facility. An ESP application may refer to a reactor's or reactors' design parameters or a PPE,
5   which is a set of values of plant design parameters that an ESP applicant expects will bound the
6   design characteristics of the reactor or reactors that might be built at a selected site; alternatively
7   an ESP may refer to a detailed reactor design. An ESP is not a license to build a nuclear power
8   plant; rather, the application for an ESP initiates a process undertaken to assess whether a
9   proposed site is a suitable location for such a plant should the applicant decide to pursue a CP
10   or COL.
11
12   Section 102 of the National Environmental Policy Act of 1969 (NEPA) (42 USC 4321) directs that
13   Federal agencies prepare an EIS for major Federal actions that significantly affect the quality of
14   the human environment. The NRC has implemented Section 102 of NEPA in 10 CFR Part 51.
15   Subpart A of Title 10 of the Code of Federal Regulations (CFR) Part 52 contains the NRC
16   regulations related to ESPs. In addition, as set forth in 10 CFR 52.18, the Commission has
17   determined that an EIS will be prepared during the review of an application for an ESP. The
18   purpose of Dominion's proposed action, issuance of the ESP, is to provide stability in the
19   licensing process by addressing site safety and environmental issues before the plants are built
20   rather than after construction is completed. Part 52 of Title 10 describes the ESP as a "partial
21   construction permit." An applicant for a CP or COL for a nuclear power plant or plants to be
22   located at a site for which an ESP has been issued can reference the ESP, and matters
23   resolved in the ESP proceeding are considered resolved in the subsequent proceeding.
24   However, issuance of either a CP (and OL) or COL to construct and operate a nuclear power
25   plant is a major Federal action that requires its own environmental review in accordance with
26   10 CFR Part 51.
27
28   Three primary issues – site safety, environmental impacts, and emergency planning – must be
29   addressed in the ESP application. Likewise, in its review of the application, the NRC assesses
30   the applicant's proposal in relation to these issues and determines if the application meets the
31   requirements of the Atomic Energy Act of 1954 and NRC regulations. Site safety and
32   emergency planning are addressed in the staff's safety evaluation report. This SDEIS addresses
33   the environmental impacts related to the changes proposed in Revision 6 of the ER. Pursuant to
34   10 CFR 52.17(a)(2), however, the applicant did not address the benefits of the proposed action
35   (e.g., the need for power). In accordance with 10 CFR 52.18, the Draft EIS and this SDEIS are
36   focused on the environmental effects of construction and operation of a reactor, or reactors, that
37   have characteristics that fall within the postulated site parameters.
38
39   The holder of an ESP, or an applicant for a CP or COL that references an ESP that includes a
40   site redress plan, may, in accordance with 10 CFR 52.25, perform the site preparation and
41   preliminary construction activities enumerated in 10 CFR 50.10(e)(1), provided that the final ESP
42   EIS concludes that the activities will not result in any significant adverse environmental impacts

1   that cannot be redressed. Dominion's application included a site redress plan that specifies how
2   the applicant would stabilize and restore the site to its preconstruction condition (or conditions
3   consistent with an alternative use) in the event these site preparation activities are performed but
4   a nuclear power plant is not constructed on the ESP site. Pursuant to 10 CFR 52.17(a)(2),
5   Dominion did not address the benefits of the proposed action (e.g., the need for power). In
6   accordance with 10 CFR 52.18, the EIS is focused on the environmental effects of construction
7   and operation of a reactor, or reactors, that have characteristics that fall within the design
8   parameters that would be specified in the ESP if it is granted.
9

10   Upon acceptance of the Dominion ESP application for docketing, the NRC began the
11   environmental review process described in 10 CFR Part 51 by publishing in the *Federal Register*
12   a Notice of Intent (68 FR 65961) to prepare an EIS and conduct scoping. The staff visited the
13   North Anna ESP site during December 2003 and held a public scoping meeting on
14   December 8, 2003, in Mineral, Virginia. Subsequent to the site visit and the scoping meeting and
15   in accordance with NEPA and 10 CFR Part 51, the staff determined and evaluated the potential
16   environmental impacts of constructing and operating two new nuclear power plants at the North
17   Anna ESP site, and stated its preliminary finding in a Draft EIS issued on December 2, 2004.
18   On December 10, 2004, the staff issued a Notice of Availability in the *Federal Register*
19   (69 FR 71854). On December 17, 2004, the U.S. Environmental Protection Agency (EPA)
20   issued a Notice of Filing (69 FR 75535), and initiated a 75-day comment period for the Draft EIS,
21   which ended March 2, 2005.
22

23   The Draft EIS set forth (1) the results of the NRC staff's preliminary analyses, which considered
24   and weighed the environmental effects of the proposed action (issuance of the ESP) and of
25   constructing and operating two new nuclear units at the ESP site; (2) mitigation measures for
26   reducing or avoiding adverse effects; (3) the environmental impacts of alternatives, and (4) the
27   staff's preliminary recommendation regarding the proposed action.
28

29   The staff conducted a public meeting on February 17, 2005, to describe the preliminary results of
30   the NRC environmental review, answer questions, and provide members of the public with
31   information to assist them in formulating comments on the Draft EIS. After the comment period,
32   the staff considered all comments received. The staff's disposition of these comments, along
33   with comments received on this SDEIS will be set forth in Appendix E of the Final EIS.
34

35   During the course of preparing this SDEIS, the staff reviewed the revised ER submitted by
36   Dominion, consulted, as necessary, with Federal, State, Tribal and local agencies; and followed
37   the guidance set forth in review standard RS-002, *Processing Applications for Early Site Permits,*
38   to conduct an independent review of the issues with respect to the changes presented in
39   ER Revision 6. The review standard draws from the previously published NUREG-0800,
40   *Standard Review Plan for the Review of Safety Analysis Reports for Nuclear Power Plants,* and
41   NUREG-1555, *Standard Review Plans for Environmental for Nuclear Power Plants.*
42

To guide its assessment of environmental impacts of a proposed action or alternative actions, the NRC has established a standard of significance for impacts using Council on Environmental Quality (CEQ) guidance (40 CFR 1508.27). Using this approach, the NRC has established three significance levels – SMALL, MODERATE, or LARGE – which are defined below:

SMALL – Environmental effects are not detectable or are so minor that they will neither destabilize nor noticeably alter any important attribute of the resource.

MODERATE – Environmental effects are sufficient to alter noticeably, but not to destabilize, important attributes of the resource.

LARGE – Environmental effects are clearly noticeable and are sufficient to destabilize important attributes of the resource.

Mitigation measures were considered for each resource area and are presented in the appropriate sections.

The staff's preliminary recommendation, in view of the environmental impacts described in the Draft EIS, and the impacts reviewed in this SDEIS in relation to the changes presented in ER Revision 6, is that the ESP for North Anna Units 3 and 4 should be issued. This recommendation is based on (1) the ER submitted by Dominion, as revised; (2) consultation with Federal, State, Tribal and local agencies; (3) the staff's independent review; (4) the assessments summarized in the Draft EIS and this SDEIS, including the potential mitigation measures identified in the ER and in both the Draft EIS and SDEIS. In addition, in making its recommendation, the staff has concluded that the alternative sites considered are not obviously superior to the proposed site. Finally, the staff concludes that the site preparation and preliminary construction activities enumerated in 10 CFR 50.10(e)(1) would not result in any significant adverse environmental impact that cannot be redressed.

Abbreviations/Acronyms

ABWR	advanced boiling water reactor
ac	acre(s)
ACE	U.S. Army Corps of Engineers
ACR-700	Advanced CANDU Reactor
ADAMS	Agency-wide Documents Access and Management System
AEC	U.S. Atomic Energy Commission
ALARA	as low as reasonably achievable
ALWR	advanced light-water reactor
ATWS	anticipated transient without scram
BEA	Bureau of Economic Analysis
BMP	best management practices
Bq	becquerel(s)
Btu	British thermal unit(s)
BWR	boiling water reactor
C	Celsius
CEDE	committed effective dose equivalent
CEQ	Council on Environmental Quality
CFR	Code of Federal Regulations
cfs	cubic feet per second
Ci	curie(s)
cm	centimeter(s)
COL	combined construction and operating license, combined license
CP	construction permit
CWA	Clean Water Act of 1977 (also known as the Federal Water Pollution Control Act)
CWIS	cooling water intake system
CZMA	Coastal Zone Management Act
d	day
DBA	design-basis accident
DEIS	draft environmental impact statement
DGIF	Department of Game and Inland Fisheries
DOE	U.S. Department of Energy
EAB	exclusion area boundary
EAC	Early Action Compact
EC	Energy Conservation (mode of cooling tower use)
EIS	environmental impact statement
EPA	U.S. Environmental Protection Agency

ER	Environmental Report
ESBWR	economic simplified boiling water reactor
ESE	east-southeast
ESP	early site permit
F	Fahrenheit
FR	*Federal Register*
ft	foot, feet
FWPCA	Federal Water Pollution Control Act (also known as the Clean Water Act of 1977)
FWS	U.S. Fish and Wildlife Service
gal	gallon(s)
GEIS	generic environmental impact statement
gpd	gallons per day
gpm	gallons per minute
GT-MHR	gas turbine-modular helium reactor
ha	hectare(s)
HLW	high-level waste
HPS	Health Physics Society
hr	hour(s)
IAEA	International Atomic Energy Agency
ICRP	International Commission on Radiological Protection
IEEE	Institute of Electrical and Electronics Engineers, Inc.
IHA	Indicators of Hydrologic Alteration
in.	inch(es)
INEEL	Idaho National Engineering and Environmental Laboratory
IRIS	international reactor innovative and secure
ISFSI	independent spent fuel storage installation
kg	kilogram(s)
km	kilometer(s)
kV	kilovolt(s)
kWh	kilowatt hour(s)
L	liter(s)
LAAC	Lake Anna Advisory Committee
lb	pound(s)
LLW	low-level waste
LOCA	loss-of-coolant accident
LOS	level-of-service

LPZ	low population zone
LWR	light-water reactor
m	meter(s)
m/sec	meter(s) per second
m³/d	cubic meter(s) per day
m³/s	cubic meter(s) per second
MBq	million Becquerel(s)
mGy/yr	milligray per year
MGD	million gallons per day
mi	mile(s)
MIT	Massachusetts Institute of Technology
mL	milliliter(s)
mph	miles per hour
mrad	millirad(s)
mrem	millirem(s)
MSL	mean sea level
mSv	millisievert(s)
MT	metric ton(s) (or tonne[s])
MTU	metric ton(s)-uranium
MW	megawatt(s)
MWC	Maximum Water Conservation (mode of cooling tower use)
MWd/MTU	megawatt-days per metric ton of uranium
MW(e)	megawatt(s)-electric
MW(t)	megawatt(s)-thermal
MWh	megawatt hour(s)
NA	not applicable
NAPS	North Anna Power Station
NCDC	National Climatic Data Center
NCHS	National Center for Health Statistics
NCRP	National Council on Radiation Protection and Measurements
NEPA	National Environmental Policy Act of 1969
NESC	National Electric Safety Code
NHP	National Historic Park
NHPA	National Historic Preservation Act
NIEHS	National Institute of Environmental Health Sciences
NNE	north-northeast
NOAA	National Oceanographic and Atmospheric Administration
NO_x	nitrogen oxide(s)
NPDES	National Pollutant Discharge Elimination System
NRC	U.S. Nuclear Regulatory Commission

NUG	non-utility generator
ODCM	Offsite Dose Calculation Manual
OL	operating license
OSHA	Occupational Safety and Health Administration
PBMR	pebble bed modular reactor
PCB	polychlorinated biphenyl
PPE	plant parameter envelope
ppm	parts per million
PWR	pressurized water reactor
RAI	Request for Additional Information
RCIC	reactor core isolation cooling
REMP	radiological environmental monitoring program
RIS	resident important (fish) species
rms	root mean square
ROI	region of interest
RRY	reference reactor-year
RSA	Rapidan Service Authority
Ryr^{-1}	per reactor year
s	second
SAIC	Science Applications International Corporation
SCDHEC	South Carolina Department of Health and Environmental Control
SCDNR	South Carolina Department of Natural Resources
SDEIS	Supplement to the Draft Environmental Impact Statement
SER	safety evaluation report
SHPO	State Historic Preservation Officer
SODI	Southern Ohio Diversification Initiative
SO$_x$	sulfur oxide(s)
SPCC	Spill Prevention Control and Countermeasure
SR	State Route
SRS	Savannah River Site
SSAR	Site Safety Analysis Report
SSE	south-southeast
Sv	sievert(s)
SWR	Service Water Reservoir
SWU	separative work units
TEDE	total effective dose equivalent
TRU	transuranic (waste)

TVA	Tennessee Valley Authority
UCO	uranium oxycarbide
UFSAR	Updated Final Safety Analysis Report
UHS	ultimate heat sink
U.S.	United States
USCB	U.S. Census Bureau
USDA	U.S. Department of Agriculture
USEC	United States Enrichment Corporation, Inc.
USGS	U.S. Geological Survey
VAC	Virginia Administrative Code
VATAX	Virginia Department of Taxation
VDCR	Virginia Department of Conservation and Recreation
VDEQ	Virginia Department of Environmental Quality
VDGIF	Virginia Department of Game and Inland Fisheries
VDH	Virginia Department of Health
VDOT	Virginia Department of Transportation
VDSS	Virginia Department of Social Services
VEC	Virginia Employment Commission
VEPCo	Virginia Electric & Power Company (Virginia Power)
VNHP	Virginia Natural Heritage Program
VPDES	Virginia Pollutant Discharge Elimination System
yd	yard(s)
yr	year(s)
WHTF	Waste Heat Treatment Facility

1.0 Introduction

This section was changed to discuss the submittal of Revision 6 to the Environmental Report and the issuance of the Supplement to the Draft Environmental Impact Statement.

On September 25, 2003, the U.S. Nuclear Regulatory Commission (NRC) received an application pursuant to Title 10 of the Code of Federal Regulations (CFR), Part 52 from Dominion Nuclear North Anna, LLC (Dominion) for an early site permit (ESP) for the North Anna ESP site located in Louisa County, Virginia, near the town of Mineral. On December 10, 2004, NRC issued a *Federal Register* notice (68 FR 65961) announcing the availability of NUREG-1811, *Draft Environmental Impact Statement for an Early Site Permit (ESP) at the North Anna ESP Site* (Draft EIS) and sought comment on the Draft EIS. In the Draft EIS, the NRC staff evaluated the impacts of constructing and operating two new nuclear power units at the North Anna ESP site, based on Revision 3 to Dominion's application (NRC 2004a). On April 13, 2006, Dominion submitted Revision 6 to its application, which included a revised Environmental Report (ER) (Dominion 2006a). This Supplement to the Draft EIS (SDEIS) evaluates the environmental impacts of changes from Revision 3 to Revision 6, including Dominion's responses to the NRC staff's request for additional information (RAI) on Revision 6. On June 21, 2006, Dominion submitted Revision 7 to its application which included the necessary information from the RAI responses (Dominion 2006b).

In Revision 6 to the North Anna ESP application, Dominion proposed (1) changing its approach for cooling the proposed Unit 3 from the once-through cooling system, as described in previous versions of the ER, to a closed-cycle system and (2) increasing the maximum power output per unit from 4300 megawatts-thermal (MW(t)) to 4500 MW(t) for each of the proposed Units 3 and 4 (referred to hereafter as Units 3 and 4). Under the revised cooling system approach, Unit 3 would use a closed-cycle, combination wet and dry cooling system. The proposed increase in power level corresponds to the revision of the maximum power of an economic simplified boiling water reactor (ESBWR), one of the reactor designs included in the plant parameter envelope (PPE) and evaluated in the Draft EIS.

The NRC staff determined that the changes to the proposed action were substantial; therefore, the staff decided to prepare a Supplement to its Draft EIS (referred to as the SDEIS) pursuant to 10 CFR 51.72. On May 16, 2006, following receipt of Dominion's ER Revision 6, the staff published a Notice of Intent to prepare a Supplement to the Draft EIS for the North Anna ESP application in the *Federal Register* (71 FR 28392). The scope of this SDEIS is limited to the environmental impacts associated with the change in the cooling system for Unit 3 and the increase in the maximum power level for both units. The evaluation presented in this SDEIS replaces the evaluation of the impacts associated with the originally proposed once-through cooling for Unit 3 and modifies the analysis of impacts related to the power level increase. These revised evaluations, along with public comments received on the analysis presented in

1 this SDEIS, will be incorporated into the Final EIS together with comments and responses
2 received concerning the original Draft EIS and the staff's consideration of such comments.
3
4 This SDEIS follows the structure of the section contents of the Draft EIS. Those sections that
5 are not affected by the changes in the revised ER are so identified. Some sections in the
6 SDEIS were not affected by the change, but the staff included them solely to provide context for
7 the reader. For example, in this chapter, the background, alternatives to the proposed action,
8 compliance and consultations, and report contents were not affected by the change but are
9 included to provide context. Sections in which the thermal or electric power level or the cooling
10 system are mentioned but not used in the evaluation for those sections will be changed in the
11 Final North Anna ESP EIS (Final EIS) to include the proper reference and are not shown here
12 as a specific change.
13

1.1 Background

15
16 *This section is not affected by changes presented in Revision 6 of the ER and is provided solely*
17 *for context.*
18
19 An ESP is a Commission approval of a site or sites for one or more nuclear power facilities.
20 Issuance of an ESP is an action separate from the issuance of a construction permit (CP) and a
21 operating license (OL) or a combined construction and operating license (combined license or
22 COL) for such a facility. The ESP application and review process makes it possible to evaluate
23 and resolve safety and environmental issues related to siting before the applicant makes a large
24 commitment of resources. If the ESP is approved, the applicant can "bank" the site for up to
25 20 years for future reactor siting. In addition, if the ESP includes a site redress plan, the ESP
26 holder can perform the site preparation and preliminary construction activities enumerated in
27 10 CFR 50.10(e)(1). An ESP does not authorize construction or operation of a nuclear power
28 plant. To construct or operate a nuclear power plant, an ESP holder must obtain a CP and an
29 OL, or a COL, which are separate major Federal actions for which EISs would be prepared in
30 accordance with 10 CFR Part 51.
31
32 As part of its evaluation of the environmental aspects of the action proposed in an ESP
33 application, the NRC prepares an EIS in accordance with 10 CFR 52.18 and 10 CFR Part 51.
34 Because site suitability encompasses construction and operational parameters, the EIS
35 addresses impacts of both construction and operation of reactors and associated facilities. In a
36 review separate from the EIS process, the NRC analyzes the safety characteristics of the
37 proposed site and emergency planning information. These latter two analyses are documented
38 in a safety evaluation report (NRC 2005a) that presents the conclusions reached by the NRC
39 regarding the following issues:
40

- whether there is reasonable assurance that a reactor or reactors, having characteristics that fall within the parameters for the site, can be constructed and operated without undue risk to the health and safety of the public

- whether there are significant impediments to the development of emergency plans

- whether site characteristics are such that adequate security plans and measures can be developed.

In addition, if the applicant proposes either major features of emergency plans or complete and integrated emergency plans, the safety evaluation report documents whether such major features are acceptable, or whether the complete and integrated emergency plans provide reasonable assurance that adequate protective measures can and will be taken in the event of a radiological emergency. Dominion has chosen to propose major features of emergency plans.

1.1.1 Plant Parameter Envelope

This section is not affected by changes presented in Revision 6 of the ER and is provided solely for context.

The applicant for an ESP need not provide a detailed design of a reactor or reactors and the associated facilities but should provide sufficient bounding parameters and characteristics of the reactor or reactors and the associated facilities so that an assessment of site suitability can be made. Consequently, the ESP application may refer to a PPE as a surrogate for a nuclear power plant and its associated facilities.

A PPE is a set of values of plant design parameters that an ESP applicant expects will bound the design characteristics of the reactor or reactors that might be constructed at a given site. The PPE values are a bounding surrogate for actual reactor design information. Analysis of environmental impacts based on a PPE approach permits an ESP applicant to defer the selection of a reactor design until the CP or COL stage. The PPE reflects bounds of the values for each parameter that it encompasses rather than the characteristics of any specific reactor design. Changes to the PPE are discussed in more detail in Section 3.2 of this SDEIS.

1.1.2 Site Preparation and Preliminary Construction Activities

This section was changed to reflect the permit conditions the staff has proposed for the site preparation and limited construction activities.

The holder of an ESP, or an applicant for a CP (10 CFR Part 50) or a COL (Subpart C of 10 CFR Part 52) that references an ESP with an approved site redress plan, may in accordance with 10 CFR 52.25(a) perform the site preparation and preliminary construction activities enumerated in 10 CFR 50.10(e)(1), provided the final ESP EIS concludes that the activities will not result in any significant adverse environmental impacts that cannot be redressed. Dominion provided a site redress plan as part of its ESP application (Dominion 2006c). Activities permitted under an ESP containing a site redress plan include preparation of the site for construction of the facility, installation of temporary construction support facilities, excavation for facility structures, construction of service facilities, and construction of certain structures, systems, and components that do not prevent or mitigate the consequences of postulated accidents (10 CFR 50.10(e)(1)). The site redress plan specifies how the applicant would stabilize and restore the site to its preconstruction condition (or conditions consistent with an alternative use) in the event these site preparation activities are performed but a nuclear power plant is not constructed on the ESP site.

Should the NRC grant the ESP and the ESP holder decides to perform the activities authorized by 10 CFR 52.25, "Extent of Activities Permitted," the ESP holder must obtain from the landowner the authority to undertake those activities on the ESP site. In obtaining such a right, the ESP holder must also obtain the corresponding right to implement the site redress plan described in the staff's Final EIS in the event that no plant is built on the ESP site. The staff proposes to include a condition in any ESP that might be issued requiring that the ESP holder obtain the right to implement the site redress plan before initiating any activities authorized by 10 CFR 52.25. In addition, Section 401 of the Clean Water Act requires that applicants for Federal permits that would allow discharges into navigable waters obtain a certification that any such discharges will comply with the Clean Water Act. As discussed in Section 1.5 of this chapter, the staff proposes to include a condition prohibiting Dominion from conducting any pre-construction activity that would result in a discharge into navigable waters without first submitting to the NRC a Virginia Water Protection Permit which (under Virginia's State Water Control Law at Virginia Code § 62.1-44.15:5(A) constitutes the certification required under Clean Water Act § 401) or a determination by the Virginia Department of Environmental Quality (VDEQ) that no certification is required.

1.1.3 ESP Application and Review

A description of the SDEIS process is added to clarify how issues that are not resolved at the
ESP stage are addressed at the COL stage and an expanded explanation of how the
information provided in the ER is used by the staff as a basic source of information.

In accordance with 10 CFR 52.17(a)(2), Dominion submitted an ER as part of its ESP
application (Dominion 2006a). The ER focuses on the environmental effects of construction
and operation of reactors with characteristics that fall within the PPE. The ER also includes an
evaluation of alternative sites to determine whether there is an obviously superior alternative to
the proposed site. The ER is not required to include, nor does it include, an assessment of the
benefits of the proposed action (e.g., the need for power) or a discussion of energy alternatives.

The NRC staff conducts its reviews of ESP applications in accordance with guidance set forth in
review standard RS-002, *Processing Applications for Early Site Permits* (NRC 2004b). The
review standard draws from the previously published NUREG-0800, *Standard Review Plan for*
the Review of Safety Analysis Reports for Nuclear Power Plants (NRC 1987), and
NUREG-1555, *Standard Review Plans for Environmental Reviews for Nuclear Power Plants*
(ESRP) (NRC 2000). RS-002 provides guidance to NRC staff reviewers to help ensure a
thorough, consistent, and disciplined review of any ESP application. As stated in RS-002, an
applicant may elect to use a PPE approach instead of supplying specific design information.
The staff's June 23, 2003, responses to comments received on draft RS-002 (in NRC's
document system [ADAMS] under the Accession Number ML031710698) provide additional
insights on the staff's expectations and potential approach to the review of an application
employing the PPE approach (NRC 2003). Specifically, the NRC staff adapted the ESRP
review guidance to the PPE concept. The findings in this EIS reflect the adaptation of the
ESRP guidance to the PPE approach.

Pursuant to 10 CFR 52.18, an EIS prepared by the NRC staff on an application for an ESP
focuses on the environmental effects of construction and operation of a reactor, or reactors, that
have characteristics that fall within the postulated site parameters. Such an EIS must also
include an evaluation of alternative sites to determine whether there is any obviously superior
alternative to the site proposed. The Commission's regulations recognize that certain matters
need not be resolved at the ESP stage (i.e., an assessment of the benefits, need for power)
and, thus, may be deferred until an applicant decides to apply for a CP or COL. Further, the
NRC staff realizes that certain information pertaining to the environmental impacts of
construction and operation of new nuclear power facilities may not be available when the NRC
staff reviews an ESP application.

Dominion's ESP application, including its ER, was submitted under oath or affirmation as part of
the application for an ESP. Applicants use the body of NRC regulatory guidance

1   (e.g., Regulatory Guides, Review Standards, and Standard Review Plans) and can take
2   advantage of approaches and methods that are acceptable to the NRC to analyze
3   environmental impacts. The staff relied upon the ER as a source of basic information about the
4   plant parameters, the site, the region, and the environment. The applicant and the NRC are not
5   required to have alternate positions on the significance of environmental impacts; nevertheless,
6   at times there are different conclusions reached based on different methods and assumptions.
7   Subsequent to the acceptance of the application, the staff visited the site; consulted with local,
8   State, Tribal and Federal agencies; and conducted its own independent review. The Draft EIS
9   and this SDEIS are the result of the staff's review and properly includes material from various
10   sources including the ER. In the end, the NRC is responsible for the reliability of all of the
11   information used in its EIS. If, as part of its independent review, the NRC determines that
12   information presented in the ER is useful and the NRC confirms its accuracy, then the NRC may
13   use the information and analyses in its EIS.
14
15   In its analysis of some issues, the staff relied on reasonable assumptions made by Dominion or
16   the staff. The NRC staff will verify the continued applicability of these assumptions at the CP or
17   COL stage to determine whether there is new and significant information from that discussed
18   herein.
19
20   In its application and in responses to requests for additional information (RAIs), Dominion did
21   not or was unable to provide information and analysis for certain issues sufficient to allow the
22   NRC staff to complete its independent analysis. The staff was unable to determine a unique
23   significance level for such issues in this SDEIS, and therefore, these issues are not resolved for
24   the North Anna ESP site. For such issues, Dominion did not offer, nor did the staff identify
25   bases for assumptions that would allow resolution.
26
27   As provided by 10 CFR 52.39(a)(2), the Commission shall treat those matters that are resolved
28   through this EIS as resolved in any later proceeding on an application for a CP or COL
29   referencing the requested North Anna ESP. However, as discussed in the NRC staff's
30   July 6, 2005, letter to Mr. A. Heymer of the Nuclear Energy Institute, a CP or COL applicant
31   must identify whether there is new and significant information on these resolved issues
32   (NRC 2005a). This complements the obligation of a COL applicant referencing an ESP to
33   provide information to resolve any significant environmental issue not considered in the
34   previous proceeding on the ESP. Inasmuch as an ESP and a COL are major Federal actions,
35   both actions require the preparation of an EIS pursuant to 10 CFR 51.20. As provided in
36   10 CFR 52.79 and under the National Environmental Policy Act of 1969 (NEPA), the CP or COL
37   environmental review will be informed by the EIS prepared at the ESP stage, and the NRC staff
38   intends to use tiering and incorporation-by-reference whenever it is appropriate to do so. The
39   CP or COL applicant must address any other issue not considered and not resolved in the EIS
40   for the ESP. Moreover, pursuant to 10 CFR 51.70(b), the NRC is required to independently
41   evaluate and be responsible for the reliability of all information used in an EIS prepared for a CP

1   or COL application, and the staff may (1) inquire into the continued validity of information
2   disclosed in an EIS for an ESP that is referenced in a COL application, and (2) look for any new
3   information that may affect the assumptions, analyses, or conclusions reached in the ESP EIS.
4
5   In addition, measures and controls to limit any adverse impact will be identified and evaluated
6   for feasibility and adequacy in limiting adverse impacts at the ESP stage, where possible, and at
7   the CP or COL stage. As a result of the staff's environmental review of the ESP application, the
8   staff may determine that conditions or limitations on the ESP may be necessary in specific
9   areas, as set forth in 10 CFR 52.24. Therefore, the staff identified in the Draft EIS when and
10   how assumptions and bounding values limit its conclusions on the environmental impacts to a
11   particular resource.
12
13   Following requirements set forth in 10 CFR Part 51 and the guidance in RS-002, the NRC
14   environmental staff (and technical experts from the Pacific Northwest National Laboratory
15   retained to assist the staff) visited the North Anna ESP site and alternative sites in
16   December 2003; January, February, September, and December 2005; and May 2006 to gather
17   information and to become familiar with the sites and their environs. During these site visits, the
18   staff and its contractor personnel met with the applicant's staff, public officials, Federal and
19   State regulators, and the public. A list of the organizations contacted is provided in Appendix B.
20   Other documents related to the North Anna ESP site were reviewed and are listed as
21   references where appropriate.
22
23   Upon acceptance of the Dominion ESP application for docketing, the NRC began the
24   environmental review process described in 10 CFR Part 51 by publishing in the *Federal*
25   *Register* a Notice of Intent (68 FR 65961) to prepare an EIS and conduct scoping. The staff
26   visited the North Anna ESP site during December 2003 and held a public scoping meeting on
27   December 8, 2003, in Mineral, Virginia. Subsequent to the site visit and the scoping meeting
28   and in accordance with NEPA and 10 CFR Part 51, the staff determined and evaluated the
29   potential environmental impacts of constructing and operating two new nuclear power plants at
30   the North Anna ESP site, and stated its preliminary finding in a Draft EIS issued on
31   December 2, 2004. On December 10, 2004, the staff issued a Notice of Availability in the
32   *Federal Register* (69 FR 71854). On December 17, 2004, the U.S. Environmental Protection
33   Agency (EPA) issued a Notice of Filing (69 FR 75535), and initiated a 75-day comment period
34   for the Draft EIS, which ended March 2, 2005.
35
36   A public meeting was conducted on February 17, 2005, at Mineral, Virginia, to describe the
37   results of the NRC environmental review, answer questions related to the review, and provide
38   members of the public with information to assist them in formulating their comments on the
39   Draft EIS.
40

Introduction

1   On April 13, 2006, Dominion submitted Revision 6 to its application (Dominion 2006a). In
2   response to the changes proposed in ER Revision 6 related to the Unit 3 cooling system and
3   the maximum power level of both Units 3 and 4, the NRC staff re-evaluated the environmental
4   impacts of these issues and has documented its conclusion in this SDEIS. The scope of the
5   SDEIS is limited to the environmental impacts associated with the changes in the ER Revision 6
6   cooling system for Unit 3 and the maximum power level of the PPE. This new evaluation will
7   replace the now obsolete evaluation of the impacts of once-through cooling for Unit 3 in the
8   Draft EIS and will modify the analysis of impacts related to the power level increase. These
9   revised evaluations, along with public comments received on the analysis presented in this
10 SDEIS, will be incorporated into the Final EIS together with comments and the staff's
11 consideration of comments received concerning the Draft EIS.
12
13 To guide its assessment of environmental impacts of a proposed action or alternative actions,
14 the NRC has established a standard of significance for impacts using Council on Environmental
15 Quality (CEQ) guidance (40 CFR 1508.27). Using this approach, the NRC has established
16 three significance levels – SMALL, MODERATE, or LARGE – which are defined below:
17
18   SMALL – Environmental effects are not detectable or are so minor that they will neither
19   destabilize nor noticeably alter any important attribute of the resource.
20
21   MODERATE – Environmental effects are sufficient to alter noticeably, but not to
22   destabilize, important attributes of the resource.
23
24   LARGE – Environmental effects are clearly noticeable and are sufficient to destabilize
25   important attributes of the resource.
26
27 The Final EIS will present the staff's analysis that considers and weighs the environmental
28 impacts of the proposed action at the North Anna ESP site, including the environmental impacts
29 associated with construction and operation of reactors at the site, the impacts of constructing
30 and operating reactors at alternative sites, the environmental impacts of alternatives to granting
31 the ESP, and mitigation measures available for reducing or avoiding adverse environmental
32 effects. The Final EIS will also provide the NRC staff's recommendation to the Commission
33 regarding the suitability of the North Anna ESP site for construction and operation of reactors
34 with characteristics that fall within the PPE.
35

1.2 The Proposed Federal Action

This section is changed to reflect that (1) the total nuclear generating capacity to be added would not exceed a total of 9000 MW(t) for the two units (4500 MW(t) each), rather than the originally proposed 8600 MW(t) (4300 MW(t) each) and (2) Unit 3 would use a closed-cycle, combination wet and dry cooling system rather than the originally proposed once-through system.

The proposed Federal action is the issuance, under the provisions of 10 CFR Part 52, of an ESP for the North Anna ESP site for nuclear power facilities with characteristics that fall within the PPE. In addition, Dominion proposes a plan for redressing the environmental effects of certain site preparation and preliminary construction activities (i.e., those activities enumerated in 10 CFR 50.10(e)(1)) performed by an ESP holder under 10 CFR 52.25. In accordance with the plan, the site would be redressed if the NRC issues the requested ESP (containing the site redress plan), the ESP holder performs these site preparation and preliminary construction activities, the ESP is not referenced in an application for a CP or COL, and no alternative use is found for the site. While Dominion is not currently proposing construction and operation of new units, this EIS analyzes the environmental impacts that could result from the construction and operation of two new nuclear units at the North Anna ESP site, or at three alternative sites. These impacts are analyzed to determine whether the proposed ESP site is suitable for the new units and whether there is an alternative site that is obviously superior to the proposed site.

The North Anna ESP site proposed by Dominion is located in Louisa County in northeastern Virginia, near the town of Mineral. It is completely within the confines of the current North Anna Power Station (NAPS) site, which is located on a peninsula on the southern shore of Lake Anna approximately 8 km (5 mi) upstream of the North Anna Dam. Lake Anna is approximately 27 km (17 mi) long with 435 km (272 mi) of shoreline. The lake was created in 1971 by the construction of a dam on the main stem of the North Anna River. Virginia Electric and Power Company, a subsidiary of Dominion Resources, Inc., owns the land above and below the lake surface and around the lake up to the expected high-water mark.

For purpose of the ESP application, no specific plant design was selected by Dominion for the ESP site; instead, a set of values of plant parameters (i.e., the PPE) has been specified for the staff's evaluation of the future development of the North Anna site. Dominion has for the purpose of preparation of a combined license application selected the Economic Simplified Boiling Water Reactor (ESBWR) (Dominion 2005a). However, for the ESP review, Dominion's application uses the PPE approach. The PPE is based on the addition of power generation from two distinct units, to be designated as North Anna Units 3 and 4. Each unit represents a portion of the total generation capacity to be added and would consist of one or more reactors or reactor modules. These multiple reactors or modules (the number of which may vary

1 depending on the reactor type selected) would be grouped into distinct operating units. The
2 total nuclear generating capacity to be added would not exceed 9000 MW(t). Cooling water for
3 Unit 3, the first of the proposed new units, was originally envisioned as being provided by
4 Lake Anna using a once-through cooling system. With the changes proposed in ER Revision 6,
5 Unit 3 would now be cooled using a closed-cycle, combination wet and dry cooling tower
6 system. Unit 4 would use dry cooling towers.
7
1.3   The Purpose and Need for the Proposed Action
9
10 *This section is not affected by changes presented in Revision 6 of the ER and is provided solely*
11 *for context.*
12
13 The purpose and need for the proposed action (i.e., ESP issuance) is to provide stability in the
14 licensing process by addressing site safety and environmental issues before the plants are built
15 rather than after construction is completed. The ESP process allows for early resolution of
16 many safety and environmental issues that may be identified for the ESP site. In the absence
17 of an ESP, safety and environmental reviews of applications for operating licenses under
18 10 CFR Part 50 would take place during plant construction. Alternatively, all safety and
19 environmental issues would have to be addressed at the time of the staff's review of a COL
20 submitted under 10 CFR Part 52 if no ESP for the site were referenced. Although actual
21 construction and operation of the facility would not take place unless and until a COL is granted,
22 certain lead-time activities, such as ordering and procuring certain components and materials
23 necessary to construct the plant, may begin before the COL is granted. As a result, without the
24 ESP review process, there could be a considerable expenditure of funds, commitment of
25 resources, and passage of time before site safety and environmental issues are finally resolved.
26
1.4   Alternatives to the Proposed Action
28
29 *This section is not affected by changes presented in Revision 6 of the ER and is provided solely*
30 *for context.*
31
32 Section 102(2)(C)(iii) of NEPA states that EISs will include a detailed statement on alternatives
33 to the proposed action. The NRC regulations for implementing Section 102(2) of NEPA provide
34 for inclusion of a chapter in an EIS that discusses the environmental impacts of the proposed
35 action and the alternatives (10 CFR Part 51, Subpart A, Appendix A). Chapter 8 of this EIS
36 discusses the environmental impacts of three categories of alternatives: (1) alternative sites,
37 (2) system design alternatives, and (3) the no-action alternative. The Commission determined
38 that evaluation of energy alternatives is not required for an ESP.
39

1 The three alternative sites that are considered in detail in this EIS include lands within
2 Dominion's Surry Power Station in Virginia, the U.S. Department of Energy Portsmouth
3 Gaseous Diffusion Plant in Ohio, and the U.S. Department of Energy Savannah River Site in
4 South Carolina. Chapter 8 also includes sections discussing (1) Dominion's region of interest
5 for identification of alternative plant sites, (2) the methodology used by Dominion to select the
6 proposed ESP site and alternative sites, and (3) generic issues that are consistent among the
7 alternative sites. Chapter 9 compares the environmental impacts at the North Anna ESP site to
8 the alternative sites and to the no-action alternative, and qualitatively determines whether any
9 one of the alternative sites considered is obviously superior to the proposed site.
10

1.5 Compliance and Consultations

12
13 *Changes to this section reflect a discussion about a proposed permit condition to govern site*
14 *preparation and limited construction activities.*
15
16 Prior to construction and operation of a new reactor or reactors, Dominion is required to hold
17 certain Federal, State, and local environmental permits, as well as meet relevant Federal and
18 State statutory requirements. In its ER, Dominion provided a list of environmental approvals
19 and consultations associated with the North Anna ESP. Because an ESP is limited to
20 establishing the acceptability of the proposed site for future development, with the exception of
21 the Clean Water Act and Coastal Zone Management Act (CZMA) certifications, the
22 authorizations Dominion will need from Federal, State, and local authorities for construction and
23 operation are not yet necessary; therefore, they have not been obtained. A National
24 Atmospheric Administration "stay of review" for the CZMA consistency concurrence review was
25 removed March 31, 2006, with Dominion's submittal of additional analyses to VDEQ
26 (Dominion 2006b).
27
28 Section 401 of the Clean Water Act specifies that "Any applicant for a Federal license or permit
29 to conduct any activity including, but not limited to, the construction or operation of facilities,
30 which may result in any discharge into the navigable waters, shall provide the licensing or
31 permitting agency a certification from the State..." (401 certification). Dominion is unable to
32 obtain 401 certification from the Commonwealth of Virginia at the ESP stage. In a letter dated
33 October 6, 2005 (Dominion 2005b), responding to an request for additional information,
34 Dominion stated:
35
36   To address the timing of this certification, the ESP should include a condition prohibiting
37   Dominion from conducting any pre-construction activity that would result in a discharge into
38   navigable waters without first submitting to the NRC a Virginia Water Protection Permit
39   (which under Virginia's State Water Control Law at Va. Code § 62.1-44.15:5(A) constitutes

1   the certification required under FWPCA § 401) or a determination by the Virginia DEQ that
2   no certification is required.
3
4   The Commonwealth of Virginia agreed to the ESP permit condition prohibiting discharges to
5   navigable waters until a 401 certification is obtained or waived by the Commonwealth
6   (VDEQ 2006). In addition, Dominion would need to obtain the other necessary authorizations in
7   order to conduct the site preparation and preliminary construction activities allowed by
8   10 CFR 52.25(a). Authorizations and consultations potentially relevant to the proposed ESP
9   are included in Appendix L.
10
11 The staff reviewed the list and contacted the appropriate Federal, State, and local agencies to
12 identify any compliance, permit, or significant environmental issues of concern to the reviewing
13 agencies that may impact the suitability of the North Anna ESP site for the construction and
14 operation of the reactors that fall within the PPE.
15

16 ## 1.6   Report Contents
17
18 The format of this SDEIS follows the format of the North Anna Draft EIS. However, in this
19 SDEIS, text is provided only in sections that have changed as a result of cooling system and
20 power increase changes described in ER Revision 6 or to provide context related to the
21 NRC staff's evaluation of those changes. The subsequent chapters of this SDEIS are
22 organized as follows:
23
24 • Chapter 2 describes the proposed use of a closed-cycle, combination wet and dry
25   cooling system and discusses the environment that would be affected by this change.
26
27 • Chapter 3 documents the characteristics of the closed-cycle, combination wet and dry
28   cooling system to be used as the basis for evaluation of the environmental impacts.
29
30 • Chapter 4 documents the staff's evaluation of the environmental impacts of construction
31   of the closed-cycle, combination wet and dry cooling system.
32
33 • Chapter 5 documents the staff's evaluation of the environmental impacts of operation of
34   the closed-cycle, combination wet and dry cooling system and of radiological doses
35   associated with the power increase.
36
37 • Chapter 6 documents the staff's evaluation of the fuel cycle, transportation, and
38   decommissioning impacts in support of the power increase.
39

- Chapter 7 documents the staff's evaluation of cumulative impacts, as defined in 40 CFR Part 1508, as revised to reflect the change to a closed-cycle, combination wet and dry cooling system for proposed Unit 3 and the power increase.

- Chapter 8 discusses once-through cooling as a system design alternative and presents the analysis of the no-action alternative and alternative sites.

- Chapter 9 sets forth the staff's comparison of the environmental impacts associated with the closed-cycle, combination wet and dry cooling system with the impacts of the alternatives.

- Chapter 10 summarizes the findings of the preceding chapters and presents the staff's evaluation of the environmental impact of the closed-cycle, combination wet and dry cooling system.

The appendices provide the following additional information:

- Appendix A - Contributors to the Environmental Impact Statement Related to Dominion Nuclear North Anna, LLC's Application for an Early Site Permit at North Anna Nuclear Plant Site

- Appendix B - Organizations Contacted

- Appendix C - Chronology of NRC Staff Environmental Review Correspondence Related to Dominion Nuclear North Anna, LLC's Application for Early Site Permit at North Anna Nuclear Plant Site

- Appendix D - Scoping Meeting Comments and Responses

- Appendix E - Draft Environmental Impact Statement Comments and Responses Volume II

- Appendix F - Key Correspondence

- Appendix G - Environmental Impacts of Transportation

- Appendix H - Supporting Documentation on Radiological Dose Assessment

- Appendix I - Plant Parameter Envelope Values

- Appendix J - Dominion Nuclear North Anna, LLC Commitments and Assumptions Relevant to the Analysis of Impact (a tabulation to be included in the Final EIS)

- Appendix K - Staff's Independent Review of Water Budget and Water Temperature Impacts

- Appendix L - Authorizations and Consultations.

1.7 References

10 CFR Part 50. Code of Federal Regulations, Title 10, *Energy,* Part 50, "Domestic Licensing of Production and Utilization Facilities."

10 CFR Part 51. Code of Federal Regulations, Title 10, *Energy,* Part 51, "Environmental Protection Regulations for Domestic Licensing and Related Regulatory Functions."

10 CFR Part 52. Code of Federal Regulations, Title 10, *Energy,* Part 52, "Early Site Permits; Standard Design Certifications; and Combined Licenses for Nuclear Power Plants."

40 CFR Part 1508. Code of Federal Regulations, Title 40, *Protection of Environment*, Part 1508, "Terminology and Index."

68 FR 65961. "Dominion Nuclear North Anna, LLC; Notice of Intent To Prepare an Environmental Impact Statement and Conduct Scoping Process." *Federal Register,* Vol. 68, No. 226. November 24, 2003.

69 FR 75535. "Environmental Impact Statements; Notice of Availability." *Federal Register*, Vol. 69, No. 242. December 17, 2004.

71 FR 28392. "Notice of Intent to prepare a supplement to the Draft EIS for the North Anna ESP Site." Vol. 71, No. 974. May 16, 2006.

Clean Water Act (also referred to as the Federal Water Pollution Control Act). 33 USC 1251,et seq.

Coastal Zone Management Act of 1972 (CZMA). 16 USC 1451, et seq.

Dominion Nuclear North Anna, LLC (Dominion). 2005a. Letter to the NRC dated November 22, 2005, Dominion's Submittal of ESP Application Schedule.

1   Dominion Nuclear North Anna, LLC (Dominion). 2005b. Letter to the NRC dated October 6,
2   2005, Dominion's Response to the Supplemental Request for Additional Information
3   (Accession No. ML052790657).
4
5   Dominion Nuclear North Anna, LLC (Dominion). 2006a. *North Anna Early Site Permit*
6   *Application – Part 3 – Environmental Report*. Revision 6, Glen Allen, Virginia.
7
8   Dominion Nuclear North Anna, LLC (Dominion). 2006b. *North Anna Early Site Permit*
9   *Application – Part 4 – Programs and Plans*. Revision 7, Glen Allen, Virginia.
10
11 Dominion Nuclear North Anna, LLC (Dominion). 2006c. *North Anna Early Site Permit*
12 *Application – Part 4 – Programs and Plans*. Revision 6, Glen Allen, Virginia.
13
14 National Environmental Policy Act of 1969 (NEPA). 42 USC 4321, et seq.
15
16 U.S. Nuclear Regulatory Commission (NRC). 1987. *Standard Review Plan for the Review of*
17 *Safety Analysis Reports for Nuclear Power Plants*. NUREG-0800, Washington, D.C.
18
19 U.S. Nuclear Regulatory Commission (NRC). 2000. *Standard Review Plan for Environmental*
20 *Reviews for Nuclear Power Plants*. NUREG-1555, Vol. 1, Washington, D.C.
21
22 U.S. Nuclear Regulatory Commission (NRC). 2003. Response to comments on Draft RS-002
23 *Processing Applications for Early Site Permits* (ML031710698).
24
25 U.S. Nuclear Regulatory Commission (NRC). 2004a. *Draft Environmental Impact Statement for*
26 *an Early Site Permit (ESP) for the North Anna ESP Site*. NUREG-1811, Washington, D.C.
27
28 U.S. Nuclear Regulatory Commission (NRC). 2004b. *Processing Applications for Early Site*
29 *Permits*. RS-002, Washington, D.C.
30
31 U.S. Nuclear Regulatory Commission (NRC). 2005a. *Safety Evaluation Report for an Early*
32 *Site Permit (ESP) at the North Anna ESP Site*. NUREG-1835, September 2005.
33
34 U.S. Nuclear Regulatory Commission (NRC). 2005b. Letter dated July 6, 2005, from
35 W. Beckner, NRC, to A. Heymer, Nuclear Energy Institute, responding to comments on the
36 scope of an NRC staff review of a combined license application.
37
38 Virginia Department of Environmental Quality (VDEQ). 2006. Letter dated June 16, 2006 from
39 Jeffery A. Steers, Regional Director, Virginia Department of Environmental Quality.
40
41

2.0 Affected Environment

This chapter was modified to incorporate background information necessary to support the evaluation of the impacts of construction and operations in later chapters as related to changes proposed in Revision 6 of the Environmental Report. Section summaries are provided for context.

The site proposed by Dominion Nuclear North Anna, LLC (Dominion) for an early site permit (ESP) is located in Louisa County, Virginia, within the existing boundaries of the currently operating North Anna Power Station (NAPS) (Dominion 2006a). Virginia Electric and Power Company (referred to as Virginia Power or VEPCo) and Dominion are wholly owned subsidiaries of Dominion Resources, Inc. The site is on the shore of Lake Anna approximately 64 km (40 mi) north-northwest of Richmond. Two operating nuclear generating units, Units 1 and 2, are currently located on the NAPS site, and a small hydroelectric power plant is located at the base of the North Anna Dam.

The environment affected by issuance of the proposed ESP, based on Dominion's Revision 3 to its ESP application (Dominion 2004), was described in the *Draft Environmental Impact Statement (EIS) for an Early Site Permit (ESP) at the North Anna ESP Site* (Draft EIS) (NRC 2004a). Since that time, Dominion revised its application, and in Revision 6, changed proposed plant parameters relating to the cooling system for proposed Unit 3 and the maximum power level for proposed Units 3 and 4 (referred to hereafter as Units 3 and 4) (Dominion 2006a).

The environment affected by the proposed ESP is described in this chapter. To provide context regarding the construction and operational impacts discussed in Chapters 4 and 5, this chapter gives abbreviated descriptions of the existing environment drawn from the Draft EIS. The description of the environment itself is not affected by the changes presented by Dominion in Revision 6 of its ER (Dominion 2006a). However, some information has been included in this chapter to update or provide additional relevant detail needed to support the evaluation of the Unit 3 closed-cycle, combination wet and dry cooling system.

2.1 Site Location

This section is not affected by changes presented in ER Revision 6 and is provided solely for context.

The proposed location for Units 3 and 4 is wholly within the NAPS site and is west of and adjacent to the existing facilities of NAPS Units 1 and 2 (Figure 2-1). Two other NAPS units received construction permits on July 26, 1974, but were not constructed. The NAPS site is located in rural Louisa County, Virginia, which had a population of about 25,000 in 2000. NAPS is located within a triangle formed by the cities of Richmond, Charlottesville, and Fredericksburg, Virginia. Figure 2-2 shows the location of NAPS in relation to the major

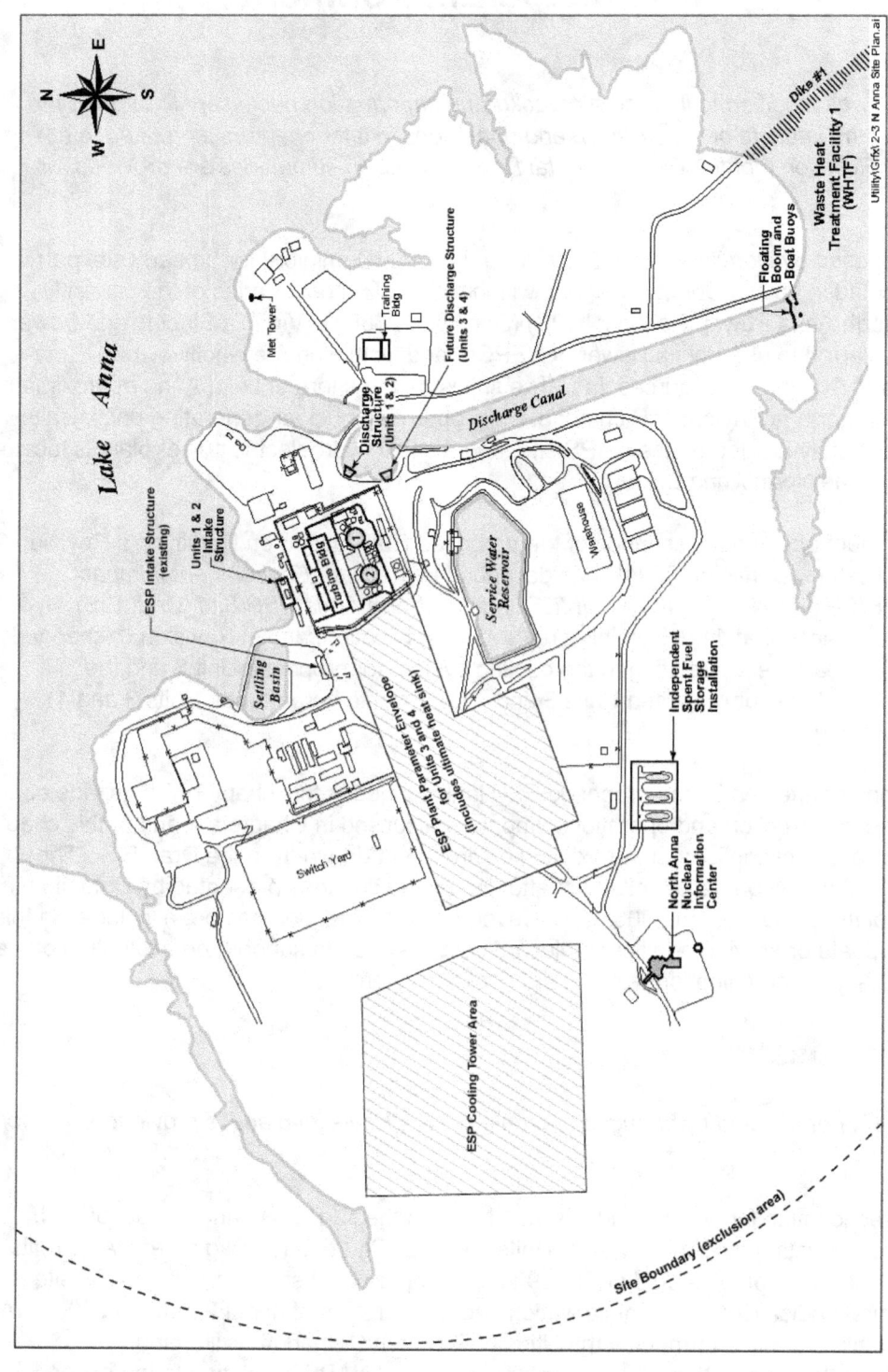

Figure 2-1. North Anna ESP Site Boundaries within the Existing NAPS Site

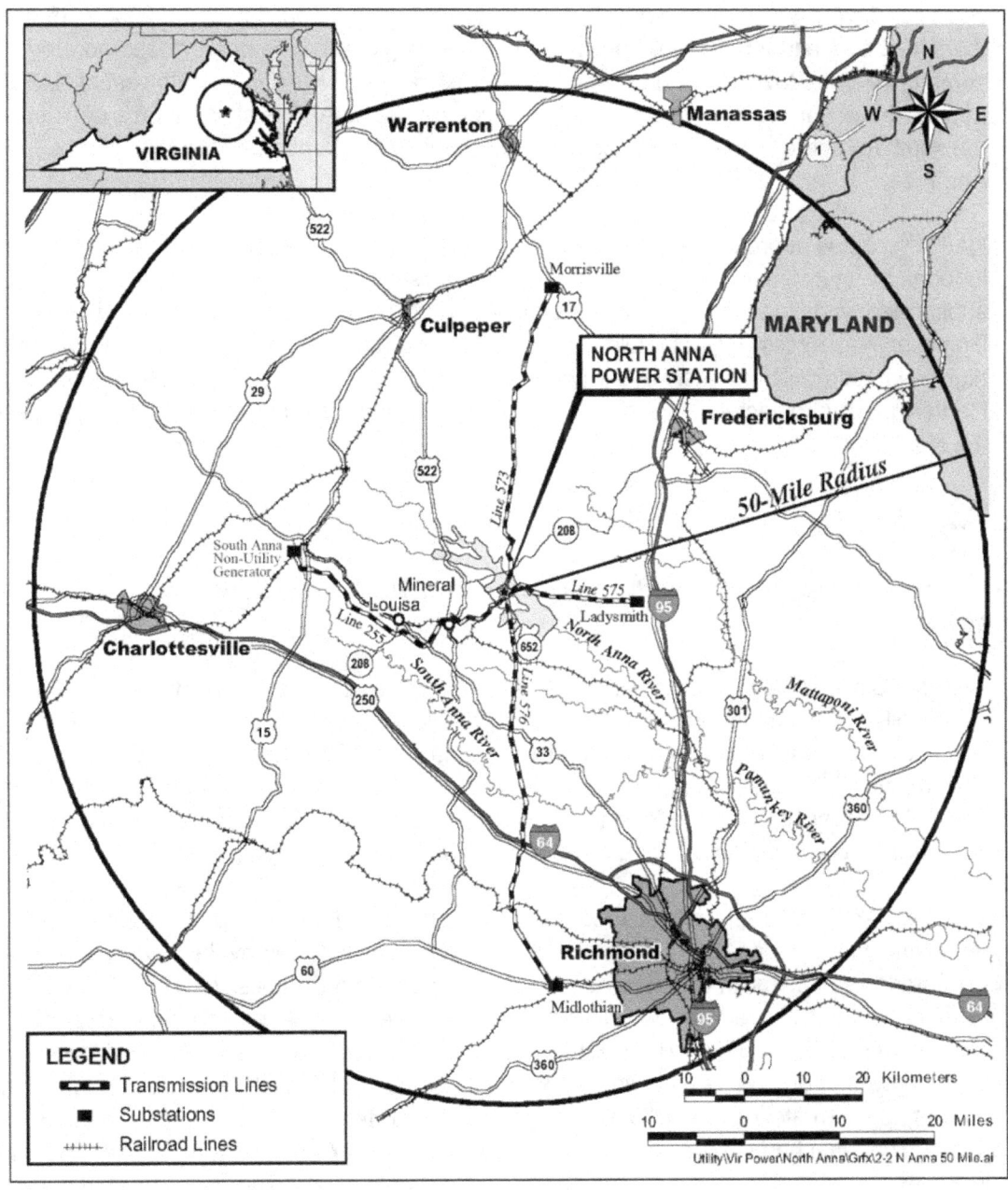

Figure 2-2. Location of North Anna Power Station, 80-km (50-mi) Region

1

1   cities and towns within an 80-km (50-mi) radius. Interstate 95 passes within 26 km (16 mi) of
2   the NAPS site, and Interstate 64 passes within 29 km (18 mi). The nearest incorporated
3   community is the town of Mineral, which is approximately 10 km (6 mi) southwest of NAPS.
4   Louisa, the county seat, is 19 km (12 mi) west of the site. NAPS is situated on a peninsula on
5   the southern shore of Lake Anna, approximately 8 km (5 mi) upstream from the North
6   Anna Dam.

7

8   NAPS occupies approximately 422 ha (1043 ac) of land. In addition, the waste heat treatment
9   lagoons cover approximately 1400 ha (3400 ac), as shown in Figure 2-3. All site land,
10   subsurface lands, and mineral rights are owned jointly by Virginia Power, a subsidiary of
11   Dominion Resources, Inc., and Old Dominion Electric Cooperative. No public or commercial
12   highways, railroads, or waterways traverse the site. Virginia Power also owns and operates the
13   North Anna Hydroelectric Project, an 855-kW-capacity hydroelectric power plant at the base of
14   the North Anna Dam.

15

2.2 Land

17

18   *This section is not affected by changes presented in ER Revision 6 and is presented solely for*
19   *context.*

20

21   The NAPS site is situated on a peninsula of Lake Anna's southern shore at the end of State
22   Route (SR) 700. Lake Anna, an artificial reservoir, was created in 1971 by Virginia Power by
23   erecting a dam on the main stem of the North Anna River. The reservoir was filled by
24   December 1972. Downstream of the dam, the North Anna River flows southeasterly, joining the
25   South Anna River to form the Pamunkey River about 43 km (27 mi) southeast of the NAPS site.
26   The earthen dam that impounds Lake Anna is about 8 km (5 mi) southeast of NAPS.

27

28   The Lake Anna reservoir (or "the reservoir") was formed by impounding the North Anna River
29   above the North Anna Dam. Construction of the dam was licensed by the Virginia State
30   Corporation Commission in 1969 (Virginia State Corporation Commission 1969). The Lake
31   Anna reservoir is divided into two distinct bodies of water, Lake Anna and the Waste Heat
32   Treatment Facility (WHTF). The WHTF is composed of three lagoons and is designated by the
33   Commonwealth of Virginia as a waste heat treatment facility. The lagoons have a total surface
34   area of approximately 1400 ha (3400 ac) and are separated from Lake Anna by a series of
35   dikes. The main body of the lake is approximately 27 km (17 mi) long with 435 km (272 mi) of
36   irregular shoreline and approximately 3900 ha (9600 ac) of water surface. The land adjacent to
37   Lake Anna is becoming increasingly residential as the area is developed. No new
38   transportation routes (roads or railroad lines) or new industrial activities are currently planned in
39   the vicinity of NAPS.

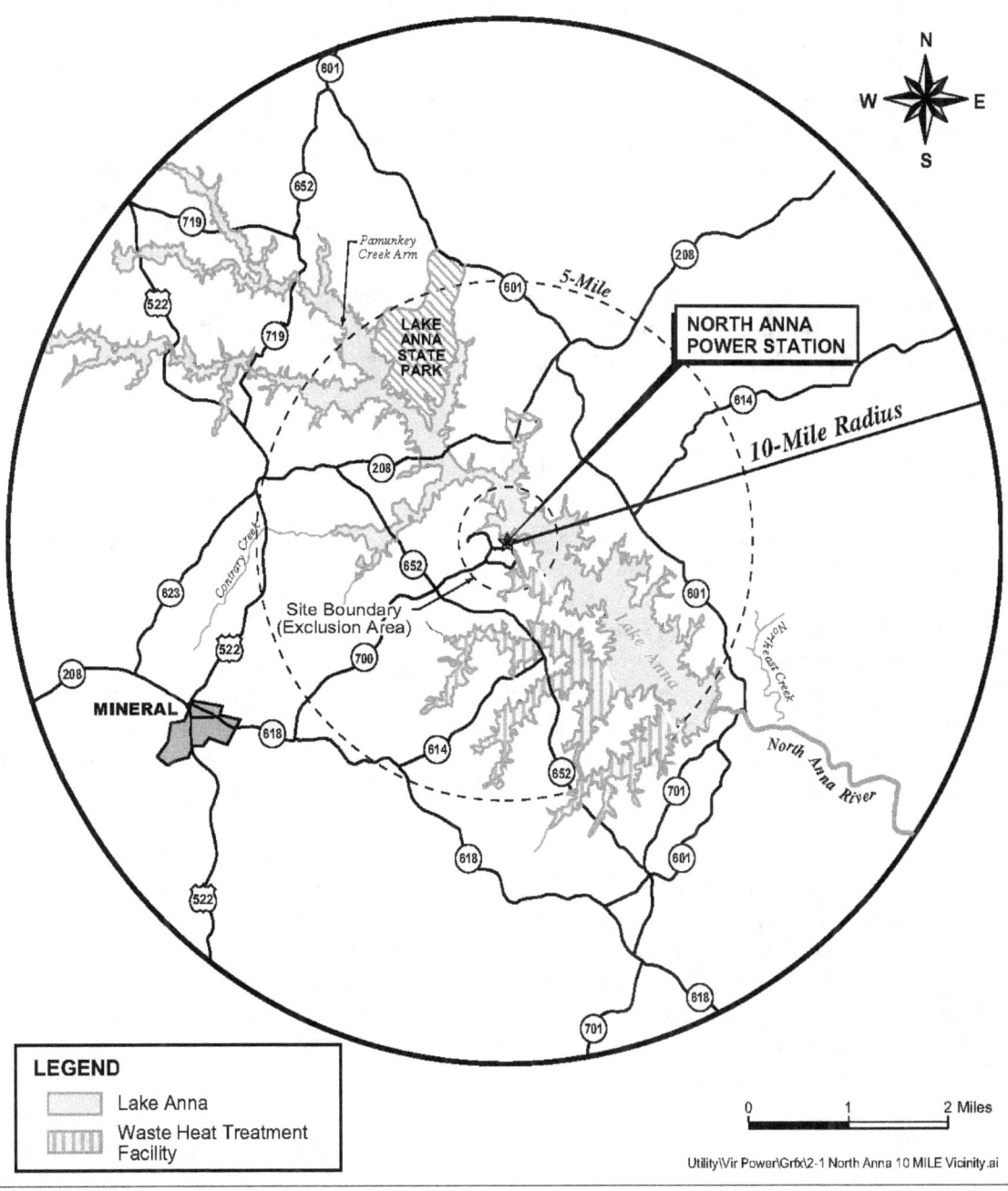

1
2   **Figure 2-3**. North Anna Power Station Vicinity Map, 16-km (10-mi) Region

1   The primary land cover on the NAPS site is pine and pine-hardwood mixed forest (70 percent).
2   Approximately 20 percent of the site is used for nuclear power station facilities and activities
3   including electricity generation, maintenance and distribution facilities, warehouses, training and
4   administration buildings, lagoons and settling basin, parking lots, roads, a railroad line,
5   information center, and the independent spent fuel storage installation (ISFSI). About
6   10 percent of the site is cleared area that includes landscaped ground, open areas, laydown
7   areas, three historic cemeteries, a weapons range used for security training, and a recreation
8   and picnic area used by employees of Dominion Resources, Inc., and its subsidiaries.
9
10 The footprint of the land use identified in Figure 2-1 is unchanged from the Draft EIS
11 (NRC 2004a) and would not change with the addition of cooling towers for Unit 3. As a result,
12 the description of land use is not affected by the changes presented in ER Revision 6
13 (Dominion 2006a).
14

15 ## 2.3 Meteorology and Air Quality

16
17 *This section is not affected by changes presented in ER Revision 6 with the exception that a*
18 *discussion of dew point temperature measurements was added.*
19
20 The ESP site is located in the Piedmont region of Virginia. The climate in this region is
21 considered continental. Summers are generally warm and humid, while winters are generally
22 mild. Based on data collected from the onsite meteorological station starting as early as 1974,
23 the prevailing winds are from the south-southwest at both the 10- and 48.4-m (33- and 159-ft)
24 levels (Dominion 2006a), although there is some seasonal variation. The average temperature
25 at the lower meteorological station level is 13.2°C (55.8°F) and is moderated by the presence of
26 Lake Anna.
27
28 Louisa County, where the North Anna ESP site is located, is within the Northeastern Virginia
29 Intrastate Air Quality Control Region (AQCR), and is classified as in-attainment for all criteria
30 pollutants for which the National Ambient Air Quality Standards have been established
31 (40 CFR 81.347). VDEQ would regulate airborne emissions at the North Anna ESP site during
32 construction activities and routine non-radiological emissions during operation. Any emissions
33 from the operation of the proposed units are not expected to jeopardize compliance with
34 requirements set forth under the current permit.
35
36 With the addition of combination wet and dry cooling towers at the site, the potential impacts
37 from fogging, icing, and salt deposition might occur both within and outside the plant site
38 boundary. The extent to which these events would occur depends on both the local
39 meteorological conditions and the design of the wet and dry cooling towers. To determine the
40 impact of the towers on the environment during operations (Chapter 5), Dominion used dew
41 point temperature measurements and an analytical computer code. The input data for this code

1 included hourly wind speed and direction data, station barometric pressure, and dry bulb and
2 dew point temperature data collected at the 10-m (33-ft) level of the primary meteorological
3 tower during the years of 1998 through 2000, and meteorological data obtained from the nearby
4 National Weather Service Stations in Richmond and at Dulles Airport in Virginia. The
5 information collected from these sources was input to an analytical code that estimated the
6 potential impacts outlined in Chapter 5.
7
8 A general characterization of onsite humidity conditions and the potential for fogging resulting
9 from increased emission of water vapor to the atmosphere can be expressed in terms of dew
10 point depression, which is the difference between dry bulb and dew point temperature. In
11 response to questions raised by the staff, summary onsite data were provided for the number of
12 hours when the dew point depression was predicted to be five degrees or less as a function of
13 season, time of day, and wind direction for the same period that was used to estimate the
14 impacts from combination wet and dry cooling tower operation (Dominion 2006b). For the
15 winter and spring seasons, the greatest occurrence when the dew point depression was five
16 degrees or less occurred when winds were from the west-northwest. During the summer
17 season, the greatest occurrence was with winds from the southwest, while in the fall the
18 greatest occurrence was with winds from the west. In all cases, the greatest occurrence was
19 during the early morning hours. For all seasons and all wind directions, the amount of time that
20 the dew point depression was five degrees or less ranged from 37 percent (winter) to
21 28 percent (spring).
22

23 ## 2.4 Geology
24
25 *This section is not affected by changes presented in ER Revision 6 and is provided solely for*
26 *context.*
27
28 The North Anna ESP site lies within the Piedmont Physiographic Province (Trapp and
29 Horn 2000). The Piedmont Province is bounded on the west by the Blue Ridge Province and on
30 the east by the Coastal Province. The boundary between the Coastal Province and the
31 Piedmont Province is the Fall Line. The Fall Line is a low, east-facing cliff paralleling the
32 Atlantic coastline from New Jersey to the Carolinas. It separates hard Paleozoic metamorphic
33 rocks of the Appalachian Piedmont to the west from the softer, gently dipping Mesozoic and
34 Tertiary sedimentary rocks of the Coastal Plain. This erosional scarp, the site of many
35 waterfalls, often represents an obstruction to upstream passage of migratory fish.
36

2.5 Radiological Environment

This section is not affected by changes presented in ER Revision 6 and is provided solely for context.

A radiological environmental monitoring program (REMP) has been conducted around the NAPS site since 1976 (NRC 1977). The REMP includes monitoring of the airborne exposure pathway, direct exposure pathway, water exposure pathway, aquatic exposure pathway from Lake Anna and the North Anna River, and ingestion exposure pathway in a 40-km (25-mi) radius of NAPS. The preoperational environmental radiation monitoring program sampled various media in the environment to establish a baseline to determine the magnitude and fluctuation of radioactivity in the environment once the existing units began operation (AEC 1973). The preoperational monitoring program included collection and analysis of samples of air particulates, precipitation, milk, crops, soil, well water, surface water, fish, and silt as well as measurement of ambient gamma radiation. After operation of NAPS Units 1 and 2 began, the monitoring program continued to assess their radiological impacts to workers, the public, and the environment. Modifications to the monitoring program are made based on changes in the area, such as milk production, agricultural uses, and changes in lake use. Radiological releases from the existing units are summarized in the annual effluent reports and radiological operating reports. Since the Draft EIS was published in November 2004, VEPco issued more recent reports (see *Radiological Environmental Operating Report* [VEPCo 2005a] and the *Annual Radioactive Effluent Release Report* [VEPCo 2005b]).

The NRC staff reviewed historical data on releases from the existing units and estimated occupational and population doses. The data and analysis showed that doses from the existing units to the maximally exposed individuals around NAPS were a small fraction of the limits specified in Federal environmental radiation standards: 10 CFR Part 20; 10 CFR Part 50, Appendix I; and 40 CFR Part 190.

2.6 Water

This section is not affected by changes presented in ER Revision 6 and is provided solely for context.

Surface Water Hydrology

The dominant hydrological feature of the NAPS site is the Lake Anna reservoir, which was formed by impounding the North Anna River above the North Anna Dam. The Lake Anna reservoir is divided into two distinct bodies of water: Lake Anna and the WHTF. The WHTF is composed of three waste heat treatment lagoons separated from the lake by a series of dikes (Figure 2-4). Lake Anna is approximately 27 km (17 mi) long with 435 km (272 mi) of irregular

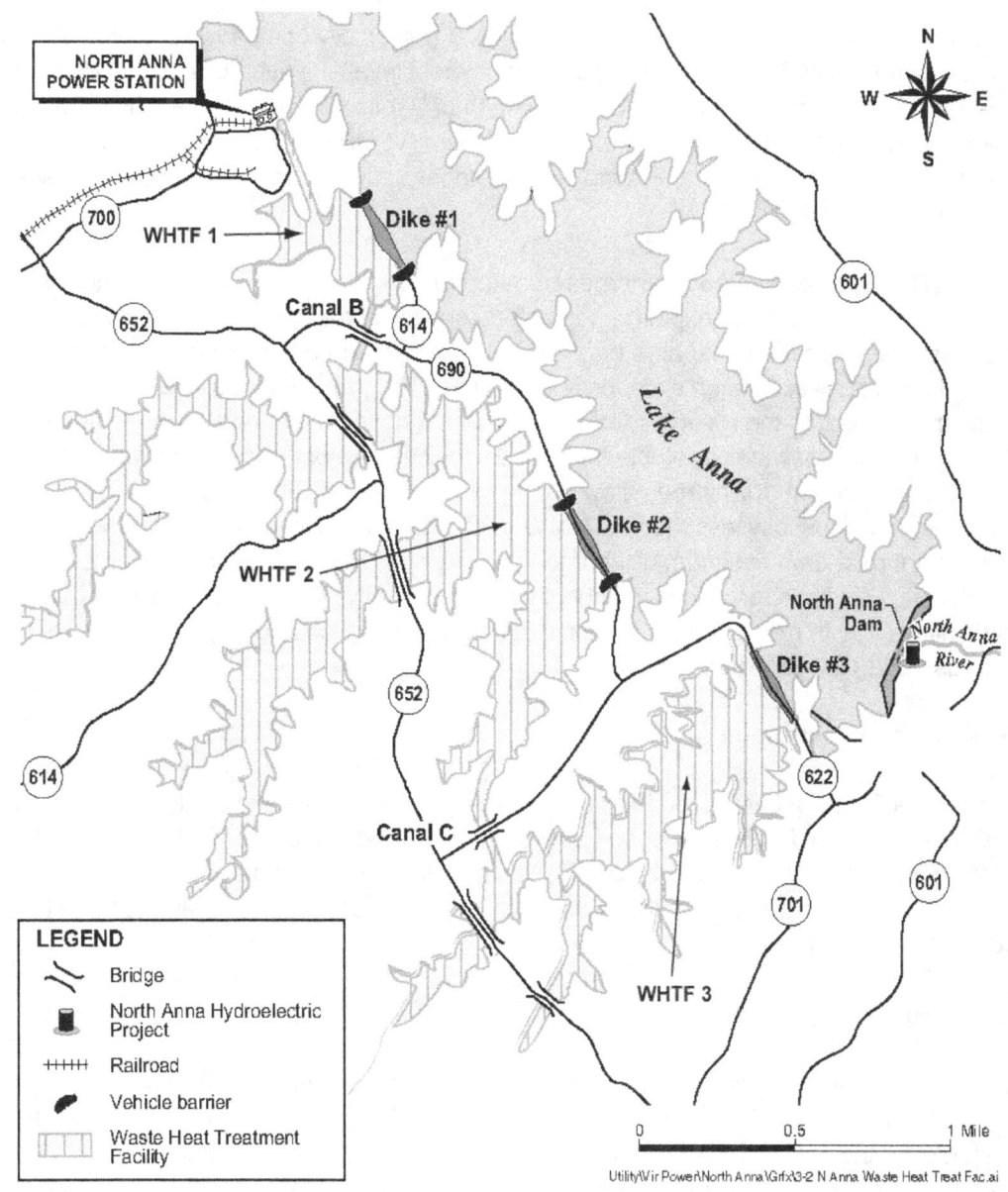

Figure 2-4. Lake Anna and the North Anna Power Station Waste Heat Treatment Facility

1   shoreline and approximately 3900 ha (9600 ac) of water surface area at the normal pool

2   elevation of 76.2 m (250 ft) above mean sea level (MSL). By comparison, the WHTF has a

3   surface area of 1400 ha (3400 ac) at the normal pool level elevation of 76.2 m (250 ft) MSL.

4   The WHTF is designated by the Commonwealth of Virginia as a waste heat treatment facility in

5   the Virginia Pollutant Discharge Elimination System (VPDES) permit (VDEQ 2001) for NAPS

6   (Figure 2-4). The gates of Lake Anna Dam are operated to maintain a steady pool elevation at

7   76.2 m (250 ft) MSL.

8

9   The WHTF receives heated discharges from the existing units. The time for water to flow

10   through the WHTF and its exposure to the atmosphere allows the WHTF to dissipate some of

11   the waste heat to the atmosphere before the water is returned to Lake Anna. In an average

12   year at the site, precipitation exceeds evaporation. The presence of the reservoir and the

13   discharge of heat to the reservoir from Units 1 and 2 have increased evaporation and reduced

14   the total quantity of water available for release downstream of the dam. However, the dam

15   provides a minimum downstream flow during low water conditions. The historical pre-dam

16   minimum flows (usually less than 0.14 m^3/s [5 cfs] during dry summer months) were less than

17   the current post-dam minimum discharges of 0.57 m^3 (20 cfs). Seasonal patterns of

18   precipitation and evaporation also impact water availability. Over an annual cycle, this seasonal

19   variability tends to result in a water deficit during July, August, and September and a water

20   surplus during the rest of the year.

21

22   **Groundwater Hydrology**

23

24   Recharge of the aquifers in the Piedmont Physiographic Province is predominately from local

25   infiltration. The hydraulic connection between the reservoir and nearby aquifers results in a rise

26   of the water table for those aquifers in proximity of the lake. Given the relatively small

27   fluctuations of lake water surface elevation, it is not expected that the water table in these

28   aquifers would vary significantly.

29

30   **Hydrological Monitoring**

31

32   Currently, Dominion collects measurements directly associated with the current site operation

33   that are required under the terms of its existing VPDES permit. Dominion also records lake

34   level elevations at the dam. Dominion was able to use this existing monitoring program as part

35   of the pre-application monitoring program for the ESP site. At the site, Dominion records data

36   from 19 groundwater wells. Nine of these wells are associated with NAPS Units 1 and 2, one

37   was installed near the ISFSI, and nine pre-ESP-application wells were installed in 2002.

38

39   At various times in the past, the U.S. Geological Survey (USGS) has maintained four streamflow

40   gauges in the vicinity of NAPS. Two gauges measured streamflows of tributaries draining into

41   Lake Anna and two measured streamflows downstream of Lake Anna Dam. Because of limited

1 inflow data, it is not possible to create a reliable water budget for Lake Anna directly from inflow
2 and discharge measurements. No water velocity measurements within Lake Anna have been
3 recorded.
4
5 **Surface Water Use**
6
7 The existing NAPS units are the largest users of surface water in the region. Because of the
8 limited projected development in the three upstream counties and policies promoting the use of
9 storm water management practices that limit the impact of impervious surfaces, upstream
10 land-use changes are not expected to appreciably alter the patterns of inflow to the reservoir.
11 However, growth in downstream demands for surface water withdrawals could result in
12 increased water conflicts, particularly during drought periods.
13
14 **Groundwater Use**
15
16 Groundwater in the vicinity of the ESP site is primarily obtained from springs and wells in either
17 the saprolite or underlying crystalline bedrock. The production of groundwater in the vicinity of
18 the ESP site is generally not sufficient to satisfy large water demands because of the relatively
19 low yield of the aquifers; therefore, the majority of groundwater development in the area is for
20 domestic and agricultural use.
21
22 **Surface Water Quality**
23
24 Localized elevated temperature in Lake Anna is the most significant surface water-quality
25 concern associated with both the existing NAPS units. In the vicinity, eight of the tributaries
26 draining into Lake Anna are on the Virginia 2004 Clean Water Act Section 303(d) list as
27 impaired for one or more of the following attributes: fecal coliform bacteria, pH, or dissolved
28 oxygen. Downstream of Lake Anna, the discharge is not listed as impaired until it reaches the
29 Chesapeake Bay. Dominion has a VPDES permit for Units 1 and 2 issued by VDEQ
30 (VDEQ 2001). Before Units 3 and 4 could begin to operate, Dominion would be required to
31 obtain a VPDES permit for discharges from these units.
32
33 **Groundwater Quality**
34
35 There are no site-specific data available for the nonradiological chemistry of the groundwater
36 underlying the ESP site. Groundwater sampling undertaken in 1992 as part of the Louisa
37 County Water Testing Program has identified coliform contamination in aquifers near the ESP
38 site. This contamination is most likely attributable to private septic systems in the area.
39

Thermal and Chemical Monitoring

The current temperature monitoring program in Lake Anna reservoir includes both continuous fixed-location temperature stations and temperature profile locations that are sampled twice per year. The VPDES permit requires monitoring of a variety of constituents including pH, chlorine, copper, nickel, chromium, zinc, suspended solids, oil and grease, and biological oxygen demand. While temperature is monitored both inside and outside the WHTF, no chemical monitoring is required outside the WHTF. Community-based monitoring of Lake Anna and WHTF water quality has been performed by volunteers from the Lake Anna Civic Association. Results from this community-based monitoring program are provided to the Commonwealth of Virginia and the U.S. Environmental Protection Agency.

Summary

The description of the existing hydrology, water quality, and use are not affected by the changes presented in ER Revision 6 (Dominion 2006a). The water resources, namely the WHTF and Lake Anna, however, would be affected by the change in the proposed cooling water system, and this change is reflected in Chapter 5 of this Supplement to the Draft EIS (SDEIS).

2.7 Ecology

This section is not affected by changes presented in the ER Revision 6 and is provided solely for context.

Forests in the Piedmont Physiographic Province are nominally characterized by oak-hickory-pine forest (Woods et al. 1999). However, the portion of northeastern Virginia that includes the North Anna ESP site has been settled since the colonial era and, therefore, no longer contains virgin forests. Vegetative cover surrounding the ESP site is an irregular patchwork of row crops, pastures, pine plantations, abandoned (old) fields, and second growth forests of hardwoods and mixed pine-hardwoods (Dominion 2006a). The Lake Anna reservoir is adjacent to the site, oriented from northwest to southeast.

Much of the proposed North Anna ESP site construction area consists of dirt roads, cleared areas, parking lots, buildings, and other areas recovering from prior disturbance. Because of past development or use, undisturbed habitats are absent from this area.

The changes in the cooling system of Unit 3 from the once-through system proposed in ER Revision 3 (Dominion 2004) to the combination wet and dry cooling system as proposed in Revision 6 (Dominion 2006a) would result in the use of more land for the cooling towers and less lake bank disturbance for the smaller intake. Although the cooling towers would use more land, the cooling tower footprint proposed in Revision 3 is sufficient for the cooling towers for

1   Units 3 and 4; therefore, no additional land disturbance is proposed. Consequently, this section
2   is not affected by the changes presented in the ER Revision 6 (Dominion 2006a).
3
4   **2.7.1 Terrestrial Ecology**
5
6   *This section is not affected by changes presented in ER Revision 6, but was updated to provide*
7   *information regarding eagles.*
8
9   Approximately 32 ha (80 ac) of the 729-ha (1803 ac) NAPS site is currently forested; most of the
10   forested portion of the site is within the area where cooling towers would be constructed.
11   Wildlife species found in the forested portions of the ESP site and surrounding areas are those
12   typically found in the forested portions of the North Anna site and in upland Piedmont forests of
13   north-central Virginia. Wildlife species in the old-field habitat of the laydown area and in the
14   transmission line rights-of-way within the ESP site would include most of those found in the
15   adjacent wooded areas.
16
17   The rolling terrain at the ESP site slopes down to the waters of Lake Anna, resulting in
18   essentially no marsh habitat along the shoreline at the site. Hydrophytic vegetation occurs
19   in a thin band approximately 0.3 to 1 m (1 to 3 ft) wide along the edge of the lake
20   (Dominion 2006a). Two intermittent streams flowing north into an unnamed arm of Lake Anna,
21   just northwest of the power-block area, bisect the area where cooling towers would be located.
22   A narrow band of wetlands is associated with each of these streams. A small isolated wetland
23   is located within the ESP site. Dominion has mapped these wetlands, and has provided the
24   information to the U.S. Army Corps of Engineers for evaluation (Dominion 2005).
25
26   Several species of resident and migratory wading birds and waterfowl use Lake Anna.
27   Waterfowl are typically most abundant at Lake Anna during the winter. Lake Anna provides
28   important habitat for migratory waterfowl on the Atlantic flyway, especially during extremely cold
29   winters when the elevated water temperature from station operation maintains a large ice-free
30   body of water.
31
32   Bald eagles (*Haliaeetus leucocephalus*), a Federal- and State-listed threatened species, are
33   occasionally observed along Lake Anna. There are no known eagle nests on the ESP site
34   (NRC 2002). The nearest known nest is approximately 4.2 km (2.6 mi) to the west. The
35   loggerhead shrike (*Lanius ludoviciana*), a State-listed threatened species, occasionally has
36   been observed in the vicinity of NAPS and is known to breed in central Virginia (VDGIF 2004),
37   but breeding loggerhead shrikes have not been recorded at the North Anna site or along the
38   transmission line rights-of-way (Dominion 2006a).
39
40   There are no known populations of any plants species listed as threatened or endangered
41   by the U.S. Fish and Wildlife Service (FWS) or the Commonwealth of Virginia on the North Anna
42   site (Dominion 2006a; NRC 2002; FWS 2004).

1   Dominion currently performs no terrestrial ecological monitoring (Dominion 2006a). However,
2   Dominion does cooperate with private organizations such as the local chapter of the Audubon
3   Society to allow informal monitoring of selected resources at and near NAPS, and has worked
4   with the Virginia Department of Conservation and Recreation (VDCR) Natural Heritage Program
5   to conduct rare plant surveys in transmission line rights-of-way.
6
7   The description of the terrestrial ecology is not affected by the changes presented in the ER
8   Revision 6 (Dominion 2006a). Discussions of the impacts on the terrestrial ecology as a result
9   of cooling tower construction and operation are presented in Chapters 4 and 5, respectively, of
10   this SDEIS.
11

12   ## 2.7.2   Aquatic Ecology
13

14   *This section is not affected by changes presented in ER Revision 6 and is provided solely for*
15   *context.*
16

17   The aquatic resources in the vicinity of the North Anna ESP site are associated with Lake Anna,
18   the WHTF, and the North Anna River (VEPCo 2001). Lake Anna reservoir is typical of many
19   shallow reservoirs found in the southern and mid-Atlantic states. It contains numerous
20   phytoplankton, zooplankton, and benthic macroinvertebrate communities. Thirty-nine species of
21   fish (representing 12 families) have been identified in the reservoir (VEPCo 1986). It appears to
22   support a greater standing crop of fish than most U.S. reservoirs with thriving populations of
23   several forage and game fish species. Non-native fish species, including striped bass, walleye,
24   threadfin shad, and blueback herring, have been stocked in Lake Anna by Virginia Department
25   of Game and Inland Fisheries (VDGIF). Striped bass, introduced during 1973, have been
26   stocked annually since 1975 to create and maintain a "put-grow-and-take" recreational fishery.
27   Professional fishing guides take clients fishing for largemouth, striped bass, black crappie, and
28   walleye on Lake Anna, but there is no commercial fishing in the lake.
29

30   Before the North Anna River was impounded, the fish community of the river downstream of the
31   Contrary Creek inflow was dominated by pollution-tolerant species. In the years following
32   impoundment (and partial reclamation of the Contrary Creek mine sites), there was a steady
33   increase in measures of abundance and diversity of fish in the reservoir. Fish counts taken
34   during 1984 and 1985 indicate that the operation of the existing units had little or no effect on
35   fish diversity downstream from the dam (Dominion 2006a).
36

37   The WHTF is the body of water into which waste heat from the existing units is discharged via a
38   canal. It is separated from Lake Anna by a series of dikes. A weir at Dike 3 allows water to flow
39   from the WHTF to the lake. The same aquatic communities occur in the WHTF and Lake Anna.
40   Fish can swim from Lake Anna into the WHTF and back. However, fish are not stocked in the
41   WHTF, and only residents who live around the WHTF have access.
42

1   The lower North Anna River below the North Anna Dam is small, approximately 23 to 46 m
2   (75 to 150 ft) wide, but it supports a diverse assemblage of stream fishes. There is no
3   commercial fishing in the North Anna River, but recreational fishing is popular. Unless stream
4   flow is unusually high, powerboats are impractical, so most anglers fish from shore or from
5   canoes and kayaks. Recreational fishermen generally seek largemouth and smallmouth bass
6   or redbreast sunfish. Bluegill and redear sunfish are present as well, but receive less attention
7   from anglers (Dominion 2006a).
8

9   Virginia Power has monitored fish populations in the Lake Anna reservoir and the North Anna
10   River for more than 25 years. No Federally or State-listed fish or mussel species has been
11   collected in any of these monitoring studies, nor has any listed species been observed in creel
12   surveys or occasional special studies conducted by Virginia Power biologists. In addition, no
13   Federally or State-listed aquatic plant species has been collected in any of the monitoring
14   studies associated with the existing NAPS Units 1 and 2, nor has any listed species been
15   observed in surveys or special studies conducted by Virginia Power biologists. VDGIF also
16   conducts aquatic ecology monitoring as part of its management of the Lake Anna fisheries.
17

18   No Federally or State-listed fish species' range includes Lake Anna or the North Anna River,
19   and none is believed to occur in counties adjacent to Lake Anna or the North Anna River
20   (i.e., Caroline, Hanover, Louisa, Orange, and Spotsylvania Counties). Two aquatic species
21   listed by the FWS as Federally endangered potentially occur in the counties adjacent to
22   Lake Anna reservoir or the North Anna River. They are dwarf wedgemussel (*Alasmidonta*
23   *heterodon*) and the James River spiny mussel (*Pleurobema collina*), neither of which have
24   been observed or collected in local streams.
25

26   No Federally or State-listed aquatic plant has a range that includes Lake Anna or the North
27   Anna River, and none is believed to occur in counties adjacent to Lake Anna or the North Anna
28   River (i.e., Caroline, Hanover, Louisa, Orange, and Spotsylvania Counties).
29

30   The description of the aquatic ecology is not affected by the changes in ER Revision 6
31   (Dominion 2006a). Discussions of the impacts on the aquatic ecology as a result of cooling
32   tower construction and operation and the influence of the cooling water system on lake levels
33   and temperatures are presented in Chapters 4 and 5, respectively, of this SDEIS.
34

2.8 Socioeconomics

This section is not affected by changes presented in ER Revision 6 and is provided solely for context.

This section presents the socioeconomic resources that potentially could be impacted by the construction, operation, and decommissioning of two new nuclear power units at the North Anna ESP site.

Demographics

The potential impact area for the analysis discussed in this section was determined by where the majority of employees of the currently operating NAPS Units 1 and 2 reside. There are approximately 720 employees currently at NAPS. Approximately 79 percent of these employees live in Henrico, Louisa, Orange, and Spotsylvania Counties and the City of Richmond (NRC 2002).

All or parts of 32 counties and five major cities are located within 80 km (50 mi) of the proposed North Anna ESP site. The largest population center within 16 km (10 mi) of the site is the town of Mineral, which is southwest of NAPS. The town of Louisa, located west of the ESP site, falls within the 32-km (20-mi) radius.

The area within 16 km (10 mi) of the ESP site is predominately rural and is characterized by farmland and wooded tracts. No significant industrial or commercial facilities are in the area, and none are anticipated. Recreational use of Lake Anna, which is the cooling water source for NAPS, is the greatest contributor to a transient population. Numerous recreational sites, consisting of boat ramps, wet slips, camping sites, picnic areas, etc., are located around the reservoir.

Migrant workers are typically members of minority or low-income populations. Given the expected small number of migrant workers, even if they were concentrated at a single location, they would remain only for a short time and would not materially change the population characteristics of any particular census tract within Louisa County.

Economy

The communities potentially impacted socioeconomically by activities at the ESP site are in Henrico, Louisa, Orange, and Spotsylvania Counties and the City of Richmond, all of which are located in central Virginia. The greatest impacts would be observed in Louisa County, where the NAPS site is located. All these counties, but not the City of Richmond, experienced steady growth in population and economic activity during the 1990s.

1 **Transportation**
2
3 There are 32 counties within the 80-km (50-mi) radius of the ESP site. One county is in
4 Maryland while the remaining counties are in Virginia. The 31-county Virginia area is served by
5 two major freeways (Interstates 95 and 64). General transportation studies have been
6 undertaken of highways in the region, and plans are in place to upgrade several highways,
7 including those in areas around Lake Anna.
8
9 **Property Taxes**
10
11 Dominion has a significant impact on the economic well-being of Louisa County, paying on
12 average about 46 percent of the total property taxes between 1995 and 2003. Louisa and
13 Spotsylvania Counties have both been impacted by Lake Anna and the economic development
14 around Lake Anna. Orange County has been impacted to a lesser extent by this development
15 because it has fewer miles of shoreline. Over time, the percentage contribution of total NAPS
16 property taxes payable to Louisa County for NAPS Units 1 and 2 will decline, assuming the
17 current rate of economic growth in the county continues.
18
19 **Aesthetics and Recreation**
20
21 Access to the NAPS site itself is provided by SR 700, a narrow, two-lane road leading to the
22 plant boundary. The terrain is gently undulating and wooded. Most of the site structures are
23 screened from public view up to the proximity of the plant boundary. Noise from plant
24 operations is not noticeable, particularly from points outside the NAPS plant boundary.
25
26 **Housing**
27
28 During refueling outages, site employment increases by as many as 700 temporary workers for
29 30 to 40 days. Each county in the area of potential impact has a comprehensive land-use plan.
30 The county showing the greatest increase in housing units over the decade of the 1990s was
31 Spotsylvania County, which would be expected given its economic growth over the decade.
32 Rental rates for reasonable housing in Louisa County are considered high for a small rural area,
33 and the availability of rental apartments and housing is limited. There is also a shortage of
34 rental housing in Orange and Culpeper Counties, and in nearby Charlottesville.
35
36 **Public Services**
37
38 Public water supply is not a constraint to growth in the vicinity of NAPS. There are supply
39 concerns in some individual municipalities and in some of the impacted counties. There are no
40 limitations on new sources of water from groundwater, and many of the treatment plants located
41 in the area of potential impact have reserve treatment capacity, especially in the larger
42 metropolitan areas. In cases where municipal systems are approaching reserve-capacity limits,

1   plans are in place to address those limitations by constructing new treatment systems or
2   expanding existing facilities.
3
4   Social services in the Commonwealth are provided in each county by the Virginia Department of
5   Social Services (VDSS), which operates offices in each county. None of the nearest three
6   counties has a hospital; however, there are major medical facilities in Fredericksburg and
7   Henrico Counties and in Richmond and Charlottesville. Emergency medical and fire services
8   are in transition from volunteer-based to full-time professional. Louisa, Orange, and
9   Spotsylvania Counties either are adding new buildings to their primary and secondary school
10   systems in the vicinity of Lake Anna or are concerned about overcrowding and are considering
11   such additions.
12

13   **Summary**
14

15   The description of the area's socioeconomic demographics and community characteristics is not
16   affected by the changes presented in ER Revision 6 (Dominion 2006a). Discussions of the
17   impacts of construction as related to the changes in ER Revision 6 are presented in Chapter 4,
18   and those of operation are presented in Chapter 5.
19

20   # 2.9   Historic and Cultural Resources
21

22   *This section is not affected by changes presented in ER Revision 6 and is provided solely for*
23   *context.*
24

25   The area around the North Anna ESP site is rich in prehistoric and historic Native American and
26   historic Euro-American resources. The prehistoric Native American occupation of the region
27   including the North Anna site spans from about 10,000 B.C. to A.D. 1600. Toward the end of
28   that period (about A.D. 1500 to 1675), initial contacts with Europeans and cultural changes
29   associated with subsequent European settlement of the area took place. European settlement
30   of the area around the North Anna site began shortly after 1700, and Louisa County was formed
31   in 1742. The earliest economy of the area was based on cultivation of tobacco in the fertile
32   lands along the North and South Anna River valleys. The area just upriver from the North Anna
33   site was the scene of intensive gold mining in the period from about 1830 to 1900.
34

35   A cultural resource assessment was conducted in 2001, and examination of historic and cultural
36   resource files at the Virginia Department of Historic Resources Archives indicated that no
37   previously recorded cultural resource sites were known to exist at NAPS (Ahlman and
38   Mullin 2001). Similarly, a review of historical documentation at the Louisa County Historical
39   Museum indicates few historic resources in the vicinity of the North Anna site. A field inspection
40   of the proposed ESP project area (Voigt 2003) concluded that much of the proposed ESP site
41   lies within previously disturbed areas, particularly in the eastern portion. However, some

1   undisturbed areas in the western sector have some potential for the presence of cultural
2   resources. Reconnaissance-level historic and archaeological investigations completed in
3   1969 and 1970 for both the North Anna site area and the lake bed area yielded few results
4   (AEC 1973). Cultural resource surveys along transmission line rights-of-way associated with
5   NAPS have largely resulted in negative findings for cultural resources (Saunders 1976;
6   MacCord 1981).

8   The proposed plant footprint for Units 3 and 4 cooling towers is unchanged in the ER
9   Revision 6. Therefore, the description of the historic and cultural resources is not affected by
10 the changes presented in ER Revision 6 (Dominion 2006a).

2.10 Environmental Justice

This section is not affected by changes presented in ER Revision 6 and is provided solely for context.

Environmental justice refers to a Federal policy under which each Federal agency identifies and
addresses, as appropriate, disproportionately high and adverse human health or environmental
effects of its programs, policies, and activities on minority[a] or low-income populations. The
memorandum accompanying Executive Order 12898 (59 FR 7629) directs Federal agencies to
consider environmental justice under the National Environmental Policy Act (NEPA). Although it
is not subject to the Executive Order, the Commission has voluntarily committed to undertake
environmental justice reviews (NRC 2004b). The staff examined the geographic distribution of
minority and low-income populations within 80 km (50 mi) of the North Anna site. Minority
populations exist in several counties within 80 km (50 mi) of NAPS, and all minority block
groups are more than 16 km (10 mi) from NAPS. Census block groups containing low-income
populations are concentrated in the City of Richmond. Also, Henrico and Chesterfield Counties,
to the southeast between approximately 65 and 80 km (40 and 50 mi) from the North Anna site,
have low-income populations. Other areas of low-income populations include Buckingham
County southwest of the site and Charlottesville.

The description of the local demography for use in the environmental justice analysis is not
affected by the changes presented in ER Revision 6 (Dominion 2006a).

(a) Minority categories are defined as American Indian or Alaskan Native; Asian; Native Hawaiian or other
Pacific Islander; or Black races; or Hispanic ethnicity; "other" may be considered a separate minority
category. The 2000 Census included multi-racial data. The staff should consider multi-racial
individuals in a separate minority category, in addition to the aggregate minority category when the
Census Bureau releases the updated information (NRC 2004b).

2.11 Related Federal Projects

This section is not affected by changes presented in ER Revision 6 and is provided solely for context.

The staff reviewed the possibility that activities of other Federal agencies might impact the issuance of an ESP to Dominion. Any such activities could result in cumulative environmental impacts and the opportunity for a Federal agency to become a cooperating agency for preparation of the EIS (10 CFR 51.10(b)(2)). After reviewing the Federal activities in the vicinity of the NAPS site, the staff determined that there were no Federal project activities that would make it desirable for another Federal agency to become a cooperating agency for preparation of the EIS. During the course of preparing the draft EIS, NRC consulted with the FWS and the National Oceanic and Atmospheric Administration's National Marine Fisheries Service as required by Section 102(2)(C) of NEPA.

The description of related Federal projects is not affected by the changes presented in ER Revision 6 (Dominion 2006a).

2.12 References

10 CFR Part 20. Code of Federal Regulations, Title 10, *Energy*, Part 20, "Standards for Protection Against Radiation."

10 CFR Part 50. Code of Federal Regulations, Title 10, *Energy*, Part 50, "Domestic Licensing of Production and Utilization Facilities."

10 CFR Part 51. Code of Federal Regulations, Title 10, *Energy*, Part 51, "Environmental Protection Regulations for Domestic Licensing and Related Regulatory Functions."

40 CFR Part 81. Code of Federal Regulations, Title 40, *Protection of Environment*, Part 81, "Designation of Areas for Air Quality Planning Purposes."

40 CFR Part 190. Code of Federal Regulations, Title 40, *Protection of Environment*, Part 190, "Environmental Radiation Protection Standards for Nuclear Power Operations."

59 FR 7629. Executive Order 12898, "Federal Actions to Address Environmental Justice in Minority and Low-Income Populations." *Federal Register,* Vol. 59, No. 32. February 16, 1994.

Ahlman T. and J. Mullin. 2001. *Cultural Resource Assessment, North Anna Power Station, Louisa County, Virginia.* The Louis Berger Group, Inc., Richmond, Virginia.

Dominion Nuclear North Anna, LLC (Dominion). 2004. *North Anna Early Site Permit Application – Part 3 – Environmental Report*. Glen Allen, Virginia.

Dominion Nuclear North Anna, LLC (Dominion). 2005. Letter to the U.S. Army Corps of Engineers regarding confirmation of wetlands delineation. November 16, 2005, Glen Allen, Virginia.

Dominion Nuclear North Anna, LLC (Dominion). 2006a. *North Anna Early Site Permit Application – Part 3 – Environmental Report*. Revision 6, Glen Allen, Virginia.

Dominion Nuclear North Anna, LLC (Dominion). 2006b. Letter from E. Grecheck (DNNA) to NRC dated May 24, 2006, submitting additional information in response to an NRC request dated May 10, 2006, (ML061510131).

MacCord H.A., Sr. 1981. *An Archaeological Reconnaissance Survey of the North Anna – Louisa 230 kV Line, Louisa County, Virginia*. The Archaeological Society of Virginia, Richmond, Virginia.

National Environmental Policy Act of 1969 (NEPA). 42 USC 4321, et seq.

Saunders J.R., Jr. 1976. *An Initial survey of the North Anna – Louisa Transmission Line, Louisa County, Virginia*. Report on file at the Virginia Department of Historic Resources, Richmond, Virginia.

Trapp H., Jr. and M.A. Horn. 2000. *Ground Water Atlas of the United States, Segment 11, Delaware, Maryland, New Jersey, North Carolina, Pennsylvania, Virginia, West Virginia*. In U.S. Geological Survey, Hydrologic Investigations Atlas 730-L, Washington, D.C.

U.S. Atomic Energy Commission (AEC). 1973. *Final Environmental Statement Related to the Continuation of Construction and the Operation of Units 1 and 2 and Construction of Units 3 and 4, North Anna Power Station*. Washington, D.C.

U.S. Fish and Wildlife Service (FWS). 2004. Threatened and Endangered Species System (TESS) Listings by State and Territory as of 03/01/2004. Virginia. Accessed at http://ecos.fws.gov/tess_public/TESSWebpageUsaLists?state=VA on March 1, 2004.

U.S. Nuclear Regulatory Commission (NRC). 1977. *Regulatory Guide 1.111 – Methods for Estimating Atmospheric Transport and Dispersion of Gaseous Effluents in Routine Releases from Light-Water-Cooled Reactors*. Regulatory Guide 1.111, Revision 1, Washington, D.C.

1   U.S. Nuclear Regulatory Commission (NRC). 2002. *Generic Environmental Impact Statement*
2   *for License Renewal of Nuclear Plants, Supplement 7, Regarding North Anna Power Station,*
3   *Units 1 and 2.* NUREG-1437, Office of Nuclear Reactor Regulation, Division of Regulatory
4   Improvement Programs, Washington, D.C.
5
6   U.S. Nuclear Regulatory Commission (NRC). 2004a. *Draft Environmental Impact Statement for*
7   *an Early Site Permit (ESP) at the North Anna ESP Site*, NUREG-1811 Draft, Office of Nuclear
8   Reactor Regulation, Division of Regulatory Improvement Programs, Washington, D.C.
9
10   U.S. Nuclear Regulatory Commission (NRC). 2004b. "Policy Statement on the Treatment of
11   Environmental Justice Matters in NRC Regulatory and Licensing Actions." *Federal Register,*
12   August 24, 2004, Washington, D.C.
13
14   Virginia Department of Environmental Quality (VDEQ). 2001. *Authorization to Discharge Under*
15   *the Virginia Pollutant Discharge Elimination System and the Virginia State Water Control Law,*
16   *Virginia Electric and Power Company, North Anna Nuclear Power Station.* Permit No.
17   VA0052451, Commonwealth of Virginia, Department of Environmental Quality, Richmond,
18   Virginia.
19
20   Virginia Department of Game and Inland Fisheries (VDGIF). 2004. Geographic Search.
21   Fish and Wildlife Information Service. Accessed at www.vafwis.org/perl/vafwis.pl/vafwis.login.
22   (Note: This is a protected website that is accessible only through VDGIF authorization.)
23
24   Virginia Electric and Power Company (VEPCo). 1986. *Section 316(a) Demonstration for North*
25   *Anna Power Station: Environmental Studies of Lake Anna and the Lower North Anna River.*
26   Virginia Power Corporate Technical Assessment, Water Quality Department, Richmond,
27   Virginia.
28
29   Virginia Electric Power Company (VEPCo). 2001. *Application for License Renewal for North*
30   *Anna Power Station, Units 1 and 2, Appendix E, Environmental Report – Operating License*
31   *Renewal Stage.* Richmond, Virginia.
32
33   Virginia Electric and Power Company (VEPCo). 2005a. *Annual Radiological Environmental*
34   *Operating Report North Anna Power Station - January 1, 2004 to December 31, 2004.*
35   Richmond Virginia.
36
37   Virginia Electric and Power Company (VEPCo). 2005b. *Annual Radiological Effluent Release*
38   *Report for the North Anna Power Station – January 1, 2004 to December 31, 2004.* Richmond,
39   Virginia.
40
41   Virginia State Corporation Commission. 1969. Order of June 12, 1969, Case No. 18869,
42   p. 353, Richmond, Virginia.

1   Voigt E. 2003. *Field Inspection, Early Site Permit (ESP) Project, North Anna Power Station,*
2   *Louisa County, Virginia.* Letter report from Louis Berger Group to Dominion Resources
3   Services, Richmond, Virginia.
4
5   Woods A.J., J.M. Omernik, and D.D. Brown. 1999. Level III Ecoregions of Delaware, Maryland,
6   Pennsylvania, Virginia, and West Virginia. National Health and Environmental Effects Research
7   Laboratory. U.S. Environmental Protection Agency. Corvallis, Oregon. Accessed at
8   http://www.epa.gov/wed/pages/ecoregions/reg3_eco.htm on March 1, 2004.
9

3.0 Site Layout and Plant Parameter Envelope

This chapter was changed to modify certain sections to reflect information necessary to support the description of the plant parameter envelope aspects related to changes proposed in Revision 6 of the Environmental Report. Section summaries are provided for context.

The proposed North Anna early site permit (ESP) site is located in Louisa County in predominately rural northeastern Virginia, and is within the current North Anna Power Station (NAPS) boundaries. The site is situated approximately 64 km (40 mi) northwest of Richmond, Virginia. The approach Dominion Nuclear North Anna, LLC (Dominion) used to identify the key plant parameters and site characteristics needed to assess the environmental impacts of the proposed action in Revision 3 of its application (Dominion 2004a) was described in the *Draft Environmental Impact Statement (EIS) for an Early Site Permit (ESP) at the North Anna ESP Site* (Draft EIS), which was published in November 2004 (NRC 2004a). In Revision 6 to the ER (Dominion 2006), Dominion proposed a change in plant parameters related to the cooling system for proposed Unit 3 and maximum power output for both proposed Units 3 and 4 (referred to hereafter as Units 3 and 4), affecting both the safety and the environmental portions of the application.

Changes to Revision 6 of the ER are reflected in this section, particularly for site layout and plant parameters affected by changes in the maximum power level and the proposed cooling system for Unit 3. This Supplement to the Draft Environmental Impact Statement (SDEIS) relies on abbreviated descriptions from the Draft EIS published in November 2004 for discussion of the site layout, existing facilities, and other portions of the Draft EIS that remain relevant.

3.1 External Appearance and Site Layout

This section was changed to reflect the addition of a combination wet and dry cooling tower system for Unit 3 and to reference the change in proposed power output as presented in Revision 6 of the ER.

The proposed North Anna ESP site, most of which has been previously disturbed, is located within the existing NAPS site in an area adjacent to the existing units (Figure 3-1). NAPS consists of two operational pressurized water reactors (PWRs) furnished by Westinghouse Electric Company, a shared turbine building, a switchyard, intake and discharge structures, and support buildings. NAPS is located on the shore of Lake Anna, an impoundment created in 1971 by constructing a dam on the main stem of the North Anna River to create a source of cooling water for NAPS. The Lake Anna reservoir is divided into Lake Anna, which serves as the cooling water source for NAPS Units 1 and 2, and the Waste Heat Treatment Facility (WHTF), which receives the heated discharge. The existing units use a spray pond for an ultimate heat sink. A radioactive waste disposal system, a fuel-handling system, an

1 independent spent fuel storage installation, auxiliary structures, and other onsite facilities
2 necessary for a complete operating nuclear power plant also exist on the NAPS site. With the
3 exception of a few support buildings that may be relocated, existing structures at the NAPS site
4 would remain unchanged with the addition of new units. The ESP site characteristics are listed
5 in Appendix I, Table I-2 of this document.
6
7 For purposes of the ESP application, a specific plant design has not been selected for the
8 proposed new Units 3 and 4; instead, a set of plant-parameter values was chosen for the staff's
9 evaluation of the development of the North Anna ESP site. This plant parameter envelope
10 (PPE) is based on the addition of two new power generating units, each of which would be a
11 stand-alone unit with its own support systems. Appendix I, Table I-2 lists the PPE values used
12 by the staff. Dominion states that the new units would share ancillary support structures such
13 as maintenance facilities, office centers, and wastewater and water treatment plants. Each new
14 unit would represent a portion of the total generation capacity to be added, and may consist of
15 one or more reactors or reactor modules. These multiple reactors or modules (the number of
16 which may vary depending on the reactor type selected) would be grouped into distinct
17 operating units. The nuclear generating capacity to be added would not exceed
18 4500 megawatts-thermal (MW(t)) per unit, or up to a total of 9000 MW(t) for two units. For the
19 cooling systems, Dominion has proposed using combination wet and dry cooling towers for
20 Unit 3 and dry cooling towers for Unit 4. The proposed location for the cooling towers is
21 illustrated in Figure 3-1.
22

23 ## 3.2 Plant Parameter Envelope
24
25 *This section was changed to reflect modifications to the ESP parameter values relating to the*
26 *higher power output and the Unit 3 cooling system as presented in Revision 6 of the ER.*
27
28 An applicant for an ESP need not provide a detailed design of a reactor or reactors and the
29 associated facilities, but should provide sufficient values for parameters for the reactor or
30 reactors and the associated facilities so that an assessment of site suitability can be made.
31 Consequently, the ESP application may refer to a PPE as a surrogate for a nuclear power plant
32 and its associated facilities.
33
34 A PPE is a set of values of plant design parameters that an ESP applicant expects would bound
35 the design characteristics of the reactor or reactors that might be constructed at a given site.
36 The PPE values are surrogates for actual reactor design information. Analysis of environmental
37 impacts based on a PPE approach permits an ESP applicant to defer the selection of a reactor
38 design until the construction permit (CP) or combined construction and operating license
39 (combined license or COL) stage. The PPE reflects the value of each parameter that it
40 encompasses rather than the characteristics of any specific reactor design.
41

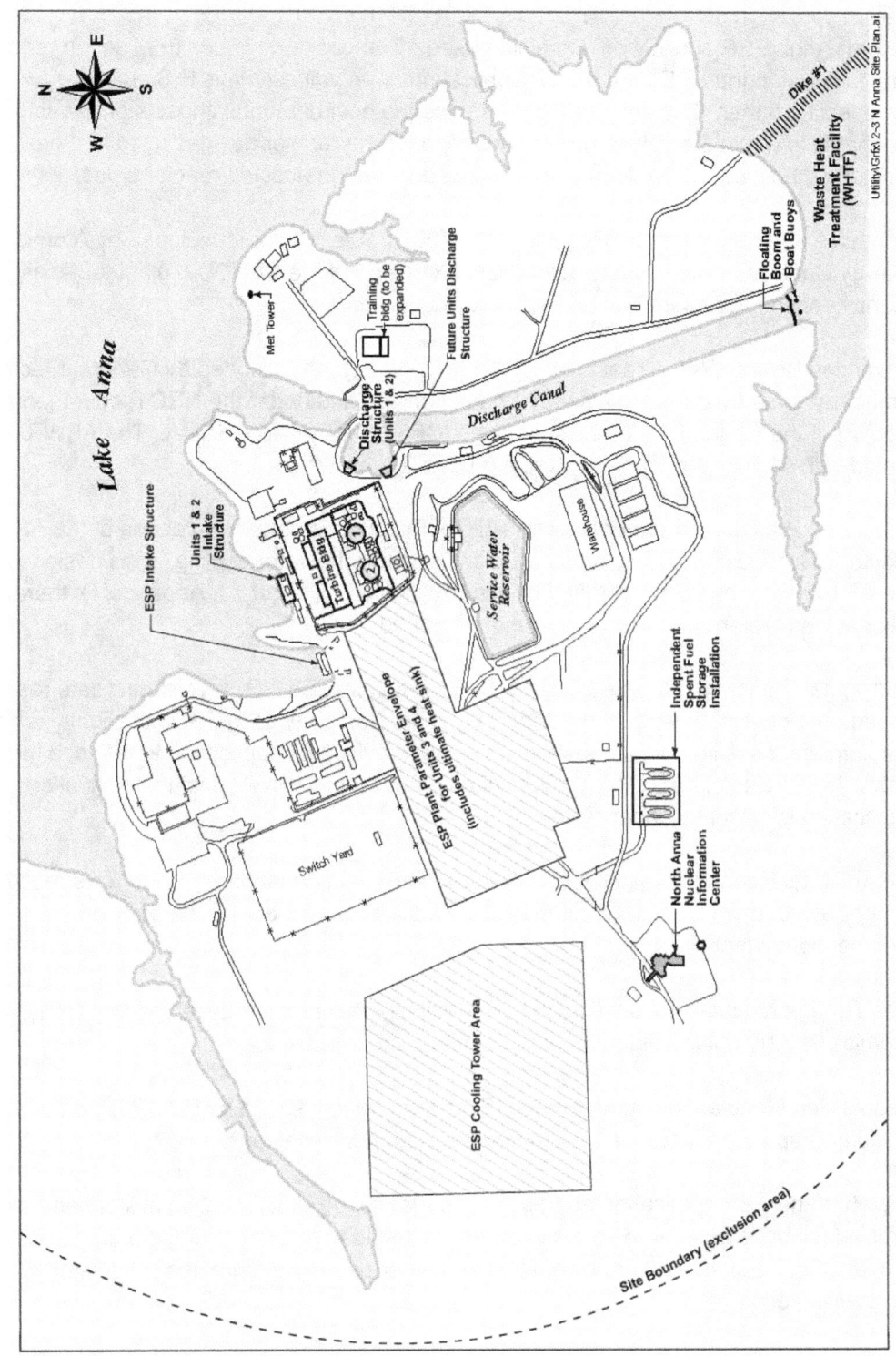

Figure 3-1. Proposed Major ESP Structures (Units 3 and 4) in Relation to Existing Units 1 and 2 Structures

1 In its North Anna ESP application, Dominion used a composite of values from seven reactor
2 designs to develop the PPE for the ESP application. The values in this EIS are not
3 design-specific; rather, they are used to determine the environmental impacts of a reactor
4 design that falls within the values used in this report. The reactor designs used to develop the
5 PPE include the following five light-water reactor and two gas-cooled reactor types:

6

7 • Canada Deuterium Uranium Reactor (ACR-700) – This reactor, developed by Atomic
8 Energy Canada Limited, is an evolutionary extension of the CANDU 6 plant that using very
9 slightly enriched uranium fuel and light-water cooling.

10

11 • Advanced Boiling Water Reactor (ABWR) – This reactor, developed by General Electric
12 Company, is a standardized plant that has been certified under the NRC requirements in
13 Title 10 of the Code of Federal Regulations (CFR) Part 52, Appendix A. The ABWR is
14 fueled with slightly enriched uranium and uses light-water cooling.

15

16 • Advanced Pressurized Water Reactor (AP1000) – This is an earlier version of the AP1000
17 reactor design, using slightly enriched uranium and light-water cooling. This design is not
18 the AP1000 that has been certified by the NRC in 10 CFR Part 52, Appendix D; therefore,
19 this design is referred to as the "surrogate AP1000."

20

21 • Surrogate Economic Simplified Boiling Water Reactor (ESBWR) – This surrogate reactor is
22 based on a design developed by General Electric Company, is fueled with slightly enriched
23 uranium and uses light-water cooling. Dominion revised its application to reflect a higher
24 power level value of 4500 MW(t) (Dominion 2006). The ESBWR design certification
25 application is currently under review by the NRC.

26

27 • International Reactor Innovative and Secure (IRIS) next generation PWR – This reactor is
28 under development by a consortium led by Westinghouse Electric Company and is a
29 modular light-water reactor.

30

31 • Gas Turbine Modular Helium Reactor (GT-MHR) – This reactor, developed by General
32 Atomics, is a modular helium-cooled, graphite-moderated reactor.

33

34 • Pebble Bed Modular Reactor (PBMR) – This reactor, developed by PBMR (Pty) Ltd., is a
35 modular graphite-moderated, helium-cooled gas turbine reactor.

36

37 Revision 6 of the ESP application addresses changes to the Unit 3 cooling system and the PPE
38 power level for Units 3 and 4. PPE values were adjusted to reflect the change to the cooling
39 system for Unit 3 and the higher power level and other parameter changes associated with the
40 power level increase.

1 Dominion would not be required to use any of these designs if it elects to proceed with a CP or
2 COL application; however, a CP or COL applicant referencing an ESP would have to address
3 whether the characteristics of the reactor ultimately selected fall within the values of the design
4 parameters specified in the ESP.
5
6 *Review Approach*
7
8 NUREG-1555, Vol. 1, *Environmental Standard Review Plan* (ESRP) (NRC 2000), and review
9 standard RS-002, *Processing Applications for Early Site Permits* (NRC 2004b), provide
10 guidance to the NRC staff to help ensure a thorough, consistent, and disciplined review of any
11 ESP application. The staff's June 23, 2003 response to comments received on draft RS-002
12 (NRC 2003) provide additional insights into the staff's approach to the review of an application
13 employing the PPE approach.
14
15 Because PPE values were used as a surrogate for design-specific values, the staff expected
16 Dominion to provide information sufficient for the staff to develop a reasonable independent
17 assessment of potential impacts to specific environmental resources. In some cases, the
18 design-specific information called for in the ESRP were not provided in the Dominion ESP
19 application because it did not exist or was not available. Therefore, the NRC staff could not
20 apply the ESRP guidance in those review areas. In such cases, the NRC staff used its
21 experience and judgment to adapt the review guidance in the ESRP and to develop
22 assumptions necessary to evaluate impacts to certain environmental resources to account for
23 this missing information. These assumptions are discussed in the appropriate sections of
24 this SDEIS.
25
26 Because the Dominion PPE values do not reflect a specific design, they were not reviewed by
27 the NRC staff for correctness. However, the NRC staff made a determination that the
28 application was sufficient to enable the staff to conduct its required environmental review and
29 that the PPE values are not unreasonable for consideration by the staff when making its finding
30 on the application in accordance with 10 CFR 52.18. During its environmental review, the staff
31 used its judgment to determine whether Dominion provided information sufficient for the staff to
32 perform its independent assessment of the environmental impacts of construction and operation
33 of a new nuclear unit or units. Dominion expects that the PPE values will bound the design
34 characteristics of a reactor or reactors that might be constructed at the North Anna ESP site.
35 At the COL stage, as required by 10 CFR 52.79, the applicant must, in addition to the
36 information and analysis otherwise required, submit information sufficient to demonstrate that
37 the design of the facility falls within the parameters specified in the ESP. If actual reactor
38 characteristics do not fall within the PPE values on which the staff based its estimate of the
39 potential environmental impacts resulting from constructing and operating one or more new
40 nuclear units at the ESP site, the staff will consider whether the difference between the actual
41 characteristics and the PPE value is significant.
42

1   Based on discussions with Dominion (NRC 2006), ER Revision 6 Table 3.1-1 provides
2   information from various reactor designs that were used to develop the bounding site-specific
3   PPE values contained in ER Table 3.1-9. So the values in ER Table 3.1-1 are generic values
4   and not site-specific values. Therefore, the site-specific values in ER Table 3.1-9 differ from the
5   generic values in ER Table 3.1-1. ER Tables 5.4-6 and 5.4-7 provide bounding PPE values for
6   the radionuclide activities. Therefore, the PPE values provided in ER tables 3.1-9, 5.4-6, and
7   5.4-7 are used in the staff's analysis unless specifically noted otherwise. The PPE values used
8   by the staff are provided in Appendix I of this SDEIS.
9
10 Throughout the North Anna ESP environmental report, Dominion (2006) provides:
11
12 (1)   Statements of plans to address certain issues in the design, construction, and operation of
13   the facility
14
15 (2)   Statements of planned compliance with current laws, regulations, and requirements
16
17 (3)   Statements of plans for future activities and actions that it will take should it decide to
18   apply for a CP or COL
19
20 (4)   Descriptions of Dominion's estimate of the environmental impacts resulting from the
21   construction and operation of a new nuclear unit or units on the North Anna ESP site
22
23 (5)   Descriptions of Dominion's estimates of future activities and actions of others and the
24   likely environmental impacts of those activities and actions that would be expected should
25   Dominion decide to apply for a CP or COL.
26
27 The activities described include, but are not limited to, such actions as:
28
29 • Considering the results of testing and monitoring during the development of a CP or
30   COL application
31
32 • Complying with NRC regulations and those of other agencies, including obtaining
33   appropriate permits from other agencies
34
35 • Taking actions to mitigate adverse environmental impacts (e.g., best management
36   practices)
37
38 • Addressing certain issues at the CP or COL stage that were not addressed in the ESP
39   application.
40

1   Some of these future actions are those that Dominion would be required to implement because
2   they are currently required by law, and others are actions that Dominion has indicated that it
3   would implement without the legal obligation to take such actions.
4
5   The staff performed its evaluation of the impacts of constructing and operating one or more new
6   nuclear units at the ESP site assuming that these activities and actions would be undertaken by
7   Dominion and others during future licensing activities. As discussed previously, the staff
8   developed assumptions necessary to evaluate impacts to certain environmental resources to
9   account for missing detailed information. In addition to other sources of information obtained
10 independently, the staff considered future activities and actions, estimates of expected
11 environmental impacts that were identified by Dominion in its ER, and the PPE values listed in
12 Appendix I when developing the inputs and assumptions used in the NRC staff's independent
13 review of the environmental impacts of constructing and operating one or more new units on the
14 North Anna ESP site. The staff has identified missing information with respect to particular
15 resources, the staff's assumptions in evaluating such resources, and any resulting limitations in
16 the staff's conclusions or the environmental impacts to particular resources, where appropriate.
17 In addition, as a result of the staff's environmental review of the Dominion ESP application, the
18 staff determined that conditions or limitations on the ESP may be necessary in specific areas, in
19 accordance with 10 CFR 52.24. Proposed permit conditions are set forth in individual EIS
20 sections for particular resources.
21
22 ## 3.2.1 Plant Water Use
23
24 *This section was changed to remove references to the once-through cooling system previously*
25 *described in the Draft EIS.*
26
27 This SDEIS assesses the impacts of plant water use based on the values of design parameters
28 provided by Dominion in ER Revision 6. At the ESP stage, the staff's review of the design
29 parameters is limited to an evaluation of whether the parameter values are not unreasonable.
30 At the COL stage, a COL applicant referencing the ESP is required to demonstrate that the
31 specific plant design would fall within the design parameters in the ESP. The following sections
32 describe both the consumptive and non-consumptive water uses of proposed Units 3 and 4 and
33 the associated plant water treatment systems.
34
35 ### 3.2.1.1 Plant Water Consumption
36
37 *This section was changed to reflect the water consumed by the closed-cycle, combination wet*
38 *and dry cooling system for Unit 3 as presented in Revision 6 of the ER.*
39
40 This section describes plant water consumption demands, excluding those demands that are
41 part of the normal and ultimate heat sink cooling system. Consumptive water demands

1   associated with the cooling systems are discussed in Section 3.2.2. Non-cooling system related
2   water demands are relatively small compared to the consumptive cooling demands of Unit 3.
3
4   Units 3 and 4 would have identical demands for potable water, demineralized water, and fire
5   protection water. In its ER Revision 6 (Table 3.3-1), Dominion states that the normal water
6   demands for these systems are 41.3 L/s (655 gpm) per unit (Dominion 2006). These demands
7   could increase up to a maximum of 210 L/s (3340 gpm) when the fire protection system is
8   operating at full capacity. Potable water would be provided from groundwater wells, whereas
9   the demineralized water and fire protection water would be supplied from Lake Anna.
10

11   **3.2.1.2 Plant Water Treatment**
12

13   *This section was not affected by changes presented in Revision 6 of the ER except to refer to*
14   *Unit 3 along with Unit 4 in relation to makeup water treatment.*
15

16   Because no specific design has been selected, the water treatment systems for the proposed
17   Units 3 and 4 are not specified. Currently, Lake Anna is the source for Units 1 and 2 condenser
18   cooling and service water. This water is not treated. Makeup water for the proposed Units 3
19   and 4 and both ultimate heat sink systems would require treatment with biocides, antiscalants,
20   and dispersants. Treatment of makeup water for ultra-pure water systems, such as the
21   condensate and primary cooling systems, would employ technologies such as reverse osmosis
22   and ultra-filtration. The water quality of effluents from any water treatment would be regulated
23   by a Virginia Pollutant Discharge Elimination System (VPDES) permit for the units.
24

25   **3.2.2 Cooling System**
26

27   *This section was changed to reflect the revision from a once-through system to a closed-cycle,*
28   *combination wet and dry cooling system for Unit 3 as presented in Revision 6 of the ER.*
29

30   The following sections provide detailed descriptions of the operational modes and the
31   components of the cooling water systems for the proposed Units 3 and 4. Non-cooling system
32   related water consumption, including potable, demineralized, and fire protection water demands
33   are discussed in Section 3.2.1.
34

35   The plant would primarily use wet towers to cool Unit 3 during periods of relative water surplus,
36   which are defined as periods when the water surface elevation of Lake Anna is at or above
37   elevation 76.2 m (250 ft) above mean sea level (MSL). In Revision 6 of the Dominion ER, this
38   cooling mode for Unit 3 is termed the Energy Conservation (EC) mode. During periods when
39   the elevation of Lake Anna is below 76.2 m (250 ft) MSL for a period of seven or more
40   consecutive days, excess heat generated by Unit 3 operation would be dissipated using a dry
41   cooling tower, assuming suitable atmospheric conditions exist at the site. If atmospheric
42   conditions were such that Unit 3 dry cooling towers could not completely cool the circulating

1   water, Dominion would employ wet towers to dissipate the remaining excess heat. The dry
2   cooling towers would be designed so that they would be capable of removing at least one-third
3   of the excess heat from Unit 3 under worst-case atmospheric conditions. Dominion terms this
4   cooling mode for Unit 3 as the Maximum Water Conservation (MWC) mode.
5
6   The two proposed units employ considerably different cooling systems with vastly different
7   water needs (Dominion 2006). The proposed Unit 3 would use a closed-cycle, combination wet
8   and dry cooling tower system. A conceptual diagram of this approach is illustrated in
9   Figure 3-2. Unit 4 would use a dry cooling system that transfers heat directly from the
10 condenser to an air cooled heat exchanger without the use of Lake Anna cooling water. There
11 was no change to the Unit 4 cooling system in Revision 6 of the ER, which was evaluated in the
12 Draft EIS. Therefore, the Unit 4 dry cooling system is not evaluated in this SDEIS, and
13 discussion of Unit 4 dry cooling is for context purposes only.
14
15 The heat from the turbine generator is transferred to the cooling water in the surface condenser.
16 The cooling water passes through the dry cooling tower and, in the MWC mode, transfers one
17 third of the heat to the atmosphere. Cooling water leaving the dry towers would then pass
18 through the wet towers to remove the balance of condenser/heat exchanger rejected heat by
19 spraying the water into a forced or induced air stream. After passing through the cooling
20 towers, the cooled water would be recirculated back to the surface condenser to complete
21 the closed-cycle cooling water loop. Make-up water to the circulating water system and
22 service water cooling system would be obtained from Lake Anna. Blowdown (recirculating
23 water removed from the cooling system to reduce the buildup of contaminants, such as
24 dissolved solid) from the cooling systems would be discharged to the existing plant WHTF
25 discharge canal.
26
27 **3.2.2.1 Description and Operational Modes**
28
29 *This section was changed to reflect the operating modes of the closed-cycle, combination wet*
30 *and dry cooling system for Unit 3 as presented in Revision 6 of the ER.*
31
32 The operating modes for the proposed Units 3 and 4 under normal operating and
33 emergency/shutdown conditions are described in the following paragraphs. In Revision 6 to the
34 ER, Dominion states that the minimum lake level for operation of the proposed units would be
35 an elevation of 73.8 m (242 ft) MSL.
36
37 **Unit 3 Normal Cooling**
38
39 Dominion states that the bounding thermal power generated by Unit 3 would be 4500 MW(t),
40 and that the bounding heat rejection rate to the environment would be 3020 MW
41 (1.03×10^{10} Btu/hr). Excess heat generated by the unit would be dissipated through the use of
42 a series of closed-cycle cooling towers that can operate in two modes: EC and MWC modes.

1

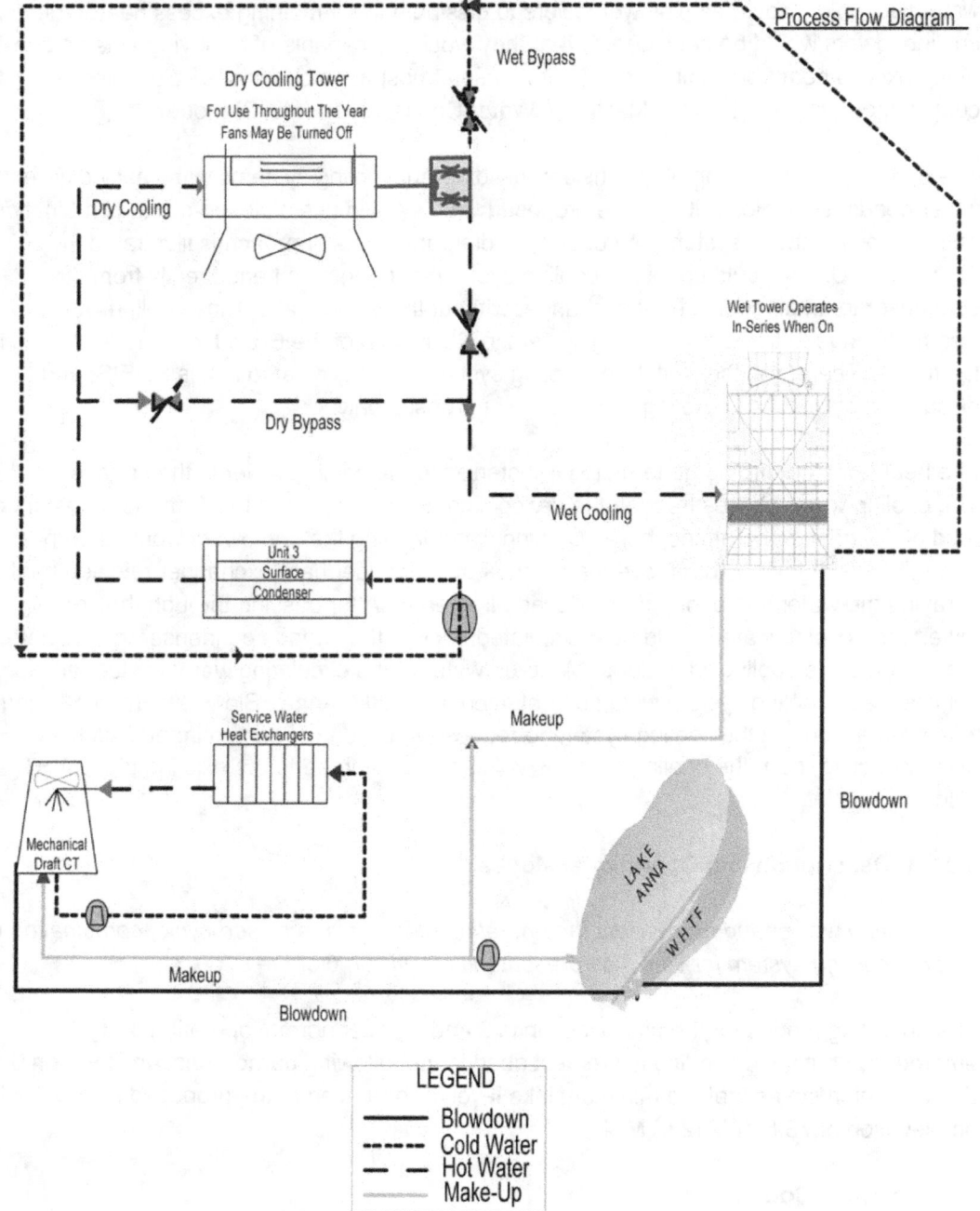

2
3 **Figure 3-2**. Conceptual Closed Loop Cooling Water Diagram
4
5

1   The EC mode of rejecting excess heat generated by Unit 3 would be employed when surplus
2   water is available from Lake Anna. Surplus water would be considered available when
3   (1) the lake level elevation of Lake Anna is at or above 76.2 m (250 ft) MSL or (2) the lake level
4   elevation has fallen below elevation 76.2 m (250 ft) MSL for a period of less than seven
5   consecutive days.
6
7   In the EC mode, excess heat generated by Unit 3 would be dissipated by closed-cycle wet
8   cooling towers. Makeup water would be supplied from Lake Anna at a maximum flow rate of
9   1405 L/s (22,268 gpm). The blowdown flow rate and the related evaporation rate associated
10 with the wet cooling towers would vary depending on thermal output from the unit and
11 environmental conditions. In its PPE, Dominion states that the maximum evaporation rate
12 would be 1053 L/s (16,695 gpm) and the maximum blowdown discharge would be 351 L/s
13 (5565 gpm) in the EC mode.
14
15 The MWC mode of rejecting excess heat generated by Unit 3 would be employed when water
16 levels in the lake drop below elevation 76.2 m (250 ft) MSL for a period of one week or more.
17 Under favorable meteorological conditions, the entire excess heat load from Unit 3 would be
18 dissipated using closed-cycle dry cooling towers. These towers would be sized so that under
19 the worst-case conditions (i.e., full power operation and a hot and humid atmosphere at tower
20 level), at least one-third of the maximum Unit 3 excess heat would be dissipated via the dry
21 tower system. The remaining excess heat would be dissipated by the wet tower system.
22 Therefore, although the MWC mode uses less water than the EC mode, it is possible that up to
23 two-thirds of the total heat load would be dissipated by wet cooling.
24
25 In the MWC mode, the maximum makeup flow rate from Lake Anna to the wet tower system
26 would be 971 L/s (15,384 gpm). The maximum blowdown discharge and evaporation rate from
27 the wet towers are 245 L/s (3844 gpm), and 728 L/s (11,532 gpm), respectively.
28
29 **Unit 4 Normal Cooling**
30
31 During normal operation, the proposed Unit 4 would use a system of closed-loop dry cooling
32 towers. The makeup water flow rate to the circulating water system would be negligible (on the
33 order of 0.06 L/s [1 gpm]). No blowdown would be generated by these towers.
34
35 **Ultimate Heat Sink**
36
37 For safety-related cooling, an ultimate heat sink (UHS) would be constructed to provide water
38 for reactor cooling and safety-related components of Units 3 and 4. The same UHS design
39 would be used for each unit. Each UHS would be composed of a mechanical draft cooling
40 tower with a 71.6 m wide by 107 m long by 15.2 m deep (235 ft wide by 350 ft long by
41 50 ft deep) engineered underground basin constructed beneath each tower (Dominion 2004b).
42 These basins would be large enough to store a water volume of 1.16×10^5 m^3 (3.06×10^7 gal),

1 which is adequate to hold a 30-day supply of emergency cooling water (Dominion 2006).
2 During periods when the ultimate heat sink cooling towers are in operation, the towers would
3 withdraw a maximum makeup flow of 110 L/s (1700 gpm) from each basin. The blowdown from
4 the UHS towers would be discharged into the WHTF.
5
6 During periods of normal plant operation, a negligible volume of makeup water would be used to
7 offset any water losses from the UHS basins. This water would originate from Lake Anna
8 (Dominion 2006).
9
10 **3.2.2.2 Component Descriptions**
11
12 *This section was changed to reflect the components of the closed-cycle, combination wet and*
13 *dry cooling system for Unit 3 as presented in Revision 6 of the ER.*
14
15 The following sections describe the intake, discharge, and heat dissipation systems for
16 proposed Units 3 and 4. Pursuant to Sections 316(a) and 316(b) of the Clean Water Act, an
17 applicant for a CP or COL referencing an ESP for the North Anna ESP site would be required to
18 obtain approval from the Commonwealth of Virginia by documenting plant design and
19 conducting site-specific analyses regarding the impacts of the thermal discharges and operation
20 of the intake systems on the Lake Anna aquatic environment.
21
22 **Intake System**
23
24 The proposed location of the intake structure for Unit 3 is shown in Figure 3-1. Any makeup
25 water required for Unit 4 could be obtained from the Unit 3 intakes. The location of the intake
26 would be in the same approximate location as the intakes planned for the two additional power
27 reactor units proposed at the time that NAPS Units 1 and 2 were licensed. The size of the
28 intake structure originally proposed for Units 3 and 4 (Dominion 2004a) was intended to support
29 once-through cooling of proposed Unit 3. In Revision 6 of the ER, Dominion reduced the size of
30 the proposed intake structure to support Unit 3 operation to 21 m (70 ft) long and 21 m (70 ft)
31 wide because of the reduced need for cooling water. The intake system proposed for ESP
32 Unit 3 would consist of a structure next to the lake with trash racks, traveling screens, and pump
33 bays, similar to the design currently in use by Units 1 and 2. As a result of the smaller intake
34 structure, Dominion expects no major modifications to the shoreline or the existing intake
35 channel. As previously proposed, the cofferdam would still be removed to allow water access
36 from Lake Anna.
37
38 **Discharge System**
39
40 Blowdown discharge from the wet towers associated with Unit 3 would enter the WHTF via the
41 discharge canal currently used by the existing units. The PPE maximum blowdown discharge
42 from Unit 3 would be 351 L/s (5565 gpm). There would be no blowdown discharge from Unit 4.

The discharge canal and WHTF canal system were designed to convey approximately 230,000 L/s (8000 cfs), and the maximum flow rate from the existing units is approximately 120,000 L/s (4300 cfs). The discharge canal and WHTF system could therefore easily accommodate the extra water discharged by the proposed units. The licensee may combine the blowdown flow from Unit 3 with the discharge from the existing NAPS units and use the current Unit 1 and 2 discharge structure, or construct a separate discharge structure in the vicinity of the partially completed discharge structure planned for the two additional power reactors proposed at the time NAPS Units 1 and 2 were licensed (see Figure 3-1).

Heat Dissipation Systems

The normal cooling needs of Unit 3 would be provided by a closed-cycle, combination wet and dry towers. The percentage of excess heat dissipated by the dry towers would depend on the availability of water from Lake Anna and ambient environmental conditions. If excess water were available, Unit 3 would be cooled entirely by use of the wet towers. Under times of relative drought and favorable meteorological conditions, the majority of the Unit 3 waste heat would be dissipated by the dry towers.

The normal cooling needs of Unit 4 would be provided solely by a closed-cycle dry tower system. Unit 4 would have a negligible consumptive water demand on Lake Anna.

Wet cooling tower systems rely primarily on evaporative heat transfer to the atmosphere to dissipate the rejected thermal load. Dry cooling tower systems rely entirely on sensible heat transfer between the fluid circulating in the condenser loop and the ambient air. Dry towers are completely closed systems and therefore use negligible amounts of makeup water and produce negligible blowdown water. Dry cooling towers use large fans to keep air flowing over their radiators, so there is an associated high energy cost that significantly reduces plant efficiency. The efficiency penalty of dry cooling towers can exceed 12 percent (EPA 2001). Dominion's combination wet and dry cooling system has an energy efficiency penalty of 1.7 to 4 percent (Dominion 2006).

For safety-related cooling, the UHS for each of the proposed Units 3 and 4 would provide water to the reactor cooling systems and safety-related components. As proposed, both plants would use the same UHS design, which would be composed of a mechanical draft cooling tower with an engineered basin constructed underground beneath it (Dominion 2006). The basin would have a storage capacity adequate to hold a 30-day supply of emergency cooling water.

3.2.3 Radioactive Waste Management System

This section is not affected by the changes presented in Revision 6 of the ER with the exception that it now reflects the revised plant parameter envelope value of 4500 MW(t) for the core thermal power level for one unit (9000 MW(t) for two units). This section is provided solely for context.

Liquid, gaseous, and solid radioactive waste management systems would be used to collect and treat the radioactive materials that are produced as a by-product of operating the proposed Units 3 and 4 on the North Anna ESP site. These systems would process radioactive liquid, gaseous, and solid effluents to maintain releases within regulatory limits and to levels as low as reasonably achievable (ALARA) before being released to the environment.

Dominion did not identify specific radioactive waste management systems for the North Anna ESP site. The PPE concept was used to provide an upper bound on liquid radioactive effluents, gaseous radioactive effluents, and solid radioactive waste releases (Dominion 2006) (See Appendix I).

Adequate design information to estimate liquid and gaseous radioactive effluents was available for four of the seven reactor designs considered in establishing PPE values. The four reactors were LWRs and included the certified ABWR, the surrogate AP1000 PWR, the ACR-700 light-water-cooled, heavy-water moderated reactor, and the surrogate ESBWR. If a different reactor design is selected at the COL stage, the applicant would have to demonstrate that the design is bounded by the PPE values.

3.2.4 Nonradioactive Waste Systems

This section is not affected by the changes presented in Revision 6 of the ER. This section is provided solely for context.

Dominion describes the nonradioactive waste systems for the proposed Units 3 and 4 in Section 3.6 of its ER (Dominion 2006). The description of the systems was not affected by Revision 6 to the ER.

3.3 Power Transmission System

This section is not affected by the changes presented in Revision 6 of the ER with the exception that it now reflects the revised plant parameter envelope value of 4500 MW(t) for the core thermal power level for one unit (9000 MW(t) for two units). This section is provided solely for context.

In its ER, Dominion indicates the existing transmission system (three 500-kV lines and one 230-kV line) has the capacity to handle the output from the existing Units 1 and 2 plus the anticipated output from the proposed Units 3 and 4 (Dominion 2006). No additional construction of transmission lines would be expected for Units 3 and 4.

3.4 References

10 CFR Part 52. Code of Federal Regulations, Title 10, *Energy*, Part 52, "Early Site Permits, Standard Design Certifications, and Combined Licenses for Nuclear Power Plants."

Clean Water Act (also referred to as the Federal Water Pollution Control Act). 33 USC 1251, et seq.

Dominion Nuclear North Anna, LLC (Dominion). 2004a. *North Anna Early Site Permit Application – Part 3 – Environmental Report*. Revision 3, Glen Allen, Virginia, July 25, 2005.

Dominion Nuclear North Anna, LLC (Dominion). 2004b. Letter Response to Request for Additional Information Regarding Safety Portion of ESP Application, No. 04-318, Glen Allen, Virginia, August 2, 2004.

Dominion Nuclear North Anna, LLC (Dominion). 2006. *North Anna Early Site Permit Applications – Part 3 – Environmental Report*. Revision 6, Glen Allen, Virginia.

U.S. Environmental Protection Agency (EPA). 2001. *Technical Development Document for Regulations Addressing Cooling Water Intake Structures for New Facilities*. EPA-821-R-01-036, Washington, D.C.

U.S. Nuclear Regulatory Commission (NRC). 2000. *Standard Review Plans for Environmental Reviews for Nuclear Power Plants*. NUREG-1555, Vol. 1, Office of Nuclear Reactor Regulation, Washington, D.C. Available at http://www.nrc.gov/reading-rm/doc-collections/nuregs/staff/sr1555/.

Site Layout and Plant Parameter Envelope

1   U.S. Nuclear Regulatory Commission (NRC). 2003. *Response to Comments on Draft RS-002,*
2   *Processing Applications for Early Site Permits*. Available at
3   http://www.nrc.gov/reading-rm/adams.html, Accession No. ML031710698.
4
5   U.S. Nuclear Regulatory Commission (NRC). 2004a. *Draft Environmental Impact Statement*
6   *(EIS) for an Early Site Permit (ESP) for the North Anna ESP Site*. NUREG-1811 Draft, Office of
7   Nuclear Reactor Regulation, Division of Regulatory Improvement Programs, Washington, D.C.
8
9   U.S. Nuclear Regulatory Commission (NRC). 2004b. *Processing Applications for Early Site*
10 *Permits*. RS-002, Washington, D.C. Available at
11 http://www.nrc.gov/reading-rm/adams.html, Accession No. ML040700236.
12
13 U.S. Nuclear Regulatory Commission (NRC). 2006. *June 7, 2006, Summary of*
14 *Teleconferences with Dominion on June 2 and June 5, 2006, Regarding the North Anna Early*
15 *Site Permit Application*. Available at http://www.nrc.gov/reading-rm/adams.html, Accession
16 No. ML061580174.
17
18

4.0 Construction Impacts at the Proposed Site

This chapter of the Supplement to the Draft Environmental Impact Statement was changed to incorporate additional information necessary to support the analysis of construction impacts related to changes proposed in Revision 6 of the Environmental Report. Section summaries are provided for context.

This chapter examines the environmental impacts of construction associated with potential site preparation activities and construction of the proposed North Anna Power Station (NAPS) Units 3 and 4 as described in the application for an early site permit (ESP) submitted by Dominion Nuclear North Anna, LLC (Dominion). As part of its application, Dominion submitted an Environmental Report (ER) and a site redress plan (Dominion 2006). The ER provides the plant parameter envelope (PPE) as the basis for the environmental review. The parameters included in the PPE and their values are listed in Appendix I. The site redress plan allows for specific site preparation activities to be conducted with approval of an ESP. These activities evaluated are those permitted by Title 10 of the Code of Federal Regulations (CFR) 52.25(a) and 10 CFR 50.10(e)(1). In the event the ESP application is approved and Dominion conducts site preparation activities but does not build the plant, Dominion would be required to implement its site redress plan.

The approach Dominion used to identify the environmental impacts of construction in Revision 3 of its application (Dominion 2004) is described in the *Draft Environmental Impact Statement (EIS) for an Early Site Permit (ESP) at the North Anna ESP Site* (Draft EIS) published in November 2004 (NRC 2004a). The changes described in Revision 6 to the application relate to two specific issues and their ramifications for other aspects of the environmental analysis. These two issues are (1) a change in the cooling system proposed for Unit 3 to a closed-cycle, combination wet and dry cooling system and (2) an increase in the thermal power level in the PPE for proposed Units 3 and 4 (hereafter referred to as Units 3 and 4). The change in the cooling system from a once-through system to a closed-cycle system is relevant to this chapter. Unlike the once-through cooling system for Unit 3 previously evaluated in the Draft EIS and which only involves intake and discharge structures, the combination wet and dry cooling system would involve the construction of an intake structure and cooling towers. The intake structure for the new system would be smaller than that of the once-through cooling system. The size of the construction workforce and construction duration to build the proposed cooling towers falls within the parameters analyzed in the Draft EIS (NRC 2004a). The change in the maximum power level does not affect this chapter.

This chapter is divided into 13 sections. Sections 4.1 through 4.9 discuss the changes in the construction impacts, if any, resulting from differences between the information in ER Revision 3 and ER Revision 6. For those sections in which there is no change in analysis as a result of ER Revision 6, brief summaries of the topics and the impact levels determined in the Draft EIS are provided solely for context. In accordance with 10 CFR Part 51, impacts have been evaluated

1   and conclusions have been made; in addition, an impact category level of potential adverse
2   impacts (i.e., SMALL, MODERATE, or LARGE) has been assigned to each resource area.
3   Negligible impacts are listed as SMALL impacts. Possible mitigation of adverse impacts, where
4   appropriate, is presented in Section 4.10, followed by a description of the site redress plan in
5   Section 4.11. A summary of construction impacts is presented in Section 4.12. Full citations for
6   the references cited in this chapter are listed in Section 4.13. Cumulative impacts of
7   construction and operation are discussed in Chapter 7. The technical analyses in this chapter
8   support the results, conclusions, and recommendations in Chapters 9 and 10.
9
10   The staff relied on the mitigation measures described in the ER and in Section 4.10 of the Draft
11   EIS in reaching its conclusions on the significance of the adverse impacts. The staff relied on
12   the infrastructure upgrades planned by the counties, cities, and towns, such as road and school
13   expansions, in evaluating the significance of the impacts. Failure to implement such
14   infrastructure upgrades could result in a larger impact in the affected resource areas.
15
16   Changes in this chapter are limited to impacts related to the construction of the proposed
17   combination wet and dry cooling system for Unit 3.
18

19   # 4.1 Land-Use Impacts
20
21   *This section is not affected by the changes presented in ER Revision 6 and is provided solely*
22   *for context.*
23
24   This section provides information regarding land-use impacts associated with site preparation
25   activities and construction of the proposed closed-cycle cooling system for Unit 3 at the North
26   Anna ESP site. This section and its subsections are not affected by the changes presented in
27   ER Revision 6.
28

29   ### 4.1.1 The Site and Vicinity
30
31   *This section is not affected by the changes presented in ER Revision 6 and is provided solely*
32   *for context.*
33
34   The ESP site is located entirely within the existing NAPS site, which is zoned for industrial use
35   by Louisa County. All construction activities for proposed Units 3 and 4, including
36   ground-disturbing activities, would occur within the existing NAPS site boundary. According to
37   Dominion (2006), approximately 52 ha (128 ac) would be affected on a long-term basis as a
38   result of permanent facilities. An additional 27.5 ha (67.9 ac) would be disturbed on a
39   short-term basis as a result of temporary activities and construction of temporary facilities and
40   laydown areas. Dominion states that it would conduct any ground-disturbing activities in

1 accordance with Federal, State and local regulatory requirements (Dominion 2006). The
2 planned power block area is relatively level. Undulating surfaces in the area of the planned
3 cooling towers would be leveled to accommodate the towers and would be contained within the
4 land area previously evaluated before the change to the Unit 3 cooling system design.
5 Dominion has submitted a site redress plan, which was evaluated in Section 4.11 of
6 the Draft EIS (NRC 2004a).
7
8 Based on the counties' comprehensive land-use plans for the surrounding vicinity, the site
9 redress plan, and its independent review, the staff concludes that the land-use impacts of
10 construction did not change. The impact level category would still be SMALL, and mitigation is
11 not warranted.
12
13 ### 4.1.2 Transmission Line Rights-of-Way and Offsite Areas
14
15 *This section is not affected by the changes presented in ER Revision 6 and is provided solely*
16 *for context.*
17
18 In the Draft EIS (NRC 2004a), the staff determined that, based on the evaluation provided by
19 Dominion, no additional electrical transmission lines or rights-of-way would be needed to
20 transmit the power generated by the proposed North Anna Units 3 and 4 to the regional power
21 grid (Dominion 2006). Even with the change in cooling system for Unit 3 and the higher power
22 level, construction would be limited to onsite work, and no additional land beyond that
23 previously evaluated would be needed to connect the new units to the grid.
24
25 Because construction would be limited to onsite work and no additional land would be needed
26 to connect the new units to the grid, the staff concludes that land-use impacts resulting from
27 construction in transmission line rights-of-way did not change. The impact level category would
28 still be SMALL, and mitigation is not warranted.
29
30 ## 4.2 Meteorological and Air Quality Impacts
31
32 *This section is not affected by the changes presented in ER Revision 6 and is provided solely*
33 *for context.*
34
35 In the Draft EIS (NRC 2004a), the staff determined that some minor air quality impacts would be
36 expected to occur during construction at the North Anna ESP site. The likely sources of air
37 quality impacts would be fugitive dust emissions from general construction activities and the
38 potential for elevated ambient air quality levels caused by emissions from the vehicles used by
39 the workforce and from construction equipment.
40

4.2.1 Construction Activities

This section is not affected by the changes presented in ER Revision 6 and is provided solely for context.

In the Draft EIS (NRC 2004a), the staff determined that all construction activities would be conducted in accordance with Virginia Administrative Codes 9 VAC 5-50 (Visible and Fugitive Dust Emissions) and 9 VAC 5-40-5680 (Emission Standard for Mobile Sources – Vehicles). Even with the change in cooling system for Unit 3, the type of general construction activities is not expected to be different from that previously evaluated.

Based on its independent evaluation of the requirements set forth in Virginia Administrative Codes and measures of dust control plans identified in the ER, the staff concludes that air quality impacts from construction, both onsite and beyond the plant boundary, would be temporary and did not change. The impact level category would still be SMALL, and further mitigation beyond the actions stated above is not warranted.

4.2.2 Transportation

This section is not affected by the changes presented in ER Revision 6 and is provided solely for context.

In the Draft EIS (NRC 2004a), the staff determined that during construction, workers traveling to and from the site could impact the local ambient air quality levels because of emissions from vehicles both during normal operation and during periods of traffic congestion when vehicles are stopped with their engines idling. Dominion indicated that it would develop and implement a construction traffic management plan to mitigate the impact of vehicular traffic on air quality (Dominion 2006). Even with the change in cooling system for Unit 3, the size of the construction workforce is not expected to be different from that previously evaluated.

Based on the mitigation identified by Dominion in its ER to develop a traffic management plan and its own independent review, the staff concludes that the impacts on the local air quality from the increase in vehicular traffic related to construction activities would be temporary and did not change. The impact level category would still be SMALL, and further mitigation beyond the actions stated above is not warranted.

4.3 Water-Related Impacts

Changes to this section reflect construction impacts on water resources from the closed-cycle, combination wet and dry cooling system.

Revision 6 to Dominion's ER proposes the use of a combination wet and dry cooling tower system to dissipate heat from Unit 3. The proposed cooling for Unit 4 remains a dry cooling tower system. For the proposed Unit 3, the makeup water flowrate from Lake Anna for the cooling tower would be considerably less than the estimated flowrate for the initially proposed once-through design. The potential impacts on water resources expected to result from constructing proposed Unit 3 are primarily from construction of the intake structures. No intake structure is required for Unit 4.

Water-related impacts involved in the construction of a nuclear power plant are similar to impacts that would be associated with any large industrial construction project. Prior to initiating construction, including any site preparation work, Dominion would be required to obtain the appropriate permits regulating alterations to the hydrological environment. These permits would likely include:

- Clean Water Act Section 404 permit. This permit would be issued by the U.S. Army Corps of Engineers (ACE), which governs impacts of construction activities on wetlands and management of dredged material.

- Clean Water Act Section 401 certification. This certification would be issued by the Commonwealth of Virginia and would ensure that the project does not conflict with water quality management programs in the Commonwealth.

- Clean Water Act Section 402(p) National Pollutant Discharge Elimination System (NPDES) construction and industrial storm water permit. This permit would regulate point source storm water discharges. The U.S. Environmental Protection Agency's (EPA) 1990 Phase 1 Storm Water regulation (40 CFR 122.26) established requirements for storm water discharges from various activities including construction activities disturbing an area of at least 2.0 ha (5.0 ac). EPA has delegated the authority for administering the National Pollutant Discharge Elimination System (NPDES) program to the Commonwealth of Virginia.

- Coastal Zone Management Act (CZMA) Section 307 Consistency Determination (and 15 CFR Part 930). Section 307(c)(3)(A) of the Coastal Zone Management Act [16 USC 1456(c)(3)(A)] requires that applicants for Federal licenses to conduct an activity in a coastal zone are to provide to the licensing agency a certification that the proposed activity complies with the enforceable policies of the State's coastal zone

1　program. While the National Oceanic and Atmospheric Administration administers the
2　CZMA, the authority to concur in or object to the consistency determination has been
3　delegated to the Virginia Department of Environmental Quality (VDEQ).
4

5　### 4.3.1　Hydrological Alterations
6

7　*Changes to this section reflect construction impacts on water resources from the closed-cycle,*
8　*combination wet and dry cooling system intake structure and intake channel.*
9

10　Unlike Dominion's initial proposal for once-through cooling for Unit 3 considered in the
11　Draft EIS, the current proposal to use a closed-cycle, combination wet and dry cooling system
12　for Unit 3 would lead to significantly less alteration of the hydrological regime. To support the
13　once-through cooling proposed for Unit 3, Dominion in Revision 3 of the ER (Dominion 2004)
14　proposed a cooling water intake structure that was approximately 46 m (150 ft) long and 61 m
15　(200 ft) wide and would have housed the trash racks, traveling screens, and intake pumps. The
16　intake channel would have extended from the intake structure toward the west slope of the
17　intake cove, and construction would have resulted in the removal or reshaping of the shoreline
18　to accommodate the intake structure and to reduce the intake approach velocity.
19

20　Dominion proposes a 21 m (70 ft) long and 21 m (70 ft) wide intake structure to support the
21　combination wet and dry cooling tower. Dominion expects no modifications to the shoreline or
22　the existing intake channel. As previously proposed, the existing cofferdam would still be
23　removed to allow water access from Lake Anna. Implementing best management practices
24　(BMPs) for dredging would minimize the sediment that would enter the lake during removal of
25　the cofferdam. Any impacts of dredging would be localized and temporary. Before initiation any
26　shoreline modification or dredging activities, Dominion would be required to obtain a 404 permit
27　from the U.S. Army Corps of Engineers (ACE).
28

29　Because the impacts of hydrological alterations resulting from construction activities would be
30　localized and temporary, the staff concludes that the impacts of hydrologic alterations would be
31　SMALL, and mitigation is not warranted.
32

33　### 4.3.2　Water-Use Impacts
34

35　*This section is not affected by the changes presented in ER Revision 6 and is provided solely*
36　*for context.*
37

38　In the Draft EIS (NRC 2004a), the staff determined that the quantity of water used for
39　construction activities at the ESP site would be similar to other large industrial construction
40　projects. Potable water supplies for the construction workforce would be necessary. Water for
41　various standard construction activities, such as dust abatement, would be provided from Lake

1   Anna. Groundwater dewatering systems may preclude the use of existing onsite wells to supply

2   water during construction, particularly potable water needs. If additional water is needed, water

3   could be imported from offsite during periods when the dewatering system is active.

4

5   Based on these considerations and because they would be localized and temporary, the staff

6   concludes that water-use impacts during construction did not change. The impact level

7   category would still be SMALL, and further mitigation beyond the actions stated above is not

8   warranted.

9

10   ### 4.3.3 Water-Quality Impacts

11

12   *This section is not affected by the changes presented in ER Revision 6 and is provided solely*

13   *for context.*

14

15   In the Draft EIS (NRC 2004a), the staff determined that water-quality impacts for the

16   construction activities would be similar to those associated with other large industrial

17   construction projects. Construction best management practices (BMP)[a] are generally used to

18   ensure that accidental spills and storm water runoff will have minimal impact on surface and

19   groundwater quality. If Dominion were to apply for and receive a construction permit (CP) or a

20   combined license (COL) referencing an ESP for the North Anna site, or if it were to conduct site

21   preparation activities under such an ESP, an NPDES permit would be required from the

22   Commonwealth of Virginia before construction activities could commence. In view of the ability

23   of the current standard engineering construction practices to limit water quality impacts and the

24   localized and temporary nature of any impacts, the staff concludes that water-quality impacts

25   caused by construction activities did not change. The impact level category would still be

26   SMALL, and that further mitigation beyond the actions stated above is not warranted.

27

28   ## 4.4 Ecological Impacts

29

30   *Changes to this section reflect construction impacts to ecological resources from the*

31   *closed-cycle, combination wet and dry cooling system and intake structure and channel.*

32

33   This section describes the potential impacts of construction of the closed-cycle, combination

34   wet and dry cooling system for Unit 3 on the ecological resources at the North Anna ESP site.

35   The section is divided into three subsections: Terrestrial Ecosystem Impacts, Aquatic

---

(a) Best management practices are recommended site management, maintenance, or monitoring activities that have been shown to work effectively to mitigate impacts. Government agencies sometimes use BMPs to specify standards of practice where a regulation may not be sufficiently descriptive.

1   Ecosystem Impacts, and Threatened and Endangered Species. Although the amount of land
2   use needed for the cooling towers would be greater than the amount of land needed for
3   once-through cooling, the amount of land identified in the PPE already incorporated the area in
4   which the cooling towers would be built.
5
6   ## 4.4.1 Terrestrial Ecosystem Impacts
7
8   *Changes to this section reflect the terrestrial impacts from the construction of the closed-cycle,*
9   *combination wet and dry cooling system.*
10
11 The total area of the North Anna ESP site is approximately 81 ha (200 ac) of which
12 approximately 49 ha (120 ac) have been developed for industrial use. No additions to the area
13 of construction (the plant footprint) in the PPE were made as a result of the change proposed
14 for Unit 3 cooling. Construction activities are not expected to have noticeable impacts on
15 ecological resources within the developed portions of the ESP site. Construction of Units 3
16 and 4 would result in the removal of approximately 32 ha (80 ac) of forested habitat within the
17 site. The North Anna ESP site does not contain any old growth timber or unique or sensitive
18 plant species or communities. In the Draft EIS (NRC 2004a), the staff determined that
19 construction activities would not noticeably reduce the local or regional diversity of plants or
20 plant communities.
21
22 There are no important terrestrial animal species or habitats (as previously evaluated by the
23 NRC [NRC 2000]) on the North Anna ESP site. A few small wetland areas and two intermittent
24 streams exist on the North Anna ESP site (Dominion 2006). Dominion has mapped these
25 wetlands, and has provided the information to the ACE for review, confirmation, and evaluation
26 of appropriate permitting. Mitigation measures will be developed as appropriate during the ACE
27 review. Watercourses and wetlands would be avoided to the extent practicable during any
28 construction. To minimize construction-related impacts to wildlife, Dominion states that it would
29 adhere to Commonwealth of Virginia permit conditions which could restrict the timing of certain
30 construction activities (Dominion 2006).
31
32 In anticipation of construction, topsoil would be removed from the construction site footprint,
33 stored, rolled, and seeded, if necessary, to minimize erosion. Some disturbed areas may be
34 graveled, paved, or compacted to prevent erosion. These and other soil preparation activities
35 would minimize impacts to the aquatic environment from earth-moving activities. When
36 construction activities are completed, areas that have been temporarily disturbed would be
37 graded and contoured, covered with topsoil, and seeded with native vegetation
38 (Dominion 2006).
39
40 Land clearing associated with construction would be conducted according to Federal and State
41 regulations, permit conditions, existing procedures, and construction and other established

1   BMPs (e.g., directed drainage ditches and silt fencing would be employed). Fugitive dust
2   emissions would be minimized by watering the access roads and construction site as
3   necessary. Therefore, impacts from dust on terrestrial ecosystems would be minimal.
4
5   The use of the combination wet and dry cooling systems introduces additional structures and,
6   therefore, the potential for avian collisions. Collisions with utility structures are not a biologically
7   significant source of mortality for thriving populations of birds with good reproductive potential
8   (EPRI 1993). The staff previously reviewed monitoring data concerning avian collisions at
9   nuclear power plants with large cooling towers and determined that the overall avian mortality is
10 low (NRC 1996). No avian collisions with existing structures at the NAPS site have been
11 reported (Dominion 2006). The number of construction-related bird collisions with onsite
12 structures is expected to be inconsequential.
13
14 The staff reviewed the potential impacts of constructing Units 3 and 4 on terrestrial ecological
15 resources, including loss of habitat, loss of wetlands, noise, dust emissions, and avian
16 collisions. Based on its independent review and the mitigation measures identified, the staff
17 concludes that the impacts of construction-related activities on terrestrial ecological resources
18 did not change. The impact level category would still be SMALL, and further mitigation beyond
19 the actions stated above is not warranted.
20
21 ## 4.4.2 Aquatic Ecosystem Impacts
22
23 *Changes to this section reflect aquatic impacts from the construction of the closed-cycle,*
24 *combination wet and dry cooling system intake structure and the intake channel.*
25
26 Construction of the new cooling water intake structure and channel for Units 3 and 4 would be
27 the primary source of construction impacts on the aquatic environment. Construction would
28 involve modifications to an existing partially completed intake structure constructed in the 1970s
29 for two power reactor units that were proposed at the time the existing NAPS Units 1 and 2 were
30 licensed. Section 3.2.2 provides a description of the proposed plant cooling water use and
31 structures including a flow diagram in Figure 3-2.
32
33 The cooling water intake structure proposed in ER Revision 3 (Dominion 2004) to support
34 once-through cooling of Unit 3 was approximately 46 m (150 ft) long and 61 m (200 ft) wide and
35 would have housed the trash racks, traveling screens, and intake pumps. In ER Revision 6
36 (Dominion 2006), Dominion proposes to reduce the size of the intake structure to 21 m (70 ft)
37 long and 21 m (70 ft) wide because of a reduced demand for water. The screen and pump
38 layout are illustrated in Figures 4-1 and 4-2. Because the structure would be smaller than
39 originally proposed, Dominion expects no modifications to the shoreline or short intake channel.
40 As previously proposed, the cofferdam would still be removed to allow water access from
41 Lake Anna.

1
2

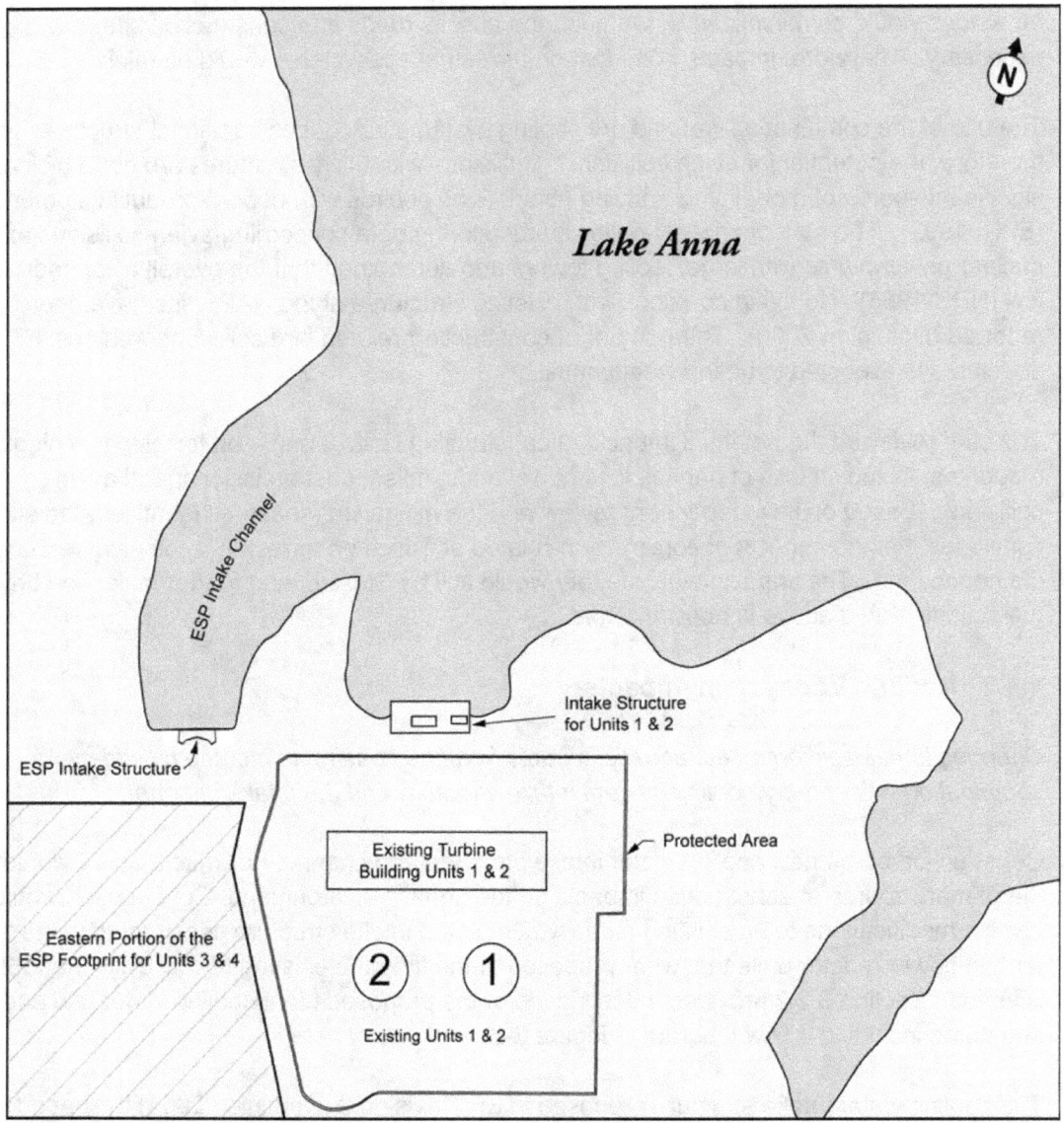

3
4

Figure 4-1. Proposed Layout of Screenwell/Pump Intake for the North Anna ESP Site

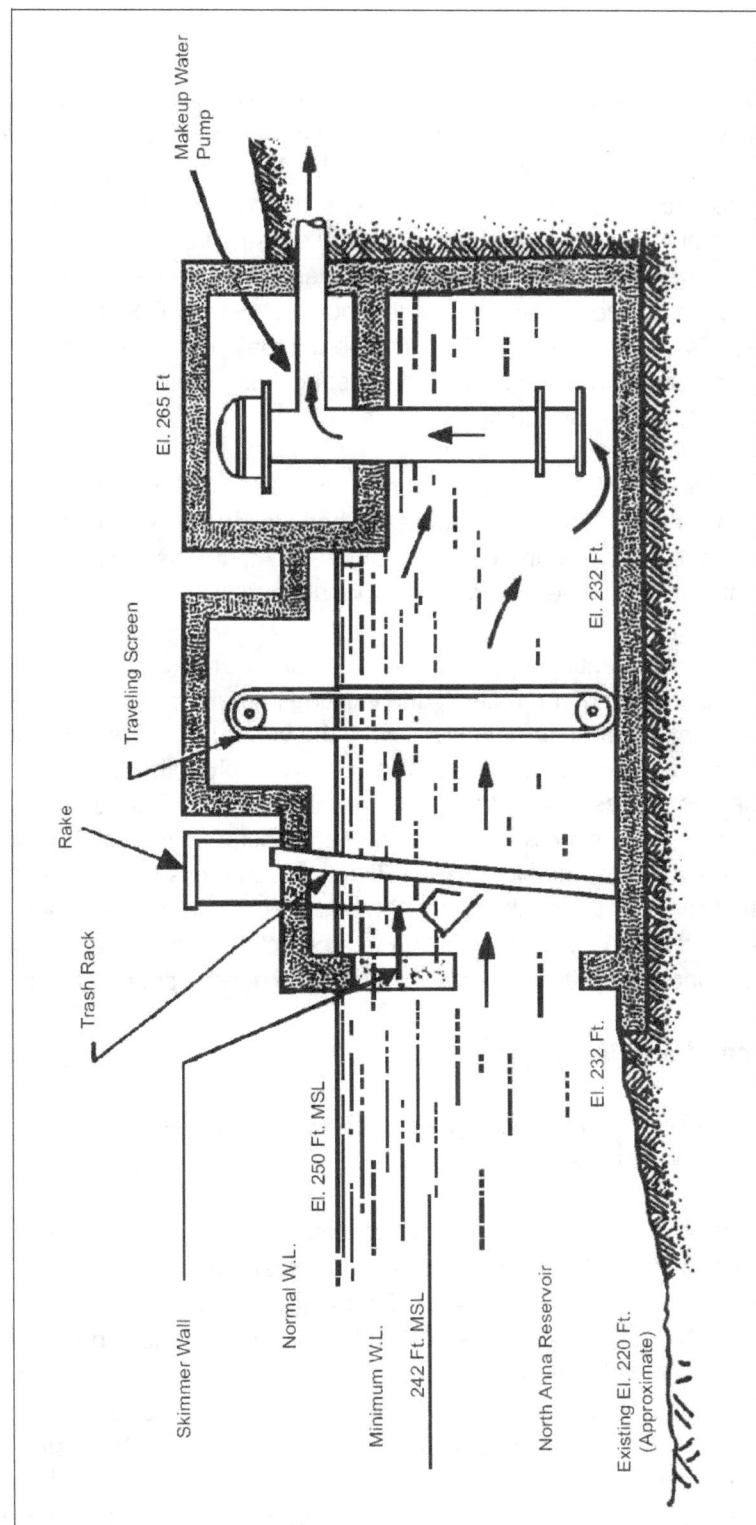

Figure 4-2. Schematic View of Pump Intake for the North Anna ESP Site

1   A temporary loss of benthic habitat and the displacement or loss of benthic organisms would be
2   expected as a result of construction activities (Dominion 2006). Fish and benthic organisms
3   inhabiting the intake channel and the lake near the intake channel may temporarily migrate from
4   the area during cofferdam removal. To minimize the impacts to benthic and fish populations in
5   Lake Anna, Dominion states it would conduct facility construction and environmental protection
6   activities in accordance with State regulations and permit requirements. Prior to any in-water
7   activities associated with the construction of the intake structure, Dominion would be required to
8   obtain a Section 404 permit from the ACE. The permit may place restrictions on any activities
9   conducted in Lake Anna during the proposed construction. These restrictions would further
10 lessen any impact to benthic or aquatic communities.

12 As a matter of practice, VDEQ would likely require that sedimentation and erosion-control BMPs
13 or effective stormwater management practices or both would be used to maintain water quality
14 and protect aquatic resources in the construction area. After construction is completed, benthic
15 and aquatic organisms would be expected to repopulate the area.

17 The staff assessed the potential impacts of construction of Units 3 and 4 on aquatic ecological
18 resources including removing or modifying the existing partially competed intake structure,
19 constructing a new intake structure, and removing the cofferdam. The applicant is expected to
20 follow sedimentation and erosion control BMPs and to comply with the VDEQ stormwater
21 management plan as well as any restrictions or requirements contained in the ACE Section 404
22 permit. No planned construction activities would be expected to impact the fisheries or any of
23 the biological communities of the North Anna River. Any impacts to the aquatic resources in
24 Lake Anna in the vicinity of the intake channel would be minor and temporary. Accordingly, the
25 staff concludes that the impact of construction-related activities on aquatic ecological resources
26 would be SMALL, and that further mitigation beyond the actions stated above is not warranted.

28 ## 4.4.3 Threatened and Endangered Species

30 *This section is not affected by the changes presented in ER Revision 6. However, additional*
31 *information is provided about eagles.*

33 No Federally listed threatened or endangered species are known to occur at or near the North
34 Anna ESP site except the bald eagle (*Haliaeetus leucocephalus*). The closest known bald
35 eagle nesting site is located more than 4 km (2.5 mi) from the North Anna ESP site. In the
36 Commonwealth of Virginia, a 0.25-mile (0.4-km) buffer zone is usually preserved to limit
37 construction activities (FWS and VDGIF 2000). Dominion follows these bald eagle nesting
38 guidelines. None of the three Federally or State-listed mussel species known to exist in the
39 region has been found in Lake Anna, the North Anna River, or other local streams. The staff
40 reviewed the potential impacts of construction of Units 3 and 4 on threatened and endangered
41 species. It is unlikely that any threatened or endangered species exist on the North Anna ESP

1 site; consequently, construction activities associated with the cooling system changes or the
2 increased power would not have an adverse effect on threatened or endangered species. The
3 staff concludes that the impacts of construction-related activities on threatened or endangered
4 species would be SMALL, and further mitigation beyond maintaining an adequate buffer zone is
5 not warranted.
6
7 4.5 Socioeconomic Impacts
8
9 *This section is not affected by the changes presented in ER Revision 6. The total number of*
10 *construction workers is not affected by changes to the cooling system or the power level*
11 *increase.*
12
13 This section evaluates the social and economic impacts to the surrounding region as a result of
14 constructing Units 3 and 4 at the North Anna ESP site. The evaluation assesses impacts of
15 construction and demands placed by the larger workforce on the surrounding region.
16 Construction activities are assumed to last up to 5 years and employ up to 5000 workers. The
17 evaluation also assesses the visual impacts of constructing the new cooling system design.
18
19 Dominion expects the workforce to be maintained for most of the construction period. This
20 construction workforce would be in addition to the 720 personnel currently employed at the site
21 (Dominion 2006). Although more extensive construction activities would be undertaken to build
22 the combination wet and dry cooling towers than was proposed for the once-through cooling
23 system, the PPE estimate of workers still bounded the workforce needed for the change to the
24 cooling system design. Therefore, the workforce would not change from that previously
25 evaluated, and the conclusion did not change as a result of the change in cooling system and
26 higher power level.
27
28 4.5.1 Physical Impacts
29
30 *This section is not affected by the changes presented in ER Revision 6 and is provided solely*
31 *for context.*
32
33 Construction activities at the North Anna ESP site may cause temporary and localized physical
34 impacts including, but not limited to, noise, odor, vehicle exhaust emissions, and fugitive dust.
35 Dominion does not expect significant vibration or shock impacts during construction because of
36 the strict control of such activities onsite (Dominion 2006). General construction activities would
37 not change appreciably as a result of the changes to the Unit 3 cooling system or the higher
38 power level.
39

1 ### 4.5.1.1 Workers and the Local Public
2
3 *This section is not affected by the changes presented in ER Revision 6 and is provided solely*
4 *for context.*
5
6 In ER Revision 6 (Dominion 2006), Dominion stated that no additional workers would be needed
7 for the additional cooling tower construction beyond the 5000 construction workers proposed in
8 ER Revision 3 (Dominion 2004), which was previously evaluated in the Draft EIS (NRC 2004a).
9
10 Dominion identified mitigation measures such as training workers, developing a fugitive dust
11 plan, and complying with the conditions specified in State and local permits. Based on its
12 review of this information, the staff concludes that the overall physical impacts to workers and
13 the local population did not change. The impact level category would still be SMALL, and
14 further mitigation beyond the actions stated above is not warranted.
15
16 ### 4.5.1.2 Buildings
17
18 *This section is not affected by the changes presented in ER Revision 6 and is provided solely*
19 *for context.*
20
21 In the Draft EIS (NRC 2004a), the staff determined that construction activities are not expected
22 to impact any offsite buildings. The buildings most exposed to shock and vibration from pile
23 driving would be those located on the NAPS site; however, Dominion has constructed the onsite
24 buildings to safely withstand any shock and vibration impacts resulting from construction
25 activities (Dominion 2006). Construction activities would not change appreciably because of the
26 changes to the Unit 3 cooling system or the higher power level.
27
28 Because the nearest offsite building is about 910 m (3000 ft) from the North Anna ESP site, the
29 staff concludes that the physical impacts to offsite buildings did not change. The impact level
30 category would still be SMALL, and mitigation is not warranted.
31
32 ### 4.5.1.3 Roads
33
34 *This section is not affected by the changes presented in ER Revision 6 and is provided solely*
35 *for context.*
36
37 In the Draft EIS (NRC 2004a), the staff determined that the transportation network in Louisa
38 County and in the ESP site vicinity is well developed. In ER Revision 6, Dominion stated that
39 no new public roads would be required as a result of construction activities, and that no public
40 roads would be altered (e.g., widened) as a result of construction activities; nevertheless, it did
41 identify several potential mitigation actions to relieve traffic congestion (Dominion 2006). While

1   Dominion stated that no public roads would need to be altered because of construction of new
2   facilities, local officials believe this would need to be evaluated prior to the start of construction.
3   Dominion states that a new access road on the NAPS site would support construction activities
4   and would be private and fully contained within the existing NAPS site boundary. The changes
5   to the cooling system for Unit 3 and the higher power level are not expected to change
6   construction impacts on roads or the need for roads in the area from those previously
7   evaluated.
8
9   Based on its independent review, the staff concludes that the overall physical impacts to local
10   roadways would be temporary and did not change. The impact level category would still be
11   SMALL, provided that mitigation actions, such as traffic control and management measures that
12   Dominion identified are undertaken.
13
14   **4.5.1.4 Aesthetics**
15
16   *Changes to this section reflect aesthetic impacts from the construction of the closed-cycle,*
17   *wet and dry cooling system.*
18
19   In the Draft EIS (2004a), the staff determined that from a visual perspective, construction
20   activities at the ESP site would generally not be visible from points outside the NAPS boundary
21   until the structures at the site approach completion. The new combination wet and dry cooling
22   towers for Units 3 and 4 are expected to be approximately 46 m (150 ft) tall, which is less than
23   the 71 m (234 ft) PPE height value for the tallest potential containment building. These
24   structures may be visible above the treeline by offsite viewers. The current North Anna
25   structures are already visible from Lake Anna and from other selected locations; it is not
26   expected that the visual impact would be appreciably different than the current visible structures
27   in those locations. Recreational users of Lake Anna would be able to observe construction
28   activities occurring on the NAPS site; however, such activities would take place on a site zoned
29   "industrial" and already containing NAPS Units 1 and 2.
30
31   Because visual impacts of construction, such as water turbidity from localized dredging and
32   fugitive dust, would be temporary and would be controlled pursuant to State regulations, and the
33   points from which they could be observed from the lake would be limited, the staff concludes
34   that the visual impacts of construction on Lake Anna and the surrounding area would be
35   SMALL, and further mitigation is not warranted.
36

1 ## 4.5.2 Demography
2
3 *This section is not affected by the changes presented in ER Revision 6 and is provided solely*
4 *for context.*
5
6 In the Draft EIS (2004a), the staff determined that increases in population directly attributable to
7 the construction workforce for Units 3 and 4 would be small because of the large workforce
8 available in the region. Some new jobs may result from the multiplier effect[a] attributable to the
9 construction workforce, but these increases, when compared to the total population base in the
10 region, would be minimal as well. Because Dominion does not expect to employ additional
11 people above the 5000-person workforce previously evaluated for the construction of Units 3
12 and 4 before the change to the Unit 3 cooling system, the staff concludes that the demographic
13 impacts did not change. The impact level category would still be SMALL, and mitigation is not
14 warranted.
15
16 ## 4.5.3 Community Characteristics
17
18 *This section is not affected by the changes presented in ER Revision 6 and is provided solely*
19 *for context.*
20
21 This section evaluates the social and economic impacts on the communities of the surrounding
22 region as a result of constructing Units 3 and 4 at the North Anna ESP site. The evaluation
23 assesses impacts of construction and demands placed by the workforce on the surrounding
24 region. Dominion does not expect to employ additional people above the 5000-person
25 workforce previously evaluated for the construction of Units 3 and 4 before the change to the
26 Unit 3 cooling system.
27
28 ### 4.5.3.1 Economy
29
30 *This section is not affected by the changes presented in ER Revision 6 and is provided solely*
31 *for context.*
32
33 In the Draft EIS (NRC 2004a), the staff determined that the impacts on the economy of
34 constructing Units 3 and 4 would generally be positive within the region. The scale of this
35 beneficial impact would vary throughout the region, with Louisa County receiving the greatest

---

(a) The multiplier effect describes the situation in which each dollar spent on goods and services by a
construction worker becomes income to the recipient who saves some but re-spends the rest on
consumption. This re-spending becomes income to someone else, who in turn saves part and
re-spends the rest. The number of times the final increase in consumption exceeds the initial dollar
spent is called the "multiplier."

1 benefit. The changes to the Unit 3 cooling system would not alter the construction impacts on
2 the economy. Based on the positive aspects of the proposed construction on the regional
3 economies and the workforce availability, the staff concludes that the impacts on the economy
4 did not change. The impact level category would still be SMALL BENEFICIAL to MODERATE
5 BENEFICIAL, and mitigation is not warranted.
6
7 **4.5.3.2 Transportation**
8
9 *This section is not affected by the changes presented in ER Revision 6 and is provided solely*
10 *for context.*
11
12 In the Draft EIS (NRC 2004a), the staff determined that the principal impacts to the
13 transportation system resulting from construction of Units 3 and 4 would be on the roads leading
14 to and from the NAPS site. The impacts could include potential congestion on some of the
15 major Federal highways and State roads leading to the NAPS site and crowding and congestion
16 at the entrance to NAPS during shift changes. To alleviate this potential problem, Dominion
17 plans to develop, in cooperation with the Virginia Department of Transportation, a traffic
18 management plan as a construction mitigation measure (Dominion 2006). Because Dominion
19 does not expect to employ additional people above the 5000-person workforce previously
20 evaluated for the construction of Units 3 and 4 before the change to the Unit 3 cooling system,
21 the staff concludes that the transportation impacts did not change. The impact level category is
22 still SMALL to MODERATE, and further mitigation beyond the actions stated above is not
23 warranted.
24
25 **4.5.3.3 Taxes**
26
27 *This section is not affected by the changes presented in ER Revision 6 and is provided solely*
28 *for context.*
29
30 In the Draft EIS (NRC 2004a), the staff reviewed the income taxes generated on wages and
31 salaries of Units 3 and 4 construction workers and Dominion corporate profits as well as sales
32 and use taxes. These taxes represent beneficial sources of income for the Commonwealth,
33 some of which would benefit the counties in the region. Property tax paid by contractors and
34 Dominion would directly benefit Louisa County. Because Dominion does not expect to employ
35 additional people above the 5000-person workforce previously evaluated for the construction of
36 Units 3 and 4 before the change to the Unit 3 cooling system, the staff concludes that the
37 impacts on income, sales and use taxes, and corporate profits did not change. The impact level
38 category is still SMALL BENEFICIAL to LARGE BENEFICIAL, and mitigation is not warranted.
39

1   **4.5.3.4 Recreation**

2

3   *This section is not affected by the changes presented in ER Revision 6 and is provided solely*
4   *for context.*

5

6   In the Draft EIS (NRC 2004a), the staff determined that construction at the North Anna ESP site
7   would result in limited visual impacts on users of Lake Anna or viewers from points outside the
8   site boundaries. Water-quality impacts of construction of a new water intake structure would be
9   subject to applicable Federal and State regulations, and any noticeable effects would be
10   transitory. Impacts on recreational users of Lake Anna as a result of these activities would be
11   minimal. Congestion on roads around Lake Anna could be exacerbated with the addition of the
12   construction workforce. Because Dominion does not expect to employ additional people above
13   the 5000-person workforce previously evaluated for the construction of Units 3 and 4 before the
14   change to the Unit 3 cooling system, the staff concludes that the recreational impacts did not
15   change. The impact level category would still be SMALL to MODERATE, and mitigation is not
16   warranted.

17

18   **4.5.3.5 Housing**

19

20   *This section is not affected by the changes presented in ER Revision 6 and is provided solely*
21   *for context.*

22

23   Dominion states in its ER that the majority of the construction workforce would come from within
24   the region (Dominion 2006). A large majority of the workforce is likely to reside already within
25   the 80-km (50-mi) radius around the NAPS site; consequently, there should be little or no impact
26   on housing. In the Draft EIS (NRC 2004a), the staff determined that if current trends persist,
27   adequate rental housing would be available within the 80-km (50-mi) radius of NAPS for those
28   workers moving into the area to establish residency. Because Dominion does not expect to
29   employ additional people above the 5000-person workforce previously evaluated for the
30   construction of Units 3 and 4 before the change to the Unit 3 cooling system, the staff concludes
31   that the impact to housing did not change. The impact level category would still be SMALL, and
32   mitigation is not warranted.

33

34   **4.5.3.6 Public Services**

35

36   *This section is not affected by the changes presented in ER Revision 6 and is provided solely*
37   *for context.*

38

39   Public services that may be affected by construction activities include water supply and waste
40   treatment facilities; police, fire, and medical facilities; and social and related services. In the
41   Draft EIS (NRC 2004a), the staff determined that public services are likely to be managed

1   based on demands and opportunities. Dominion does not expect to employ additional people
2   above the 5000-person workforce previously evaluated for the construction of Units 3 and 4
3   before the change to the Unit 3 cooling system. Based on the current availability of water and
4   waste disposal, medical, and social services and additional taxes that would likely compensate
5   for the possible need for additional services in some areas, the staff concludes that the impact
6   on the demand for public and related services as a result of construction did not change. The
7   impact level category would still be SMALL, and mitigation is not warranted.
8
9   **4.5.3.7 Education**
10
11   *This section is not affected by the changes presented in ER Revision 6 and is provided solely*
12   *for context.*
13
14   A large majority of the workforce is likely to reside already within the 80-km (50-mi) radius of the
15   NAPS site. In the Draft EIS (NRC 2004a), the staff determined that the workers moving into the
16   area to establish residency are likely to locate to the larger population areas because of
17   localized shortages of available housing. Dominion does not expect to employ additional
18   people above the 5000-person workforce previously evaluated for the construction of Units 3
19   and 4 before the change to the Unit 3 cooling system. Based on the availability of educational
20   facilities in Henrico, Spotsylvania, Orange, and Louisa Counties and the City of Richmond and
21   assuming that the housing pattern follows past experience, the staff concludes that the impact
22   of construction on educational resources did not change. The impact level category would still
23   be SMALL, and mitigation is not warranted.
24

4.6   Historic and Cultural Resources

26
27   *This section is not affected by the changes presented in ER Revision 6 and is provided solely*
28   *for context.*
29
30   All construction activities for proposed Units 3 and 4, including ground-disturbing activities,
31   would occur within the existing NAPS site boundary. According to Dominion, the area that
32   would be affected on a long-term basis as a result of permanent facilities is approximately 52 ha
33   (128 ac). An additional 27.5 ha (67.9 ac) would be disturbed on a short-term basis as a result of
34   temporary activities and facilities and laydown areas (Dominion 2006). Dominion states that it
35   would conduct any ground-disturbing activities in accordance with Federal, State and local
36   regulatory requirements (Dominion 2006). The planned power block area is relatively level.
37   Undulating surfaces in the area of the planned cooling towers would be leveled to
38   accommodate the towers.
39
40   Dominion states that it does not expect to disturb additional land beyond that previously
41   evaluated in the construction of Units 3 and 4 before the change to the cooling system. In the

1 Draft EIS (NRC 2004a), the staff determined that there is a well-managed cultural resources
2 program at the NAPS site, which includes the existence of written procedures to provide
3 immediate reaction and notification in the event of inadvertent discovery of historic and cultural
4 resources, and its cultural resource analysis and consultation. Based on the foregoing, the staff
5 concludes that the potential impacts on historic and cultural resources did not change. The
6 impact level category would still be SMALL, and further mitigation beyond the existing practice
7 is not warranted.
8
4.7 Environmental Justice Impacts
10
11 *This section is not affected by the changes presented in ER Revision 6 and is provided solely*
12 *for context.*
13
14 Environmental justice refers to a Federal policy under which each Federal agency identifies and
15 addresses, as appropriate, disproportionately high and adverse human health or environmental
16 effects of its programs, policies, and activities on minority[a] or low-income populations. In the
17 Draft EIS (NRC 2004a), the staff found no unusual resource dependencies or practices through
18 which the population could be disproportionally impacted by construction of Units 3 and 4 that
19 would result in those population being aversely affected. Dominion states that it does not
20 expect to employ additional people above the 5000-person workforce previously evaluated for
21 the construction of Units 3 and 4 before the change to the cooling system. Based on its
22 independent review, the staff concludes that offsite impacts of construction of proposed Units 3
23 and 4 at the NAPS site to minority and low-income populations did not change. The impact
24 level category would still be SMALL, and mitigation is not warranted.
25
4.8 Nonradiological Health Impacts
27
28 *This section was not affected by the changes presented in ER Revision 6 and is provided solely*
29 *for context.*
30
31 The changes described in ER Revision 6 include the construction of additional cooling towers
32 (Dominion 2006). This additional construction activity is still expected to be completed within
33 the construction period previously assumed before the change to the cooling system.

(a) The NRC Guidance for performing environmental justice reviews defines "minority" as American
    Indian or Alaskan Native; Asian; Native Hawaiian or other Pacific Islander; or Black races; or Hispanic
    ethnicity ("other" may be considered a separate minority category.) The 2000 census included multi-
    racial data. The staff should consider multi-racial individuals in a separate minority category, in
    addition to the aggregate minority category (NRC 2004b).

4.8.1 Public Health

This section is not affected by the changes presented in ER Revision 6 and is provided solely for context.

Dominion expects that individuals living near the North Anna ESP site would not experience any physical impacts greater than those that would be considered an annoyance or nuisance. In the Draft EIS (NRC 2004a), the staff determined that prior public notification would be provided in advance of atypical or noisy construction activities and that measures to minimize fugitive dust and odors would be implemented. Even with the change in cooling system for Unit 3, the type of construction activities and duration of construction are not expected to be different from that previously evaluated. Based on these mitigation measures, the State and local permits and authorization, and its independent review, the staff concludes that the nonradiological health impacts to the local population from construction did not change. The impact level category would still be SMALL, and further mitigation beyond the actions stated above is not warranted.

4.8.2 Occupational Health

This section is not affected by the changes presented in ER Revision 6 and is provided solely for context.

In the Draft EIS (NRC 2004a), the staff previously evaluated impacts of 5000 construction workers per year, assumed to work for 2080 hours per year per worker for the 5 to 7 year span of construction. Dominion does not expect to employ additional people above the 5000-person workforce previously evaluated for the construction of Units 3 and 4 before the change to the Unit 3 cooling system. In addition, the additional construction activity is still expected to be completed within the construction period previously assumed. Even with the change in cooling system for Unit 3, the type of construction activities and duration of construction are not expected to be different from that previously evaluated.

In general, human health risks for construction workers and personnel working onsite would be expected to be dominated by occupational injuries (e.g., falls, electrocution, asphyxiation). Historically, actual injury and fatality rates at nuclear reactor facilities have been lower than the average U.S. industrial rates. In the Draft EIS (NRC 2004a), the staff determined that Dominion would implement mitigation measures that include training and use of personal protective equipment and practices to minimize fugitive dust and odors. The staff assumes adherence to NRC, Occupational Safety and Health Administration, and State safety standards, practices, and procedures during construction activities. Based on the mitigation measures identified above, State and local permits and authorizations, and its independent review, the staff

1 concludes that the overall nonradiological impacts to workers from construction activities did not
2 change. The impact level category would still be SMALL, and further mitigation beyond the
3 actions stated above is not warranted.
4
5 ### 4.8.3 Noise Impacts
6
7 *This section is not affected by the changes presented in ER Revision 6 and is provided solely*
8 *for context.*
9
10 Large construction projects involve many noise-generating activities. Regulations governing
11 noise from construction activities are generally limited to worker health and safety. Even with
12 the change in cooling system for Unit 3, the type of construction activities and duration of
13 construction are not expected to be different from that previously evaluated. In the Draft EIS
14 (NRC 2004a), the staff determined that activities associated with construction of Units 3 and 4
15 at the North Anna ESP site would generate noise levels typical of larger construction projects.
16 Mitigation measures to reduce noise impacts include equipment inspection and maintenance,
17 conducting certain activities during the daytime, and responding to local concerns. Considering
18 the temporary nature of construction activities and the remote location of the North Anna ESP
19 site, the staff concludes that the noise impacts from construction did not change. The impact
20 level category would still be SMALL, and further mitigation beyond the actions stated above is
21 not warranted.
22
23 ### 4.8.4 Summary of Nonradiological Health Impacts
24
25 *This section is not affected by the changes presented in ER Revision 6 and is provided solely*
26 *for context.*
27
28 In the Draft EIS (NRC 2004a), the staff evaluated health impacts to construction workers and
29 the public. It is expected that health risks to workers would be dominated by occupational
30 injuries at rates below the average U.S. industrial rates.
31
32 Based on the mitigating actions, including operating the construction equipment within local
33 noise and air quality limits and implementing a dust control plan, and its independent review, the
34 staff concludes that the impacts of construction on nonradiological health did not change. The
35 impact level category would still be SMALL, and further mitigation beyond the actions stated
36 above is not warranted.
37

4.9   Radiological Health Impacts

This section is not affected by the changes presented in ER Revision 6 and is provided solely for context.

The changes described in ER Revision 6 include the construction of additional cooling towers (Dominion 2006). This additional construction activity is still expected to be completed within the construction period previously assumed before the change to the cooling system.

4.9.1   Direct Radiation Exposures

This section is not affected by the changes presented in ER Revision 6 and is provided solely for context.

In the Draft EIS (NRC 2004a), the staff determined that the method used to estimate the dose from direct exposure was acceptable. The two principal sources of direct radiation exposure at the construction site from NAPS Units 1 and 2 are (1) the boron recovery tank and (2) the low-level contaminated storage area, both of which are located south of the currently operating Units 1 and 2. Another source of direct radiation is the independent spent fuel storage installation (ISFSI), which is located south of the construction site. The staff reviewed the potential locations for exposures and recent records of dose rates, the locations of thermoluminescent dosimeters, the method to estimate doses to members of the public in controlled areas, and other recent data. The direct radiation exposure to workers would not be expected to change because of changes to the Unit 3 cooling system design and its construction.

4.9.2   Radiation Exposures from Gaseous Effluents

This section is not affected by the changes presented in ER Revision 6 and is provided solely for context.

In the Draft EIS (NRC 2004a), the staff reviewed data from the North Anna Units 1 and 2 radioactive effluent reports for recent years and determined that the method to estimate dose to workers from gaseous effluents was acceptable. The staff also reviewed data for the Annual Radiological Effluent Report for 2005 (VEPCo 2006) and finds that the releases previously evaluated are still representative of the typical releases for the operating units. The radiation exposure to workers from gaseous effluents would not be expected to change because of changes to the Unit 3 cooling system design and its construction.

4.9.3 Radiation Exposures from Liquid Effluents

This section is not affected by the changes presented in ER Revision 6 and is provided solely for context.

In the Draft EIS (NRC 2004a) the staff reviewed data from the North Anna Units 1 and 2 radioactive effluent reports for recent years and determined that the method to estimate dose to workers from liquid effluents was acceptable. The staff also reviewed data from the Annual Radiological Effluent Report for 2005 (VEPCo 2006) and finds that the releases previously evaluated are still representative of the typical releases from the operating units. The radiation exposure to workers from liquid effluents would not be expected to change because of changes to the Unit 3 cooling system design and its construction.

4.9.4 Total Dose to Workers

This section is not affected by the changes presented in ER Revision 6 and is provided solely for context.

In the Draft EIS (NRC 2004a), the staff determined that the method used to estimate the total dose to workers from the three pathways (direct radiation, gaseous effluents, and liquid effluents) was acceptable. The estimated annual dose to a construction worker of 0.24 mSv (24 mrem) is primarily from the direct exposure pathway, with doses from liquid and gaseous effluents as a small component of the total dose. This estimate is well within both the dose limits to individual members of the public found in 10 CFR 20.1301 of 1 mSv (100 mrem) in a year and occupational dose limits to workers found in 10 CFR 20.1201 of 0.05 Sv (5 rem). The total dose to workers would not be expected to change because of the changes to the Unit 3 cooling system design and its construction.

4.9.5 Summary of Radiological Health Impacts

This section is not affected by the changes presented in ER Revision 6 and is provided solely for context.

Based on its independent review, the staff found that the doses to the public and to construction workers to be well within NRC exposure limits designed to protect the public health, even if workers exceed the 2080 hr/yr occupancy factor, and concludes that the impacts of radiological exposures to the public and to construction workers did not change. The impact level category would still be SMALL, and mitigation is not warranted.

4.10 Measures and Controls to Limit Adverse Impacts During Construction Activities

This section is not affected by the changes presented in ER Revision 6 and is provided solely for context.

In its evaluation of environmental impacts during construction activities for the proposed new North Anna units, the staff relied on the mitigation measures identified in Section 4.10 of the Draft EIS (NRC 2004a). This is not expected to change because of the changes to the Unit 3 cooling system design and its construction.

4.11 Site Redress Plan

This section is not affected by the changes presented in ER Revision 6 and is provided solely for context.

The changes described in ER Revision 6 are related to the change in the cooling system proposed for NAPS Unit 3 to a closed-cycle, combination wet and dry cooling system and to an increase in the thermal power level. The increase in the thermal power level has no bearing on site preparation activities. The change in the cooling system, while it can result in the construction of additional structures as part of site preparation, is already included within the scope of the site preparation activities originally proposed and previously evaluated by the staff in the Draft EIS (NRC 2004a). Therefore, if Dominion receives an ESP and builds the additional cooling towers or if no applicant for a CP or COL references the ESP, or does not receive a CP or COL, Dominion would have to redress the area including these facilities. Consequently, this section is not affected by the changes presented in ER Revision 6.

4.12 Summary of Construction Impacts

Changes to this section reflect the summary of construction impacts from the proposed project presented in ER Revision 6.

Impact level categories denoted in Table 4-1 as SMALL, MODERATE, or LARGE were assigned to each resource area based on the staff's evaluation and conclusions regarding expected adverse environmental impacts, if any. A brief statement explains the basis for the impact level. Some impacts, such as the addition of tax revenue from Dominion for the local economies, are likely to be beneficial impacts to the community, and are noted as such.

Table 4-1. Characterization of Impacts from Construction of the Closed-Cycle Cooling System for Unit 3 at the North Anna ESP Site

Category	Comments	Impact Level
Land-use impacts		--
The site and vicinity	Construction activities would take place within existing site boundaries.	SMALL
Transmission line rights-of-way	No new transmission line rights-of-way would be needed.	SMALL
Air quality impacts	Construction activities would be conducted in accordance with applicable Virginia administrative codes, and dust and emissions would be minimized through a dust control plan.	SMALL
Water-related impacts		--
Hydrological alterations	Impacts would be localized and temporary. Construction activities would be conducted in accordance with applicable VDEQ administrative codes and ACE permit processes; hydrological impacts would be minimized though application of best management practices.	SMALL
Water use	Minimal water usage during construction.	SMALL
Water quality	Construction would be conducted using best management practices to control spills and storm water runoff.	SMALL
Ecological impacts		--
Terrestrial ecosystems	No important terrestrial species would be affected by construction at the NAPS site.	SMALL
Aquatic ecosystems	Construction impacts to benthic habitats would be temporary.	SMALL
Threatened and endangered species	There are no Federally listed species in the vicinity. Impacts to State-listed species would be minor.	SMALL
Socioeconomic impacts		--
Physical impacts		
Workers/local public	Construction takes place within existing plant boundaries, so impacts to the public would be minimal. Impacts to workers would be mitigated with training and protective equipment.	SMALL

Table 4-1. (contd)

Category	Comments	Impact Level
Buildings	Construction would not affect any offsite buildings, and onsite buildings were constructed to withstand vibration from construction activities.	SMALL
Roads	Growth would put pressure on local road systems, but traffic control and management measures would protect any local roads during construction.	SMALL
Aesthetics	Construction activities would be temporary, and observation points would be limited because of site location.	SMALL
Demography	Percentage of construction workers relocating to the region would be small. Most would already live within the region.	SMALL
Community characteristics		
Economy	Economic impacts of construction overall are beneficial to local economies, in this case ranging from small to moderately beneficial.	SMALL BENEFICIAL to MODERATE BENEFICIAL
Transportation	Planned upgrades and traffic management plans would reduce temporary construction transportation impacts. Impacts could be moderate in some areas without planned upgrades.	SMALL to MODERATE
Taxes	Depends on residence location; generally, impacts are beneficial, especially for property taxes and employment, ranging from small to large (Louisa County).	SMALL BENEFICIAL to LARGE BENEFICIAL
Recreation	Visual impacts of construction would be limited and temporary. Recreational use of Lake Anna would be expected to increase, and traffic mitigation would keep impacts small. Impacts could be moderate if mitigation measures are not undertaken.	SMALL to MODERATE

Table 4-1. (contd)

	Category	Comments	Impact Level
1	Housing	Adequate housing is available in Henrico and Spotsylvania Counties and in the City of Richmond to handle construction workers. If more construction workers than expected locate in Orange and Louisa Counties, the impact could be moderate.	SMALL
2	Public services	Public services are adequate for any temporary influx of workers resulting from construction at the NAPS site.	SMALL
3	Education	If Louisa County builds new schools to accommodate the temporary influx of construction workers, then all counties would have room for additional students. If no additional school capacity is added then the impact in Louisa County could be moderate.	SMALL
4	Historic and cultural resources	Proposed construction area is previously disturbed, and Dominion has a well-managed cultural resource program in place at NAPS.	SMALL
5	Environmental justice	No unusual resource dependencies in the area.	SMALL
6	Nonradiological health impacts	Emission controls and remote location of the NAPS site would keep nonradiological health impact small.	SMALL
7	Radiological health impacts	Exposures to workers would be below annual occupational and public dose limits.	SMALL

4.13 References

10 CFR Part 20. Code of Federal Regulations, Title 10, *Energy*, Part 20, "Standards for Protection Against Radiation."

10 CFR Part 50. Code of Federal Regulations, Title 10, *Energy*, Part 50, "Domestic Licensing of Production and Utilization Facilities."

10 CFR Part 51. Code of Federal Regulations, Title 10, *Energy*, Part 51, "Environmental Protection Regulations for Domestic Licensing and Related Regulatory Functions."

1   10 CFR Part 52. Code of Federal Regulations, Title 10, *Energy*, Part 52, "Early Site Permits,
2   Standard Design Certifications, and Combined Licenses for Nuclear Power Plants."
3
4   15 CFR Part 930. Code of Federal Regulations, Title 15, *Commerce and Foreign Trade*, Part
5   930, "Federal Consistency with Approved Coastal Management Programs."
6
7   40 CFR Part 122. Code of Federal Regulations, Title 40, *Protection of Environment*, Part 122,
8   "EPA Administered Permit Programs: The National Pollutant Discharge Elimination System."
9
10   Clean Water Act (also referred to as the Federal Water Pollution Control Act). 33 USC 1251,
11   et seq.
12
13   Coastal Zone Management Act of 1972 (CZMA). 16 USC 1451, et seq.
14
15   Dominion Nuclear North Anna, LLC (Dominion). 2004. *North Anna Early Site Permit*
16   *Applications – Part 3 – Environmental Report*. Revision 3, Glen Allen, Virginia.
17
18   Dominion Nuclear North Anna, LLC (Dominion). 2005a. *North Anna Early Site Permit*
19   *Applications – Part 3 – Environmental Report*. Revision 4, Glen Allen, Virginia.
20
21   Dominion Nuclear North Anna, LLC (Dominion). 2005b. *North Anna Early Site Permit*
22   *Applications – Part 3 – Environmental Report*. Revision 5, Glen Allen, Virginia.
23
24   Dominion Nuclear North Anna, LLC (Dominion). 2006. *North Anna Early Site Permit*
25   *Applications – Part 3 – Environmental Report*. Revision 6, Glen Allen, Virginia.
26
27   Electric Power Research Institute (EPRI). 1993. *Proceedings: Avian Interactions with Utility*
28   *Structures. International Workshop*. EPRI TR-103268, EPRI, Palo Alto, California.
29
30   U.S. Fish and Wildlife Service and Virginia Department of Game and Inland Fisheries (FWS and
31   VDGIF). 2000. Bald Eagle Protection Guidelines for Virginia. Accessed at
32   http://www.dgif.state.va.us/wildlife/publications/EagleGuidelines.pdf on March 23, 2004.
33
34   U.S. Nuclear Regulatory Commission (NRC). 1996. *Generic Environmental Impact Statement*
35   *for License Renewal of Nuclear Plants*. NUREG-1437, Volumes 1 and 2. Washington, D.C.
36
37   U.S. Nuclear Regulatory Commission (NRC). 2000. *Environmental Standard Review Plan:*
38   *Standard Review Plans for Environmental Reviews for Nuclear Power Plants*. NUREG-1555.
39   Vol. 1. Office of Nuclear Reactor Regulation, Washington, D.C.
40

1   U.S. Nuclear Regulatory Commission (NRC). 2004a. *Draft Environmental Impact Statement for*
2   *an Early Site Permit (ESP) at the North Anna ESP Site.* NUREG-1811, Draft, Office of Nuclear
3   Reactor Regulation, Division of Regulatory Improvement Programs, Washington D.C.
4
5   U.S. Nuclear Regulatory Commission (NRC). 2004b. Office of Nuclear Reactor Regulation
6   Office Instruction Change Notice. LIC-203, Revision 1, Procedural Guidance for Preparing
7   Environmental Assessments and Considering Environmental Issues, May 24, 2004,
8   Appendix D, Environmental Justice Guidance and Flow Chart, Washington, D.C.
9
10   Virginia Administrative Code (VAC) 9 VAC 5-50 (Visible and Fugitive Dust Emissions).
11
12   Virginia Administrative Code (VAC) 9 VAC 5-40-5680 (Emission Standards from Mobile
13   Sources - Vehicles).
14
15   Virginia Electric and Power Company (VEPCo). 2006. Annual Radioactive Effluent Release
16   Report for the North Anna Power Station – January 1, 2005 through December 31, 2005.
17   Richmond, Virginia.
18
19

5.0 Operational Impacts at the Proposed Site

This chapter of the Supplement to the Draft Environmental Impact Statement was changed to incorporate new information and analysis of operational impacts related to changes proposed in Revision 6 of the Environmental Report. Section summaries are provided for context.

This chapter examines the environmental impacts of operations associated with changes to the early site permit (ESP) application as submitted by Dominion Nuclear North Anna, LLC (Dominion) in its April 2006 Environmental Report (ER) Revision 6 (Dominion 2006a). The changes described in Revision 6 to the ER relate to two specific issues and their ramifications for other aspects of the environmental analysis. These two issues are (1) a proposed change in the cooling system for proposed Unit 3 and (2) a higher power level for proposed Units 3 and 4 (hereafter referred to as Units 3 and 4). Both changes are relevant to this chapter.

The *Draft Environmental Impact Statement (EIS) for an Early Site Permit (ESP) at the North Anna ESP Site* (Draft EIS) evaluated the environmental impacts of the ESP application in which Dominion initially proposed using once-through cooling for Unit 3 (NRC 2004a). In its ER Revision 6, Dominion proposes replacing the once-through cooling system with a closed-cycle combination wet and dry cooling system capable of functioning in two modes: (1) an Energy Conservation (EC) mode in which cooling is primarily through the use of wet towers and power output is conserved, and (2) a Maximum Water Conservation (MWC) mode in which water is conserved but more power is expended to support part of the cooling using the dry towers. As described in the Draft EIS, the proposed cooling system for Unit 4 is a dry cooling system, which reduces water consumption but increases energy consumption for operation.

The second change is an increase in reactor power level from a maximum of 4300 MW(t) to 4500 MW(t) per unit. This change is intended to align with the new maximum power designated for the economic simplified boiling water reactor (ESBWR) by the manufacturer of this reactor design (GE Nuclear 2005).

This chapter is divided into 13 sections. Sections 5.1 through 5.11 discuss the changes in the operational impacts during the 40-year operating period, if any, resulting from differences between the information in ER Revision 3 (Dominion 2004a) and ER Revision 6 (Dominion 2006a). For those sections in which there is no change in analysis as a result of the revisions in ER Revision 6, brief discussions of the topics and the impact levels determined in the Draft EIS are provided solely for context. In accordance with Title 10 of the Code of Federal Regulations (10 CFR) Part 51, impacts have been evaluated and conclusions have been made; in addition, an impact category level of potential adverse impacts (i.e., SMALL, MODERATE or LARGE) has been assigned to each affected resource area. Negligible impacts are listed as SMALL impacts. The staff's determination of the significance of the impacts is based on the assumption that the mitigation measures identified in the ER or activities planned by various

State and county governments, such as infrastructure upgrades, as discussed throughout this chapter, are implemented. Failure to implement these measures or activities could result in a change to the impacts considered by the staff. A summary of these impacts is presented in Section 5.12. The references cited in this chapter are listed in Section 5.13. The technical analyses in this chapter support the results, conclusions, and recommendations discussed in Chapters 9 and 10.

Changes in this chapter of the Supplement to the Draft EIS (SDEIS) are limited to impacts related to the operation of the proposed combination wet and dry cooling system and changes to the radiological health and postulated accidents from the proposed higher power level.

5.1 Land-Use Impacts

This introductory section is not affected by the changes presented in ER Revision 6 and is provided solely for context.

This section provides information regarding land-use impacts associated with operation of the proposed Units 3 and 4 at the North Anna ESP site. This section and its subsections are not affected by the changes presented in ER Revision 6.

5.1.1 The Site and Vicinity

This section is not affected by the changes presented in ER Revision 6 and is provided solely for context.

In the Draft EIS (NRC 2004a), the staff determined that some offsite land-use changes can be expected as a result of operational activities for Units 3 and 4. Possible changes include the conversion of some land in surrounding areas to housing developments (e.g., apartment buildings, single family condominiums and homes, and manufactured home parks) and retail development to serve plant workers. Property tax revenue from the new plants could also lead to additional growth and land conversions in Louisa County as a result of infrastructure improvements (e.g., new roads and utility services). However, any growth could be managed because all counties surrounding the North Anna ESP site have comprehensive land-use plans in place as required by the Code of Virginia § 15.2-2223. The change in Unit 3 cooling system and higher power level for Units 3 and 4 would not change expected offsite land-use previously evaluated.

Based on the foregoing, the staff concludes that the land-use impacts of operation did not change. The impact level category would still be SMALL, and further mitigation is not warranted.

5.1.2 Transmission Line Rights-of-Way and Offsite Areas

This section is not affected by the changes presented in ER Revision 6 and is provided solely for context.

In the Draft EIS (NRC 2004a), the staff determined that any two of three existing 500-kV transmission lines along with an existing 230 kV line would be expected to have sufficient capacity to carry the total output of both existing units and two new units. The existing lines are expected to have sufficient capacity to handle any output change resulting from the proposed higher power level. Dominion indicated that it would perform a system study (load flow) modeling these lines with the new units' power contribution at the construction permit (CP) or combined license (COL) stage (Dominion 2006a).

Because no additional electrical transmission lines or rights-of-way would be needed, the staff concludes that land-use impacts to other offsite areas did not change. The impact level category would still be SMALL, and mitigation is not warranted.

5.2 Meteorological and Air Quality Impacts

Changes to this section reflect meteorological and air quality impacts from the closed-cycle, combination wet and dry cooling system.

Dominion's proposed change from once-through cooling for Unit 3 to a closed-cycle, combination wet and dry cooling system for Unit 3 results in changes to certain aspects of the analysis of meteorological impacts. Lake Anna would provide the makeup water to the wet cooling towers. Because warm, moist air would be emitted to the atmosphere from the operation of the wet cooling towers, elevated plumes would at times extend above the cooling towers and be visible off site. There would also be the potential for fogging and icing at ground level as the plume loses buoyancy and for drift deposition on the local surroundings. In addition, there is the potential for ice buildup on the transmission lines and other structures within the plant boundary. The greatest impacts would occur during conditions of high humidity and low ambient temperature when the Unit 3 circulating water system is operating in the EC mode. Micro-climatic impacts would include an increase in humidity in the vicinity of the towers as well as a slightly reduced level of solar radiation in areas in the shadow of the plume, consistent with wind direction frequency.

The SACTI (Seasonal and Annual Cooling Tower Impacts) system of computer programs, initially written and assembled by the Argonne National Laboratory (ANL) (ANL 1984) for the Electric Power Research Institute was used to estimate the impact of operating the cooling towers. The version used by Dominion is dated November 1, 1990. A brief description of the application and limitations of the SACTI code regarding aesthetic aspects of the cooling tower

1   plume is included Section 5.5.1.4. The input meteorological data used to estimate the impacts
2   encompassed the period of 1998 through 2000. It included data collected onsite as well as site
3   representative data collected at the National Weather Service sites in Richmond, Virginia, and
4   Dulles Airport in Northern Virginia. For this analysis, the cooling towers were assumed to be
5   operating in the EC mode, which results in the greatest evaporation rates from the towers and,
6   therefore, the greatest level of impacts (Dominion 2006a).
7
8   The results of the staff's independent analysis indicated that for all seasons, the plume would
9   extend to a maximum height of 980 m (3200 ft) and to a length of 4900 m (16,000 ft) from the
10   tower. The annual duration of plume fogging (i.e., the plume remaining at the ground level)
11   would be about 70 hr (excluding hours of natural fog), with a majority of fogging occurring at
12   about 300 m (1000 ft) to the south-southeast from the cooling towers. Fogging would, however,
13   occur as far as 1600 m (5200 ft) from the tower. Fogging is estimated to occur during all
14   seasons except summer. The analysis indicates that icing is unlikely to occur in conjunction
15   with ground-level fogging (Dominion 2006a).
16
17   Deposition of salts from cooling tower drift would occur in all directions from the towers out to
18   1525 m (5000 ft), but would occur predominately in the areas to the north through northeast as
19   well as to the south through southeast of the towers. The maximum estimated amount of
20   deposition would be 12.6 kg/km^2/month at 175 m (575 ft) north-northeast of the cooling towers.
21   The vast majority of the drift deposition would occur within 300 m (1000 ft) of the towers.
22   Significant chemical interaction of the cooling tower plume and pollutants emitted onsite or in
23   the vicinity of the plant is not anticipated. Generally, the approach to minimize the potential for
24   contact with cooling tower drift is to limit parking or work activities in the vicinity of the cooling
25   towers. The impacts of salt deposition on terrestrial resources are discussed in Section 5.4.1.1.
26
27   In the Draft EIS (NRC 2004a), the staff determined that air quality impacts from routine releases
28   other than the cooling system would be limited to nonradiological pollutants emitted during the
29   operation of auxiliary boilers and emergency generators, and emissions from onsite service
30   vehicles. Impacts of transmission lines on air quality were reviewed elsewhere by the NRC in
31   the *Generic Environmental Impact Statement of License Renewal at Nuclear Plants*
32   (NUREG-1437) (NRC 1996). With regard to air quality impacts for criteria pollutants, given the
33   relatively large distance from the Prevention of Significant Deterioration Class I areas (see the
34   Clean Air Act, Section 169A, and 40 CFR Part 51, Subpart P) and the short time duration of any
35   emissions, the resulting impacts on local ambient air quality levels or visibility in the Class I
36   areas are estimated to be negligible.
37
38   In the Draft EIS (NRC 2004a), the staff determined that the impacts would be SMALL from other
39   potential sources of air quality impacts. The staff concludes that the potential impacts of
40   releases from vehicles, auxiliary boilers, emergency generators, cooling systems, and
41   energized transmission lines would still be SMALL, and mitigation beyond those normally taken
42   in the operation of plant equipment is not warranted.

5.3 Water-Related Impacts

Changes to this section reflect water resource impacts from the closed-cycle, combination wet and dry cooling system.

Dominion's proposed change from once-through cooling for Unit 3 to a closed-cycle, combination wet and dry cooling system for Unit 3 results in changes to the analysis of water-related issues. The dry closed-cycle cooling system for dissipation of heat for Unit 4 was unchanged from the original proposal. In the Draft EIS (NRC 2004a), the staff determined that the Unit 4 cooling system would use a maximum of 0.06 L/s (1 gpm) of water and, therefore, would have negligible water-related impacts on Lake Anna, the Waste Heat Treatment Facility (WHTF), or the North Anna River. Therefore, only the water-related impacts of proposed Unit 3 on Lake Anna, the WHTF, and the North Anna River are considered in the following sections.

After the Draft EIS was issued, VEPCo modified procedures for the existing Units 1 and 2. Previously, the procedures called for plant shutdown when the lake level falls to 74.4 m (244 ft) above mean sea level (MSL). The new limit for plant shutdown is 73.8 m (242 ft) MSL (Dominion 2006a).

During normal operation at full power, and based on Dominion's Plant Parameter Envelope (PPE) values, the primary cooling system for each proposed unit would reject 3020 MW (1.03×10^{10} Btu/hr) to the environment. Unit 3 would reject this heat load to the atmosphere via closed-cycle, combination wet and dry cooling towers. Unit 3 would employ a cooling tower system that can function in different modes which consume different amounts of water depending on the meteorological and water supply conditions. During times of relative water abundance, the Unit 3 cooling system would operate in the EC mode, which increases water consumption while decreasing energy consumption. During times of limited water availability, i.e. whenever the lake level elevation of Lake Anna falls below 76.2 m (250 ft) MSL for a period of seven or more consecutive days, the Unit 3 cooling system would operate in the MWC mode, which reduces water consumption while increasing energy consumption. The maximum water withdrawal rates in EC and MWC modes are 1405 and 971 L/s (22,268 and 15,384 gpm), respectively. During full load operation, the maximum blowdown rates in EC and MWC modes are 351 and 245 L/s (5565 and 3844 gpm), respectively. (Blowdown is the removal of recirculating water from the cooling system to reduce the buildup of contaminants, such as dissolved solids.)

Management of water resources involves balancing the tradeoffs among various and often conflicting uses. The water uses at Lake Anna and the North Anna River downstream of Lake Anna include recreation, visual aesthetics, fishery management, and a variety of consumptive uses of water, such as municipal water supplies and industrial uses (e.g., cooling water for power generation). The U.S. Environmental Protection Agency (EPA), the U.S. Army Corps of Engineers (ACE), and the Commonwealth of Virginia have jurisdiction for regulating water use

1   and water quality through Federal and State laws. Water resource management incorporates
2   the uncertainty of projections of the future supply and demand for water that results from natural
3   climate variability and man-made demands. The ability to manipulate the water supply to
4   balance periods of excess supply with periods of excess demand is limited by the available
5   water infrastructure. While the water supply is regularly being replenished by precipitation,
6   conflicts over water resources typically grow along with population.
7
8   Both Dominion and the staff independently analyzed changes in the water supply available from
9   Lake Anna that would result from operating proposed Unit 3 at the North Anna ESP site. In
10   performing their respective analyses, Dominion and the staff employed different approaches
11   and relied on different data sources. These approaches are briefly described below; however, a
12   more complete description of the Dominion analysis can be found in Sections 5.2.2 and 5.3.2 of
13   the ER (Dominion 2006a). A more complete description of the staff's analysis can be found in
14   Appendix K of this SDEIS.
15
16   The staff has reviewed long-term precipitation and evaporation data from Richmond, Virginia, to
17   characterize typical-year and critical-year conditions. Based on annual values, precipitation
18   exceeds evaporation during typical-year conditions. Using average monthly estimates,
19   however, evaporation exceeds precipitation by more than 20 percent in June. Over a typical
20   12-month period, runoff from areas draining into Lake Anna offset any decreases in the lake
21   level elevation resulting from natural evaporation. However, even minimum releases of 1.1 m^3/s
22   (40 cfs) from Lake Anna would result in decreases in lake level elevation during the months of
23   July, August, and September. Therefore, Lake Anna lake level elevations would decline during
24   both typical- and critical-years during those months. Historical summer flows in the North Anna
25   River near Partlow, Virginia, before construction of North Anna Dam were much smaller than
26   even the minimum release of 0.57 m^3/s (20 cfs) established under the Lake Level Contingency
27   Plan. Because the inflows typically exceed the regulated outflows to Lake Anna, it is therefore
28   reasonable to expect that Lake Anna would experience lake level elevation decreases during
29   the late summer months.
30
31   During the period from October 2001 through December 2002, an extreme drought occurred in
32   the region extending from Georgia to northern Virginia. As a result of this climatic anomaly,
33   Lake Anna experienced the lowest lake level elevations and lowest estimated inflows in its
34   history. During this period of drought, Dominion implemented the Lake Level Contingency Plan
35   (a condition of the North Anna Power Station (NAPS) Virginia Pollutant Discharge Elimination
36   System [VPDES] permit issued by Virginia Department of Environmental Quality [VDEQ]), and
37   releases from Lake Anna Dam were reduced from the normal minimum of 1.1 m^3/s (40 cfs) to
38   0.57 m^3/s (20 cfs). Low water conditions were quickly alleviated when normal precipitation
39   levels returned to the region. This period of extreme drought was considered as the critical
40   period in the analyses of both the applicant and the staff.
41

1   Both the staff's assessment and Dominion's water budget model of Lake Anna are based on a
2   simplified representation of the conservation of mass for the lake. The principle of conservation
3   of mass can be restated specifically for water as the change in storage of water at any time is
4   equal to the water inflow less the water outflow. In both water budget assessments, changes in
5   lake storage over time were equal to the differences between the inflows and the outflows.
6   Inflows included the drainage from the basin upstream of the lake and the precipitation
7   occurring directly on the lake. Outflows included both natural and induced (i.e., forced because
8   of operation of Units 1, 2, and 3) evaporation and releases from the dam. Groundwater can
9   either flow from the aquifer into Lake Anna, or Lake Anna water can recharge the aquifer.
10   Based on groundwater elevation measurements, the only time Lake Anna is expected to
11   recharge the adjacent aquifer would be after refilling the lake following an extended period of
12   low lake elevations. The change in storage is reflected by a change in the pool elevation.
13
14   The staff and Dominion made different assumptions to estimate the inflow to Lake Anna.
15   Because of the limited record of tributary flow measurements, there is no direct means to
16   estimate the total inflow into Lake Anna from its tributaries. The outflow from Lake Anna Dam
17   was estimated by Dominion from the U.S. Geological Survey (USGS) gauge downstream from
18   the dam at Partlow, Virginia. Dominion did not use precipitation data in its water budget
19   analysis as it assumed that the sum of precipitation, groundwater, and tributary inflows offset
20   the imbalance between the estimated evaporative losses and dam releases and the changes in
21   lake water volume. The change in lake water volume was based directly on observed records
22   of lake level elevation.
23
24   Dominion's historical evaporation estimates were based on calculations using a lake
25   temperature model developed by Massachusetts Institute of Technology (Ho and Adams 1984).
26   The staff found that relatively small errors in the pool elevation measurements using this model
27   can result in significant errors in the precipitation, groundwater, and tributary inflow estimate.
28   For example, an error of only 2.5 cm (1 in.) between daily lake elevation measurements
29   translates into an error of about 14 m^3/s (500 cfs); this can also result in negative inflow
30   estimates that are inconsistent with conservation of mass principles. The occurrence of
31   negative inflow estimates was reduced by Dominion by smoothing (i.e., using weekly averages
32   instead of daily values). Dominion's discussions and conclusions are based on the weekly
33   averaged results.
34
35   The staff estimated inflows from the watershed upstream of Lake Anna using data from the
36   adjacent Little River drainage basin adjusted for differences in the size of the drainage areas.
37   The rationale for using an adjacent drainage basin is that too few of the tributaries flowing into
38   Lake Anna are gauged for the observed data set to be useful in constructing an inflow
39   sequence. The staff also determined that it would not use the North Anna River discharges
40   downstream from North Anna Dam to directly estimate the inflows to Lake Anna because they
41   are heavily influenced by consumptive losses from Units 1 and 2 and flow regulation resulting
42   from unrecorded dam operations. The Little River drainage is 277 km^2 (107 mi^2) and is adjacent

1　to the North Anna drainage; measurements from Little River span from October 1961 to the
2　present. Based on a review of streamflow records from USGS Gauge 01671100 (Little River
3　near Doswell, Virginia), the staff selected the period from June 2000 through April 2003 as the
4　critical water period. The direct precipitation to the lake was based on precipitation records from
5　the National Weather Service meteorological station at the Richmond, Virginia, airport.
6
7　The staff estimated lake outflows based on the current operating rules for Lake Anna Dam.
8　Releases were generally performed to maintain a lake level elevation of 76.2 m (250 ft) MSL.
9　Under this condition, the staff calculated flow over the dam based on lake level. When the lake
10　level elevation dropped below 76.2 m (250 ft) MSL because inflows were inadequate to offset
11　the natural and induced evaporative losses, the release was maintained at the normal minimum
12　flow of 1.1 m^3/s (40 cfs). If the lake level elevation declined below 75.6 m (248 ft) MSL,
13　releases were assumed to decrease to 0.57 m^3/s (20 cfs) immediately. Once inflows and
14　outflows were calculated, the staff calculated the rate of evaporation from the Lake Anna
15　reservoir, factoring in the difference between the flows.
16
17　Because makeup water for ultimate heat sink (UHS) cooling towers for Units 3 and 4 would be
18　stored in an engineered basin and is much less than the water demand during normal
19　operation, the water demand when operating in UHS mode was considered to be bounded by
20　the water demand for normal operations.
21

5.3.1 Hydrological Alterations

23
24　*Changes to this section reflect water-related impacts from the closed-cycle, combination wet*
25　*and dry cooling system.*
26
27　The Unit 3 operational activity identified by the staff that would result in a detectable
28　hydrological alteration in Lake Anna is when the lake elevation is below 76.2 m (250 ft) MSL
29　and the wet cooling towers are operating. Discharges to the North Anna River downstream of
30　the North Anna Dam could also be affected by operation of the wet cooling towers, which would
31　increase the duration of reduced discharges, that is 1.1 m^3/s (40 cfs) when the lake elevation is
32　between 75.6 m (248 ft) and 76.2 m (250 ft) MSL, and 0.57 m^3/s (20 cfs) when the lake is below
33　75.6 m (248 ft) MSL).
34
35　When the lake elevation is above 76.2 m (250 ft) MSL, no hydrological effect would be
36　detectable in the lake because water available for the Unit 3 cooling system would have
37　otherwise been discharged from the North Anna Dam. The operation of Unit 3 would also result
38　in a net decrease of water available to the North Anna River equal to the consumptive water
39　loss (see also Section 5.3.2).
40
41　The staff's independent water budget analysis assumed the NAPS Units 1 and 2 and the
42　proposed Unit 3 would operate continuously. In non-drought years, the projected incremental

1 decline of the lake level attributable to Unit 3 was relatively minor. The staff determined that the
2 operation of Unit 3 would decrease the fraction of time that the lake level elevation was above
3 75.6 m (248 ft) MSL by 5 percent, from 94 percent to 89 percent of the time. With the operation
4 of Unit 3, the fraction of time the lake would be at or below elevation 75.0 m (246 ft) MSL would
5 increase by 0.9 percent, from 1.1 percent to 2.0 percent. The staff also analyzed the
6 differences in lake level elevation between the baseline (Units 1 and 2 in operation) and
7 proposed (addition of the ESP Unit 3) scenarios to examine the impacts of Unit 3 on
8 downstream flows. Considering the entire simulation period, including the critical drought
9 period, the incremental decline in lake level elevation resulting from the operation of Unit 3 was
10 less than 7.6 cm (3 in.) 70 percent of the time, less than 15 cm (6 in.) 86 percent of the time,
11 and less than 30 cm (1 ft) 94 percent of the time.
12
13 The lowest lake level elevations and greatest incremental decrease are projected to occur
14 during the month of October. When modeling lake level elevations during the critical period of
15 record, specifically targeting the minimum elevation occurring during early October 2002, the
16 staff analyzed the minimum lake level elevations for the following scenarios:
17
18 ● Units 1 and 2 (baseline conditions):   74.74 m (245.2 ft)
19 ● Units 1 and 2 plus Unit 3 (proposed conditions): 74.22 m (243.5 ft)
20
21 While the addition of Unit 3 would cause further declines in the level of Lake Anna, the staff's
22 analysis of long-term conditions using the water budget model indicates that the lake level
23 elevation would not drop below 75.6 m (248 ft) MSL during periods of normal or above average
24 precipitation. During low-water conditions, the existing NAPS Units 1 and 2 and the proposed
25 Unit 3 would be allowed to operate until the elevation of the lake reaches 73.8 m (242 ft) MSL.
26 Both the staff and Dominion estimate that during the critical period (June 2000 through
27 April 2003), the elevation of the lake would have remained above 73.8 m (242 ft) MSL had Unit
28 3 been operating.
29
30 Dominion also evaluated the impacts of raising normal operating lake level 15 to 30 cm
31 (6 to 12 in.) above 76.2 m (250 ft) MSL on shoreline areas, if VDEQ elects to consider such
32 actions to mitigate impacts on down-river flows. Increasing the lake level by approximately
33 18 cm (7 in.) would eliminate changes in the frequency of the 0.57 m^3/s (20 cfs) minimum
34 instream flow (Dominion 2006a). The staff's independent assessment (described in
35 Appendix K) estimated that the frequency of 0.75 m^3/s (20 cfs) flows would be unchanged if the
36 normal lake level were raised 25 cm (10 in.). Dominion conducted map reconnaissance,
37 helicopter flyovers, and ground-truthing from boats and concluded that there would be some
38 shifting of wetland areas, particularly in gradually sloping uplake tributary areas if lake levels
39 were raised. In addition, Dominion concluded that raising the lake level could increase localized
40 flooding potential and downstream flows, and would likely affect use of some residential and
41 marina boat ramps and docks, including those at Lake Anna State Park.
42

1   Information on operational practices and procedures was not provided in the ESP application.
2   The operation of the cooling system presented in the application was not unreasonable for
3   analysis purposes to assess hydrologic impacts. The actual procedures controlling the
4   operation of the cooling system will be determined by the Commonwealth of Virginia in the
5   required Clean Water Act, National Pollution Discharge Elimination System (NPDES) permit,
6   which is not needed until the COL stage. Based on the staff's independent assessment
7   described above and detailed in Appendix K, the staff concludes that the impacts of operation
8   on hydrological effects would be SMALL, and that mitigation is not warranted.
9

10 ## 5.3.2 Water-Use Impacts
11

12 *Changes to this section reflect water-use impacts from the closed-cycle, combination wet and*
13 *dry cooling system.*
14

15 Lake Anna, which was created as a source of cooling water for NAPS, has become a popular
16 recreation area, and the dam provides downstream flood control. The lake is not used as a
17 source of potable or industrial water, except for NAPS Units 1 and 2. The existing NAPS units
18 are the largest users of water in the region, and the addition of a third unit would add to this use.
19 Most of the water used at NAPS for Units 1 and 2 is drawn directly from Lake Anna for
20 condenser cooling. This use is non-consumptive, and the water is entirely returned to the lake
21 albeit at a warmer temperature. Although there is no consumptive use of water between the
22 intake and discharge, the elevated discharge temperature induces increased evaporative losses
23 from the remainder of the WHTF and Lake Anna resulting in a consumptive use of water.
24

25 Hanover County, one of four downstream counties, has identified a need for additional water
26 (Hanover County Department of Public Utilities 2004). The downstream users identified by
27 Hanover County are the county itself, the Doswell Limited Partnership Power Plant,
28 Paramount's King's Dominion Amusement Park, and the Bear Island Paper Company. To meet
29 its future projected demand, Hanover County proposes to withdraw 1.3 m^3/s (46 cfs) from the
30 North Anna River (Dominion 2005a). However, this diversion target withdrawal exceeds the
31 1.1 m^3/s (40 cfs) minimum discharge limit currently specified in the North Anna Lake Level
32 Contingency Plan (the "Lake Level Contingency Plan") for minimum releases during normal
33 conditions. The Lake Level Contingency Plan allows the flow to be further reduced to a
34 minimum of 0.57 m^3/s (20 cfs) during drought conditions. This potential conflict over water use
35 (which exists regardless of whether Unit 3 is ever constructed) falls within the regulatory
36 authority of the Commonwealth of Virginia.
37

38 Unlike the existing NAPS units, the majority of the water withdrawn from Lake Anna for Unit 3
39 condenser cooling would be consumed by the wet towers. Although there is some blowdown
40 from the wet towers, the discharge rate to the WHTF would be small (the full load maximum
41 would be 351 L/s (5565 gpm). Consumption of water by the wet towers would reduce the
42 overall volume of water in the lake, thereby impacting the quantity of water released at North
43 Anna Dam.

Because the North Anna Dam discharge rate is directly related to the Lake Anna surface level elevation, the lake level elevation analysis discussed above was used to estimate the impact on downstream flows in the North Anna River. The net total discharge from North Anna Dam would be reduced if Unit 3 operates as proposed. The staff determined that the fraction of time the dam would discharge 0.57 m³/s (20 cfs) increased from approximately 6 percent (Units 1 and 2 only) to 11 percent.

Because water supply generally exceeds demand, as indicated above, the staff concludes that the water supply provided by Lake Anna is adequate to meet Unit 3 and current downstream water demands except during periods of severe drought. Operation of Unit 3 would approximately double the duration of periods during drought conditions when the Lake Level Contingency Plan would be applied (i.e., when the lake level elevation would be below 75.6 m [248 ft] MSL).

Based on the foregoing, the staff concludes that during normal water years the water use impacts, including impacts on downstream users, would be SMALL, and mitigation is not warranted. During severe droughts, however, the impact to the water level could be temporarily MODERATE. Given the infrequent and temporary nature of the severe drought conditions, the fact that the minimum operational lake level elevation is 73.8 m (242 ft) MSL, and that lake level would return to normal with normal precipitation; further mitigation other than ceasing or derating operation is not warranted.

Although the staff concludes that the impact of proposed Unit 3 operation on downstream water users would be SMALL for most years and MODERATE during drought years, the staff considered mitigation in the form of increasing the reservoir storage capacity.

The staff evaluated changing the normal elevation of Lake Anna to mitigate the impact of consumptive water use associated with operation of Unit 3 on downstream flows during drought periods. The staff determined that raising the normal lake level by 25 cm (10 in.) would result in the same frequency of occurrence of 0.57 m³/s (20 cfs) discharge flows from the Lake Anna Dam as the current normal lake elevation of 76.2 m (250 ft) MSL with only NAPS Units 1 and 2 operating. Any decisions to change the normal lake elevation would be made by VDEQ.

5.3.3 Water Quality Impacts

Changes to this section reflect water treatment processes for the closed-cycle, combination wet and dry cooling system.

The discharged waste heat from operation of Unit 3 is not expected to appreciably change the water temperature of Lake Anna because the maximum blowdown flow rate (i.e., 351 L/s [5565 gpm] in EC mode and 245 L/s [3844 gpm] in MWC mode) is insignificant relative to the

1   combined discharges from Units 1 and 2 of 123,000 L/s (1,934,300 gpm). Because the Unit 3
2   cooling tower would consume water, the volume of water in Lake Anna would be reduced
3   (compared to operation of only Units 1 and 2 alone) when the lake level elevation is below
4   76.2 m (250 ft) MSL. However, assuming the heat rejection rate from operations of Units 1
5   and 2 remains constant, the reduced volume of water in the lake caused by Unit 3 operations
6   would result in an increase of average lake water temperature. Dominion addressed the rise in
7   lake temperatures caused by Unit 3 operations in section 5.2.2.1.3 of the ER, and found that the
8   average temperature rise in Lake Anna would be less than 0.06°C (0.1°F). The staff concurs
9   with the assessment.
10
11 The thermal impacts of Units 3 and 4 would be negligible because a temperature increase of
12 0.06°C (0.1°F) is insignificant. Based on the foregoing, the staff concludes that the thermal
13 impacts of the proposed new units is SMALL, and that mitigation is not warranted.
14
15 Because a specific design has not been selected, the ultimate water treatment systems for
16 proposed Units 3 and 4 have not been specified. Currently, raw cooling water from Lake Anna
17 used for condenser cooling and service water at NAPS Units 1 and 2 is not treated. Makeup
18 water for Unit 3, and the UHS systems for both Units 3 and 4 would be treated with biocides,
19 antiscalants, and dispersants. Treatment of makeup water for ultrapure water systems, such as
20 the condensate and primary cooling systems, would employ technologies such as reverse
21 osmosis and ultrafiltration.
22
23 The agency responsible for regulating the impacts on water quality of discharges into Lake
24 Anna is VDEQ. The water quality impact of effluents from Units 1 and 2 is regulated by a
25 VPDES permit that minimizes the impact on Lake Anna's water quality. Although Dominion
26 provided a chemical composition of Unit 3 blowdown in its PPE (Appendix I), the concentrations
27 of other waste streams that would discharge to the WHTF from operation of Unit 3 were not
28 defined.
29
30 An applicant for a CP or COL referencing an ESP for the North Anna ESP site would need to
31 provide information on the chemical effluents to the NRC. Based on its review, the staff
32 concludes that the issue of water quality impacts at the North Anna ESP site is not resolved.
33

34 ## 5.4   Ecological Impacts

35
36 *Changes to this section reflect ecological impacts from the Unit 3 closed-cycle, combination wet*
37 *and dry cooling system.*
38
39 Dominion has proposed that Unit 3 would be cooled using a closed-cycle, combination wet and
40 dry cooling system, and Unit 4 would be cooled using a closed-cycle system with dry cooling
41 towers. The potential impacts of new operating units on the hydrology of Lake Anna, shoreline

1   vegetation, habitats, and the associated terrestrial and aquatic ecology both of Lake Anna and
2   downstream are addressed in the following sections.
3

4   **5.4.1 Terrestrial Ecosystems**
5

6   *Changes to this section reflect the terrestrial ecological impacts from the closed-cycle,*
7   *combination wet and dry cooling system.*
8

9   This section discusses the impacts of the terrestrial ecosystems in the ESP site vicinity and
10   along the NAPS transmission line rights-of-way from the cooling systems associated with
11   operating the proposed new units at the North Anna ESP site. Closed-cycle heat dissipation
12   systems associated with nuclear power plants have the potential to impact terrestrial ecosystem
13   resources through salt drift, vapor plumes, icing, noise, and avian collisions with tall structures
14   (e.g., cooling towers). Each of these topics is discussed below.
15

16   **5.4.1.1 Cooling Tower Impacts on Terrestrial Ecological Resources**
17

18   *Changes to this section reflect terrestrial ecological impacts from the closed-cycle, combination*
19   *wet and dry cooling system.*
20

21   *Salt Drift*
22

23   Salt deposition can cause vegetation stress, either directly by deposition of salts onto foliage or
24   indirectly from accumulation of salts in the soil. An order-of-magnitude approach is typically
25   used to evaluate salt deposition on plants, because plant species sensitivities vary and
26   tolerance levels are not well documented. In this approach, deposition of sodium chloride at
27   rates up to 1 to 2 kg/ha/mo is not considered to be damaging to plants, while deposition rates
28   approaching or exceeding 10 kg/ha/mo during the growing season could cause leaf damage in
29   many species (NRC 2000a). All of the predicted deposition rates, both within and beyond the
30   site boundaries, are less than 1 kg/ha/mo.
31

32   No important terrestrial species or habitats are known to exist within the vicinity of the proposed
33   cooling towers. Important species as defined by the NRC (1999) include Federally or
34   State-listed threatened or endangered species, commercially or recreationally valuable species,
35   species essential to the maintenance and survival of rare or commercially valuable species, and
36   those that perform critical ecological functions or are biological indicators of ecosystem health.
37   Important habitats include any wildlife sanctuaries, refuges, preserves, or habitats identified by
38   State or Federal agencies as unique, rare, or of priority for protection; wetlands and floodplains;
39   and land areas identified as critical habitat for species listed by the U.S. Fish and Wildlife
40   Service (FWS) as threatened or endangered.
41

Operational Impacts at the Proposed Site

1   Because salt deposition rates are estimated to be less than 1 kg/ha/mo at all directions and
2   distances from the towers and there are no important terrestrial species or habitats likely to be
3   affected by salt deposition, the staff concludes that salt deposition impacts would be SMALL,
4   and mitigation is not warranted .
5
6   *Vapor Plumes and Icing*
7
8   The environmental impact of the operation of the wet cooling towers was evaluated by Dominion
9   using the SACTI computer model (ANL 1984), a suite of analytical tools developed by Argonne
10 National Laboratory to describe fogging, icing, salt deposition, and visible plumes from
11 traditional (e.g., non plume-abated) wet cooling towers. The model was developed specifically
12 for the Electric Power Research Institute (EPRI) for use in licensing power plants with
13 mechanical- or natural-draft cooling tower systems, has been verified with field data, and has
14 been used for many years. The SACTI program calculates the fogging, icing, salt deposition,
15 and plume height and length without consideration of water-saving techniques or features that
16 could be part of the design of the towers and result in a reduction of the size of the vapor plume.
17 Using a combination of atmospheric data from the NAPS site and National Weather Service
18 data from Richmond, Virginia, for the period 1998 to 2000, Dominion used the SACTI model to
19 calculate seasonal cooling tower plume characteristics (Dominion 2006a, b).
20
21 Dominion modeled vapor plumes and icing potential based on hourly temperature and relative
22 humidity data recorded from 1996 to 2000 in Richmond, Virginia; this modeling predicted that
23 operation of the new cooling towers would result in approximately 70 additional hours of fogging
24 per year and no additional icing. Vapor plumes (i.e., fog) produced by the cooling system would
25 have a minimal impact on the vision of flying birds, and would be unlikely to adversely affect
26 vegetation. Based on the foregoing, the staff concludes that the impacts of vapor plumes on
27 terrestrial resources would be SMALL, and mitigation is not warranted. Similarly, because no
28 icing was predicted by the model, the staff concludes that the impacts of icing would be SMALL,
29 and mitigation is not warranted.
30
31
32
33
34
35
36

5.4.1.2 Noise

Changes to this section reflect noise impacts from the closed-cycle, combination wet and dry cooling system.

Maximum noise levels from the operation of the reactors and dry cooling towers would be similar to current noise levels to which local species are adapted. Current noise levels at NAPS are occasionally as high as 100 decibels (measured at the security fence during outages), but they are typically less than 80 to 85 decibels, which is the threshold at which birds and small mammals are startled or frightened (Golden et al. 1980). Even with all combinations of wet and dry cooling towers in operation, noise levels from the cooling towers would be less than 65 decibels at the exclusion area boundary (Dominion 2006a). There are no important terrestrial species or important habitats in the vicinity of the site. Based on the foregoing, the staff concludes that the noise impacts to terrestrial ecological resources would be SMALL, and no mitigation is warranted.

5.4.1.3 Avian Collisions

Changes to this section reflect impacts to birds from collisions with the structures comprising the closed-cycle, combination wet and dry cooling system.

As discussed in the Draft EIS (NRC 2004a), the dry cooling towers proposed for Unit 4 heat dissipation are expected to be approximately 46 m (150 ft) tall, which is considerably less than the 71 m (234 ft) maximum value for the tallest building in the power block. The mechanical draft towers that may be used in the combination wet and dry cooling system for Unit 3 would be approximately 24 m (80 ft) tall. Nevertheless, for purposes of analyzing environmental impacts, Dominion's evaluation of avian collisions were based on a maximum structure height of 55 m (180 ft) for the cooling tower and 70 m (230 ft) for the maximum structure. No avian collisions with existing NAPS structures have been recorded, and the cooling towers would produce operational noise and air movements that would further decrease the likelihood of bird collisions. In view of the above, it is likely that bird collisions with the new towers would be rare. The North Anna ESP site is not within a major migratory bird concentration area along the Atlantic flyway (VDCR 2004). Dominion maintains a migratory bird protection program, including protection of nests and reporting of bird (especially raptor) strikes and other events (Dominion 2001a). Based on the foregoing, the staff concludes that impacts to birds from collisions with heat dissipation structures would be SMALL, and mitigation is not warranted.

5.4.1.4 Shoreline and Riparian Habitat

Changes to this section reflect shoreline ecological impacts related to the closed-cycle, combination wet and dry cooling system.

The increased water use and evaporation resulting from the addition of one new unit using wet cooling towers could increase the amount of shoreline exposed along Lake Anna or affect the length of time that the additional shoreline is exposed, as discussed in Section 5.3, above. This increased shoreline exposure could lead to alterations of the shoreline vegetation or enhance the introduction or spread of undesirable vegetation.

The staff evaluated the potential impacts of station operation on wetlands along the shoreline and upper reaches of Lake Anna using a hydrological assessment as discussed in Section 5.3.2, above, and in Appendix K. The maximum annual drawdown in most years would not differ greatly from that resulting from the operation of the existing units alone. The fraction of time that lake level would be at or below 75.0 m (246 ft) MSL would increase from 1 percent with two units operating to approximately 2 percent of the time with the addition of Unit 3. The surface elevation would be above 75.6 m (248 ft) MSL approximately 88 percent of the time with three operating units compared with approximately 94 percent of the time with the existing two operating units. The normal pool elevation is 76.2 (250 ft) MSL. The staff determined that the difference between the lake level with and without Unit 3 would be less than 7.6 cm (3 in.) approximately 69 percent of the time, less than 15 cm (6 in.) approximately 85 percent of the time, and less than 30 cm (1 ft) approximately 94 percent of the time. All of the periods for which the difference in the lake surface elevation with and without Unit 3 was predicted to be greater than 30 cm (1 ft) would have occurred during the two major drought events of 1980 to 1981 and 2001 to 2002.

Differences in surface elevations that fluctuate between 0 and 15 cm (0 and 6 in.) are likely to have no discernable effect on shoreline vegetation or wetlands. During the occasional periods when there are greater differences in the surface elevation, there could be noticeable temporary changes in the shoreline and wetland vegetation. Upper areas may dry out, and lower, normally inundated areas may develop stands of wetland vegetation over time. However, the increased drawdown is expected to be temporary, and even if the additional drawdown lasts for a year or more, any observable changes would not be detectable within a relatively short time after the water level returns to normal. Riparian and wetland vegetation is adapted to survive fluctuating water levels and periodic drought conditions without detectable long-term effects. Therefore, the staff concludes that the impacts to shoreline vegetation and habitats would be SMALL, and mitigation is not warranted.

The VDEQ identified the possibility of raising the lake level 15 to 30 cm (6 to 12 in.) to mitigate the impact on North Anna River downstream flows (Dominion 2006a). Dominion evaluated this potential option in Revision 6 of the ER. Dominion stated:

1   Dominion evaluated shoreline areas in an effort to assess, in general, various
2   impacts of potentially raising normal operating lake level 6 inches to 12 inches
3   above 250 ft. MSL, in the event a Virginia permitting agency process determined
4   the need for such an action. [Note: Raising normal operating lake level is not
5   being proposed to demonstrate site suitability. And though not currently
6   proposed, Virginia DEQ could require an increase in lake level to mitigate
7   impacts on down-river flows. Increasing the lake level by approximately 7 inches
8   would eliminate changes in the frequency and duration of the 20 cfs minimum
9   instream flow.]
10
11   On May 3, 2006, the staff toured Lake Anna with the applicant (NRC 2006b) and discussed the
12   option of raising the lake level between 15 to 30 cm (6 to 12 in.) to mitigate the impacts on
13   downstream flows. If the lake level were raised 15 to 30 cm (6 to 12 in.), it could impact dock
14   owners and could affect near-shore wetlands, especially the upper reaches of the lake where
15   the tributary streams are enter the North Anna River and in the areas uplake of the North Anna
16   Dam. In areas of relatively steep banks, there would be little affect on wetlands. In the area
17   below the State Road 208 bridge, the change to the wetlands would be most evident due to the
18   gradual slope of the shoreline. The net effect of raising the lake level would be to shift the
19   wetlands, but it would not result in a significant gain or loss of wetlands. The authority to raise
20   the lake level resides with the Commonwealth of Virginia.
21
22   Evaporative losses resulting from the operation of the wet cooling system for Unit 3 could cause
23   decreased flows downriver. Reduced flows could alter the riparian vegetation and habitat for
24   riparian and wetland species along the North Anna River. The staff's hydrological analysis
25   demonstrates that the fraction of time that weekly average outflow from the North Anna Dam
26   would be at or below 1.1 m^3/s (40 cfs) would increase from approximately 63 percent of the time
27   with NAPS Units 1 and 2 operating to 66 percent with the addition of Unit 3 (Dominion 2006a).
28   The analysis also predicted that the fraction of time that the outflow would be at 0.57 m^3/s
29   (20 cfs) would increase from 6 percent under current two-unit operations to approximately
30   11 percent of the time with the addition of Unit 3. Under the scenario with just NAPS Units 1
31   and 2 operating, the model predicted two periods (1998, and 2001 to 2002) when the weekly
32   average outflow would have dropped to 0.57 m^3/s (20 cfs). With the addition of Unit 3, the
33   model predicted an additional seven such periods during the 1978 to 2003 modeling period. In
34   almost all cases, the 0.57 m^3/s (20 cfs) average weekly outflow conditions commenced in
35   October, lasted for approximately two weeks to several months, and then returned to higher
36   outflow levels by the end of January.
37
38   In 1981 and 1999, the low outflow period would have commenced in early to mid-August. There
39   would have been low flow 0.57 m^3/s (20 cfs) conditions for over 14 months during the 2002
40   drought under both of the modeled conditions (i.e., the baseline with NAPS Units 1 and 2
41   operating and the proposal with Unit 3 operating as well). Low flow would have commenced in
42   early October 2001 with three units operating rather than in late October with two units

operating, and would have lasted until mid- to late December 2002 with three units operating rather than early December with just two units. Therefore, although low outflow conditions in the North Anna River were modeled to occur at a noticeably higher fraction of the time with the addition of Unit 3 compared to current conditions, in all but two cases these low-flow periods occurred during portions of the year when the riparian vegetation would have either stopped growing or would already be dormant and is, therefore, not likely to adversely affect growth or reproduction.

Therefore, it is not likely that a period of reduced outflow approximately every few years would noticeably affect the riparian vegetation downstream. The staff's analysis identified 2 out of 25 years when the 0.57 m^3/s (20 cfs) outflow conditions would have commenced during August. Low outflow during the growing season could have short-term effects on riparian vegetation; however, riparian vegetation is adapted to survive periodic fluctuations in water level and drought conditions without detectable long-term effects. The staff's analysis determined that there would be periods of reduced i.e., (0.57 m^3/s [20 cfs]) outflow during the growing season approximately once a decade. Therefore, the changes in the flow regime are not expected to noticeably change the quantity, distribution, or characteristics of the riparian or wetland vegetation and habitats along the North Anna River between the North Anna Dam and the confluence with the South Anna River. Therefore, the staff concludes that impacts of the additional units on downstream riparian habitats would be SMALL, and mitigation is not warranted.

5.4.1.5 Transmission Line Rights-of-Way

This section is not affected by the changes presented in ER Revision 6 and is provided solely for context.

In the Draft EIS (NRC 2004a), the staff previously evaluated Dominion's procedures to ensure that Dominion staff could identify and avoid rare and sensitive plant species in the NAPS transmission line rights-of-way or would modify mechanical and herbicide treatment practices used to avoid adverse impacts. In its analysis of the application to renew the operating licenses for NAPS Units 1 and 2, the staff determined that continued operation and maintenance of the transmission lines rights-of-way would not adversely impact terrestrial resources (NRC 2002b). Based on the foregoing and because there would be no new lines or alterations of the existing rights-of-way, the staff concludes that the operational impacts of Units 3 and 4 on terrestrial ecological resources did not change. The impact level category would still be SMALL, and mitigation is not warranted.

5.4.1.6 Summary of Terrestrial Ecosystems Impacts

Changes to this section reflect terrestrial ecological impacts from the closed-cycle, combination wet and dry cooling system.

The staff considered potential impacts to terrestrial ecological resources of operating the proposed Units 3 and 4, including salt drift; fogging; icing; noise; avian collisions; changes to shoreline, riparian, and wetland habitat; and transmission line rights-of-way. Based on its analysis and independent review, the staff concludes that the operational impacts of Units 3 and 4 operations on terrestrial ecological resources would be SMALL, and mitigation is not warranted.

5.4.2 Aquatic Impacts

Changes to this section reflect aquatic ecological impacts from the closed-cycle, combination wet and dry cooling system.

This section discusses the impacts on the Lake Anna and the North Anna River aquatic ecosystems from the cooling systems associated with operating the proposed new units at the North Anna ESP site. The potential impacts to the aquatic environment are expected to be related solely to the operation of Unit 3. Unit 4 is expected to use a closed-cycle, dry cooling system that uses almost no cooling water. Therefore, this analysis focuses on Unit 3 operational impacts.

5.4.2.1 Intake System

Changes to this section reflect intake structure for the closed-cycle, combination wet and dry cooling system and its associated impacts on the aquatic ecology in Lake Anna.

The existing cooling water system for NAPS Units 1 and 2 is a once-through design that withdraws water from the Lake Anna reservoir. At maximum capacity, Units 1 and 2 withdraw 122,000 L/s (1,934,300 gpm), or about 2.8 percent of the total Lake Anna volume per day (3.76×10^8 m^3 at 76.2 m MSL [305,000 ac ft at 250 ft MSL]). Each unit uses four circulating water pumps to withdraw condenser cooling water from Lake Anna, and the water is withdrawn through screens located in a cove north of the station. Each screen well contains four individual bays, and each bay is equipped with a trash rack, a traveling screen, and a vertical, motor-driven, circulating water pump. The trash racks consist of 1.3-cm (0.5-in.) wide by 8.9-cm (3.5-in.) thick vertical bars spaced 10.2 cm (4.0 in.) on center. Water flows through the trash racks at about 0.2 m/s (0.69 ft/s) (VEPCo 1985). Traveling screens in the front of the cooling water pumps filter the water and protect the pumps from damage and clogging. The traveling screens, constructed of 14-gauge wire with 9.5-mm (0.37-in.) square openings are designed to rotate once every 24 hr or whenever a predetermined differential pressure exists across the

1   screens. Debris collected at the trash racks is removed by mechanical rakes. Debris and fish
2   collected in the wire baskets are disposed of as solid waste (VEPCo 1985). After passing
3   through the plant, water is returned to the WHTF, which is separated from the main part of the
4   lake by a series of dikes.
5
6   The cooling water intake system can affect aquatic communities by either impingement or
7   entrainment. Impingement occurs when swimming organisms are not strong enough to escape
8   the cooling water intake current and are caught or stuck on the screens (i.e., impinged).
9   Impinged organisms are generally fish, but can occasionally include other semi-aquatic animals
10   such as amphibians (e.g., frogs, turtles, and salamanders), waterfowl (e.g., ducks and coots), or
11   mammals (e.g., muskrats). The screens are periodically cleaned using a spray wash system to
12   remove impinged organisms. Impingement mortality varies with species, but is considered to be
13   100 percent because NAPS Units 1 and 2 do not have a fish return system.
14
15   Entrainment is the passage of organisms through the traveling screens into the cooling water
16   system. Entrained organisms are generally small and include phytoplankton, zooplankton, and
17   fish eggs and larvae. As these entrained organisms pass through the cooling water system,
18   they are subjected to a variety of stresses that may result in mortality. Impacts to the entrained
19   organisms include physical damage from contact with pumps, pipes, and condensers;
20   pressure-related damage from passage through pumps; damage from shear associated with
21   complex water flows; damage from exposure to elevated temperatures in the condenser
22   passage; and potential exposure to toxic chemicals added to the cooling water system.
23   Entrainment mortality varies by species, but is considered to be 100 percent for closed-cycle
24   cooling systems similar to the one proposed for Unit 3.
25
26   Dominion originally proposed a once-through cooling system for Unit 3, with a cooling water
27   intake structure approximately 46 m (150 ft) long and 61 m (200 ft) wide to house the trash
28   racks, traveling screens, and intake pumps (Dominion 2004a). Shoreline contouring and
29   channel dredging were expected to occur and would have posed a potential source of
30   temporary impacts on the reservoir ecosystem in the immediate vicinity during construction
31   activities. In ER Revision 6 (Dominion 2006a), Dominion proposes to reduce the size of the
32   intake structure to support the closed-cycle, combination wet and dry cooling system to 21 m
33   (70 ft) long and 21 m (70 ft) wide because of a reduced demand for water. Assessments of
34   impact described in this SDEIS assume that Unit 3 withdraws water from a new intake structure
35   and that the maximum water withdrawal associated with Unit 3 would be 1400 L/s (49.6 cfs).
36   This represents a 1.4 percent increase in water withdrawal from Lake Anna compared to the
37   current water withdrawal of about 120,000 L/s (4300 cfs) to support Units 1 and 2. Because
38   Unit 4 employs a closed-cycle, dry cooling system, water use is negligible compared to the
39   other three units and is not considered in this impact assessment.
40

1 ### 5.4.2.2 Impingement
2
3 *Changes to this section reflect impingement losses resulting from the reduced circulating water*
4 *flow rate of the closed-cycle, combination wet and dry cooling system.*
5
6 In 1985, Virginia Power (VEPCo) published *Impingement and Entrainment Studies for North*
7 *Anna Power Station, 1978-1983* (VEPCo 1985). This document described the study design and
8 results associated with work conducted under Section 316(b) of the Clean Water Act in
9 compliance with the NAPS Environmental Technical Specifications and the then existing
10 VPDES permit under Special Conditions: Environmental Studies ("the Section 316(b) study").
11 The objective of the Section 316(b) study was to examine the effects of impingement and
12 entrainment associated with the cooling water intake system supplying Units 1 and 2 to
13 determine whether NAPS operations adversely affect fish populations in the Lake Anna
14 reservoir.
15
16 During the study years (1979 to 1983), an average of just over 47,400 fish representing
17 34 species were collected annually for each full year. Results for 1978 were not included in the
18 analysis because sampling occurred only from April to December. During each sampling
19 episode, traveling screens were washed to ensure all fish were collected, and decayed fish or
20 fish assumed to have been dead for more than 24 h were discarded. The remaining fish were
21 identified by species and counted, and up to 50 individuals of each species were weighed and
22 measured. By relating the number of fish impinged to the sampling duration and measured
23 intake flow, it was possible to estimate daily, monthly, and yearly impingement values for
24 species of interest.
25
26 The Section 316(b) study results were based on the operating conditions that existed at that
27 time and the intake configurations and specifications described in ER Revision 6 and the Draft
28 EIS (NRC 2004a). The study found that six fish species accounted for 99 percent of all fish
29 impingements (Table 5-1). The gizzard shad, *Dorosoma cepedianum,* was the species most
30 commonly impinged, and accounted for 61 percent of the observed impingements during the
31 study period (Table 5-1). More than 80 percent of the total impingements occurred from
32 January to April (Table 5-2).
33
34 Concurrent with the study, cove rotenone sampling in Lake Anna was conducted from 1979 to
35 1983 to determine the impact of the estimated impingements on species biomass. This cove
36 rotenone sampling determined that gizzard shad impingements associated with the
37 once-through cooling system used for Units 1 and 2 represented 0.32 percent of the total lake
38 biomass of that species. Impingement impacts for the other representative important fish
39 species (RIS) expressed as the biomass lost to impingement ranged from 0.02 percent for the
40 bluegill *Lepomis macrochirus,* to 3.8 percent for the black crappie, *Pomoxis nigromaculatus,*
41 (Table 5-1).
42
43

1
2
3

Table 5-1. Fish Species Most Commonly Impinged at Existing Units 1 and 2 (1979 to 1983) (Dominion 2006a)

Common Name	Scientific Name	Percent of Total Impingement	Estimated Percent of Lake Anna Total Biomass by Species
Gizzard shad	*Dorosoma cepedianum*	61	0.32
Black crappie	*Pomoxis nigromaculatus*	16	3.8
Yellow perch	*Perca flavescens*	16	1.4
Bluegill	*Lepomis macrochirus*	4	0.02
White perch	*Morone americana*	1	0.1
Striped bass	*Morone saxatilis*	1	no data

4
5
6
7
8
9
10
11
12
13

Table 5-2. Estimated Mean Monthly Impingement of RIS

14

Month	Estimated Mean Impingement per Month (all species)	
	Existing Units[a] (1979-1983)	Unit 3 Combination Wet and Dry Cooling System[a]
January	16,012	310
February	30,873	811
March	93,955	3258
April	15,702	480
May	4364	117
June	1560	37
July	1034	20
August	1680	30
September	2166	37
October	4454	101
November	5360	123
December	5280	116
Yearly Total	182,440	5440

(a) ER Revision 6 (Dominion 2006)

15
16
17
18
19
20
21
22
23
24
25
26
27
28
29
30
31

32 To assess the impacts of impingement associated with the closed-cycle, combination wet and
33 dry cooling system proposed for Unit 3, the staff compared estimates of impingement
34 associated with existing Units 1 and 2 derived from the Section 316(b) study to the losses
35 predicted for Unit 3 proposed combination wet and dry cooling system design
36 (Dominion 2006a). To assess impingement losses for Unit 3, Dominion used an intake flow rate
37 of 1723 L/s (27,309 gpm) in their calculations. This represents a maximum flow rate through the
38 intake and results in a conservative (environmentally protective) estimate of losses. Typically,

1 normal plant cooling tower makeup water needs are expected to be 971 L/s (34.3 cfs). In
2 developing its estimate of impingement losses for Unit 3, Dominion assumed the current
3 fish distribution and composition would be the same as that observed during the Section 316(b)
4 study, that a new cooling water system would operate at 100 percent pumping capacity, and
5 that the intake screen mesh size and flow velocity for Unit 3 would be the same as that of the
6 existing units.
7
8 Because the water use associated with the closed-cycle combination wet and dry cooling
9 system is much less than that of NAPS Units 1 and 2, impingement rates resulting from Unit 3
10 would be significantly less than current impingement rates (Table 5-2). Based on the
11 information provided by Dominion in ER Revision 6, adding the Unit 3 combination wet and dry
12 cooling system to the existing once-through cooling system for Units 1 and 2 employed at NAPS
13 would increase average yearly impingement from Lake Anna from 182,440 to 187,880
14 individuals, or approximately 1 percent. Moreover, these estimates are probably conservative
15 because of the flow estimates used for Unit 3 intake flow. Accordingly, the staff concludes that
16 there would be a negligible decrease in fish biomass in Lake Anna.
17
18 The Section 316(b) study conducted at NAPS from 1978 to 1983 concluded that impingement
19 associated with the once-through cooling system employed by Units 1 and 2 had not resulted in
20 significant impacts to the fish communities of Lake Anna. Subsequent monitoring since that
21 time (VEPCo 2002) has shown that fish populations in the reservoir are healthy and diverse.
22 Because the closed-cycle, combination wet and dry cooling system proposed for Unit 3
23 increases the total cooling water withdrawal from Lake Anna by approximately 1 percent and
24 would result in a minimal increase in impingement and fish biomass loss relative to the current
25 Unit 1 and 2 operations, the staff concludes that the impacts of impingement would be SMALL.
26 Although the staff concludes that the impacts of impingement to the Lake Anna fishery would be
27 small, further mitigation could reduce losses if the intake through screen flow velocity were
28 designed to be less than 0.15 m/s (0.5 ft/s), as recommended by the EPA Section 316(b).
29
30 **5.4.2.3 Entrainment**
31
32 *Changes to this section reflect entrainment losses resulting from the reduced circulating water*
33 *flow rate of the closed-cycle, combination wet and dry cooling system.*
34
35 The Section 316(b) study described above also considered entrainment. Entrainment samples
36 were collected once a week in front of the intake forebays from 1979 to 1983. Sampling was
37 conducted from March through July of each year, which encompassed the spawning periods of
38 certain reservoir fish species (e.g., the bluegill, *L. macrochirus*; the yellow perch, *Perca*
39 *flavescens*; the black crappie, *P. nigromaculatus*; the white perch, *Morone americana*; and the
40 largemouth bass, *Micropterus salmoides* (VEPCo 1985). An average of 1318 fish larvae were
41 collected annually in the entrainment samples. Entrainment samples did not contain fish eggs
42 because most of the species in the reservoir produce demersal (near the lake bottom) adhesive

1   eggs that are not generally entrained. For purposes of the Section 316(b) study and as a
2   conservative estimate of impact, Dominion assumed 100 percent mortality for all larval fish
3   entrained (VEPCo 1985).
4
5   As previously described in Section 5.4.2.2, the staff analysis was based on the Section 316(b)
6   study results and the intake configurations and specifications described in ER Revision 6 and
7   the Draft EIS. The Section 316(b) study results determined that the larvae of five species
8   accounted for the majority of larval entrainment, with the largest entrainment abundances
9   associated with the larvae of gizzard shad (*D. cepedianum*) at 65 percent (Table 5-3). Larvae
10 of white perch (*M. americana*) accounted for 15 percent of the entrainment abundances during
11 the study, and larvae of sunfish (family Centrachidae) accounted for 13.3 percent of the
12 entrainment abundances. Larvae of the yellow perch and black crappie accounted for less than
13 6 percent of entrainment abundances (Table 5-3). Based on the duration of entrainment
14 sampling, sampling gear used, and the measured flow, total larval entrainment was calculated
15 by month and by year for each species during the study period (Table 5-4).
16
17 Because larval abundance in Lake Anna is not known, it was not possible to determine the
18 percentage of larvae entrained based on the actual abundance in the lake. The adult equivalent
19 model of Goodyear (1978) was used with the following assumptions to assess the population
20 impacts caused by the loss of fish larvae from entrainment by the existing NAPS units:
21 (1) 100 percent mortality of entrained larvae; (2) stock populations are at equilibrium, and the
22 total lifetime fecundity produces two adults; (3) no compensatory mechanisms are operating;
23 and (4) 75 percent of the eggs produced by the entrained species survive to the larval stage.
24 The model estimated the number of adult fish that would have resulted from the larvae had they
25 not been lost to entrainment, and also provided an estimate of the potential percent reduction in
26 the adult fish population as a consequence of entrainment. Predicted reductions in fish
27 populations ranged from 0.01 percent for black crappie in 1978 and 1979 and sunfish in 1982,
28 to 4.13 percent for gizzard shad in 1980. Dominion concluded that reductions of this magnitude
29 would not be expected to have a significant adverse effect on the reservoir fish populations for
30 those species, especially when viewed in concert with other population mechanisms such as
31 natural compensation (VEPCo 1985). The analysis from the adult equivalent model provided a
32 conservative estimate of entrainment impact by the existing units, primarily as a result of
33 assumptions used in the analysis (VEPCo 1985).
34
35 To assess the impacts of entrainment for the proposed closed-cycle, combination wet and dry
36 cooling system, the staff compared estimates of entrainment from the Section 316(b) study for
37 the existing units to calculate entrainment losses predicted for Unit 3. As described above,
38 Dominion used an intake flow rate of 1723 L/s (27,309 gpm; 60.8 cfs) to estimate entrainment
39 for Unit 3. This represents a maximum flow rate through the intake and results in a
40 conservative (environmentally protective) estimate of losses. Typically, normal plant cooling

Table 5-3. Larval Fish Species Most Commonly Entrained at Existing Units (1978 to 1983)
(Dominion 2006a)

Common Name	Scientific Name	Estimated Percent of Total Entrainment
Gizzard shad	*Dorosoma cepedianum*	65.7
White perch	*Morone americana*	15.0
Sunfishes	*Lepomis* sp.	13.3
Yellow perch	*Perca flavescens*	4.9
Black crappie	*Pomoxis nigromaculatus*	1.0

Table 5-4. Estimated Mean Monthly Entrainment of Larvae of Common Fish Species

	Estimated Mean Entrainment per Month (all species)	
Month	Existing Units[a] (1978-1983)	Unit 3 Combination Wet and Dry[a]
March	223,513	5251
April	10,600,874	272,335
May	71,160,116	1,644,107
June	55,855,069	1,204,313
July	11,521,058	230,416
Yearly Total	149,390,630	3,354,224

(a) ER Revision 6 (Dominion 2006)

tower makeup water needs are 971 L/s (34.3 cfs). Dominion assumes the fish larvae
distribution and composition has remained the same as in the Section 316(b) study, that a new
cooling water system would operate at 100 percent efficiency, and that the intake screen mesh
size and flow velocity for the new unit would be the same as that of the existing units.

Because the water use associated with the closed-cycle, combination wet and dry cooling
system is much less than that of the once-through cooling system for Units 1 and 2, entrainment
at Unit 3 would be significantly less than at the existing units (Table 5-4). Based on the
information provided by Dominion in ER Version 6 (Dominion 2006a), adding the Unit 3 cooling
system to the existing once-through cooling system for Units 1 and 2 would increase yearly
entrainment from about 149,400 larvae to about 152,800, or approximately 1 percent.
Moreover, these estimates are likely conservative because of the high intake flow estimates
used for Unit 3. Accordingly, the staff concludes that there would be a negligible loss of larval
fish in the North Anna fish communities.

1   Based on the results of the Section 316(b) study and a comparison of entrainment abundances
2   expected to occur using the Unit 3 closed-cycle combination wet and dry cooling system, as
3   described above, the staff concludes that the impacts of entrainment of Unit 3 operations
4   superimposed on Units 1 and 2 would be negligible. The fish populations most susceptible to
5   larval entrainment represent a balanced community in Lake Anna. Over the years fishery
6   management of the reservoir has matured and changed to meet the demands for public fishing
7   through species additions (i.e., threadfin shad, *Dorosoma petenense*, and annual stockings of
8   striped bass, *M. saxatilis*, and walleye *Stizostedion vitreum*). Overall, the abundance and
9   quality of the fisheries have remained healthy and balanced despite increased fishing pressure
10   and shoreline property development. Because of the thriving populations of game fish and the
11   forage species that support them, the staff concludes that the additional entrainment resulting
12   from the operation of the Unit 3 closed-cycle, combination wet and dry cooling system would
13   represent a minor increase in entrainment, and the impacts on aquatic species in Lake Anna
14   would be SMALL.
15
16   Although the staff concludes the impacts of entrainment for Unit 3 would be SMALL, it
17   considered further mitigation by employing 1.0-mm (0.04-in.) mesh screening on the traveling
18   screen intakes associated with Unit 3. Replacing the existing 9.5-mm (0.37-in.) mesh screening
19   with 1.0-mm (0.04-in.) mesh would physically exclude the entrainment of most larvae and eggs
20   from Lake Anna. However, these organisms would be impinged on the screens because they
21   have little or no motility and cannot avoid the intake. These life forms would experience
22   mortality rates of close to 100 percent because they are fragile and are unable to escape from
23   the surface of the screens. The use of fine mesh screen technology to reduce entrainment is
24   typically employed in riverine environments where a sweeping current is present because of
25   downstream flow. A sweeping current naturally removes surface debris and impinged
26   organisms. Such sweeping flows are not present in the North Anna reservoir. Therefore, the
27   staff concludes that the use of fine-mesh screening would not significantly reduce the already
28   small entrainment losses predicted for Unit 3 operation, and mitigation is not warranted.
29
30   **5.4.2.4 Aquatic Thermal Impacts**
31
32   *Changes to this section reflect aquatic impacts from thermal discharge from the closed-cycle,*
33   *combination wet and dry cooling system.*
34
35   This section discusses the potential thermal impacts to the aquatic resources of Lake Anna from
36   the discharge of heated blowdown water from the proposed Unit 3 combination wet and dry
37   cooling system. This water would enter the existing canal, mix with the water released from the
38   NAPS Units 1 and 2 cooling system, pass into the WHTF, and ultimately flow back into Lake
39   Anna and the North Anna River. Fish and other aquatic flora and fauna could be affected if
40   there are rapid changes in water temperatures above or below their tolerance range. The staff
41   evaluated the thermal impacts on the lake's ecosystem and described the water-use impacts of
42   the cooling system for an additional unit. Except where more detailed data were available, the

1   design parameter values from the PPE (Appendix I) were used as the basis for the analysis and
2   evaluation of the Unit 3 discharge system. The staff described the physical attributes of the
3   discharge system in Section 3.2.2.
4
5   Currently, cooling water from NAPS Units 1 and 2 is discharged at a rate of 120 m^3/s (4246 cfs)
6   into the WHTF at a temperature of approximately 8°C (14°F) above intake ambient conditions.
7   The water flows through a series of lagoons and connecting canals and returns to Lake Anna at
8   Dike 3, which is just above the North Anna Dam. Waste heat is transferred to the atmosphere
9   mostly by evaporation, conduction, and back radiation, and only a small fraction of waste heat is
10   transported downstream via the North Anna Dam. Dominion estimates that with both NAPS
11   Units 1 and 2 operating, the cooling water residence time in the WHTF is approximately 7 days,
12   and that about one-half of the waste heat is dissipated in that time. Virtually all of the remaining
13   heat is dissipated to the atmosphere from the surface of Lake Anna either by evaporation or
14   radiation. The general characteristics of Lake Anna include a more riverine environment
15   upstream to a lacustrine environment downstream. Thus, the middle and lower portions of the
16   lake are generally stratified during the summer and mixed or weakly stratified in winter.
17
18   *Cold Shock*
19
20   Cold shock occurs when aquatic organisms that have been acclimated to warm water, such as
21   fish in a power plant's discharge canal, are exposed to a sudden temperature decrease. This is
22   more likely to occur if a single-unit power plant shuts down suddenly in winter. It is less likely to
23   occur at a multiple-unit plant, because a sudden temperature decrease is moderated by the
24   heated discharge from the remaining unit or units that continue to operate. Cold shock
25   mortalities at U.S. nuclear power plants are relatively rare and typically involve small numbers of
26   fish (NRC 1996).
27
28   Winter kills of fish have occurred in Lake Anna associated with cold weather and unusually cold
29   water temperatures, but plant operations were not a factor (NRC 2004a). During February and
30   March 1979, large numbers of gizzard shad were killed or stunned when Lake Anna water
31   temperatures fell below 2.2°C (36°F) (VEPCo 1985). These fish drifted into the existing units'
32   intake and were observed in impingement samples. The susceptibility of gizzard shad and
33   threadfin shad to winter kills is well known, and limited threadfin shad kills have occurred during
34   severe winters. The threadfin shad is native to the Gulf slope of the United States, peninsular
35   Florida, and Central America, and was introduced as a forage fish to a number of Virginia
36   impoundments in the 1950s, 1960s, and 1970s (Jenkins and Burkhead 1994). Because this
37   species is subject to cold kills when water temperatures drop below 8.9°C (48°F), it is able to
38   overwinter in northern latitude impoundments only when waters are heated by power plant
39   effluents (Olmsted and Clugston 1986).
40
41   As noted above, incidents of cold shock in receiving waters of nuclear power plants are
42   infrequent, and even more infrequent at multiple-unit plants. Because the maximum blowdown

1   discharge temperature is 37.8°C (100°F), and the maximum blowdown flow rate is 351 L/s
2   (12.4 cfs), the addition of the Unit 3 closed-cycle, combination wet and dry cooling system
3   represents a minor contribution of water and heat to the existing discharges associated with
4   NAPS Units 1 and 2. Thus, the presence of the Unit 3 closed-cycle, combination wet and dry
5   cooling system would have little or no impact on increasing the number of fish acclimated to
6   elevated water temperature. Moreover, the presence of multiple units at a site reduces the
7   effect of cold shock because it is unlikely that all units would cease operation simultaneously.
8   Based on the foregoing, the staff concludes that the impacts of operation from cold shock on
9   aquatic resources would be SMALL, and mitigation is not warranted
10
11   *Heat Stress*
12
13   As described in the Draft EIS, the thermal tolerance for aquatic organisms is defined in different
14   ways (NRC 2004a). Some definitions relate to the temperature that causes fish to avoid the
15   thermal plume, other definitions relate to the temperature that fish prefer for spawning, and
16   others relate to the temperatures (upper and lower) that may kill individual fishes. Some of
17   these tolerances are termed preferred temperatures, upper avoidance temperatures, and lethal
18   temperatures. A list of these tolerances for several important species found in the reservoir was
19   compiled in ER Revision 3 (Dominion 2004a) and is discussed in the Draft EIS (NRC 2004a).
20
21   In Section 5.3, the staff describes its independent assessment of the incremental impacts of the
22   proposed Unit 3 on the water temperatures within Lake Anna. The negligible increase in flow
23   and heat load from Unit 3 relative to the existing flow and heat load from Units 1 and 2 would
24   result in a negligible increase in the temperature and associated heat stress that fish in Lake
25   Anna would experience. Although temperature-related fish kills have been reported in the lake
26   and may continue to occur, it is unlikely that the operation of the Unit 3 closed-cycle,
27   combination wet and dry cooling system would result in increased mortality beyond that
28   observed during two unit operation, given the minor flow and thermal inputs to the WHTF and
29   lake. Additionally, many fish found in the lake are prolific, exhibit a high reproductive potential,
30   and compensate to offset losses. Based on the foregoing, the staff concludes the thermal
31   impacts on the fish populations of the discharge of waste heat from Unit 3 into Lake Anna would
32   be SMALL, and mitigation is not warranted.
33
34   **5.4.2.5 Striped Bass**
35
36   *Changes to this section reflect impacts to the striped bass population from thermal discharge*
37   *from the closed-cycle, combination wet and dry cooling system.*
38
39   Striped bass, *M. saxatilis*, are well adapted to residence in fresh water, and are often chosen for
40   the development of a recreational fishery in inland reservoirs. Considered a cool-water species,
41   striped bass become less mobile as the water warms and seek out thermal refuges where water
42   temperature is less than 26°C (79°F) and dissolved oxygen levels are at least 3 or 4 mg/L

1   (Cheek et al. 1985). Because striped bass are sensitive to changes in temperature and
2   vulnerable to thermal stress, some reservoir populations are susceptible to summer die-offs
3   because of elevated lake temperatures and diminished dissolved oxygen levels. In reservoirs
4   similar to Lake Anna in the southern United States, striped bass are generally found in deeper,
5   colder water during the summer months, and tend to congregate near the thermocline. The
6   striped bass fishery in Lake Anna is supported by stocking; there is no evidence that this
7   species spawns in Lake Anna or the North Anna River. Spawning has been documented in the
8   Pamunkey River during the months of April and May.
9
10 *Impacts to Striped Bass Populations in Lake Anna*
11
12 Evidence suggests that unusually high air temperatures and low rainfall in summer (e.g., the
13 drought conditions that occurred in central Virginia from 2001 to 2002) can reduce striped bass
14 habitat in some portions of Lake Anna. The closed-cycle, combination wet and dry cooling
15 system proposed for Unit 3 would contribute less than 1 percent to the current discharge flow of
16 heated water to the discharge canal and WHTF, and the maximum temperature of the
17 blowdown water (38°C [100°F]) would be within the range of temperatures currently observed in
18 the discharge canal in July and August of 37.0° to 39.1°C (98.6° to 102.4°F) (Dominion 2006a).
19 Unit 4 would use dry cooling only and would not have thermal impacts on the lake.
20
21 Based on the above, the staff determined that waste heat input to Lake Anna from the Unit 3
22 closed-cycle, combination wet and dry cooling system would not appreciably contribute to the
23 thermal heating that already occurs in Lake Anna because of natural and anthropogenic
24 (derived from human activities) inputs. Lake Anna currently is stratified during the summer and
25 is mixed or weakly stratified in winter; this pattern would not change with the operation of Unit 3.
26 Although NAPS Units 1 and 2 operations during extended drought periods decrease striped
27 bass habitat, there is no evidence indicating that suitable habitat would be eliminated in the
28 mid-lake and upper-lake areas. Because Unit 3 would not significantly contribute to the current
29 thermal inputs to Lake Anna, the staff concludes that the thermal impacts to striped bass from
30 the operation of the Unit 3 closed-cycle, combination wet and dry cooling system and the Unit 4
31 dry cooling system would be SMALL, and mitigation is not warranted.
32
33 *Impacts to Other Striped Bass Populations in the Vicinity of the ESP Site*
34
35 Striped bass are known to occur in the North Anna River downstream of the dam, but these fish
36 are believed to have passed through or over the dam from Lake Anna. Striped bass are known
37 to occur and spawn successfully in the Pamunkey River, but they are unlikely to venture above
38 the fall line during their spawning migrations (Jenkins and Burkhead 1994). Because of the
39 distance downstream from the dam to the spawning area on the Pamunkey River (river km 119
40 [mi 74]), the waste heat discharged from Units 1 and 2 is not detectable and does not affect
41 anadromous striped bass populations or known spawning habitat downstream of the dam.
42 Because the Unit 3 closed-cycle, combination wet and dry cooling system and the Unit 4 dry

1 cooling system would not significantly contribute to the current thermal inputs to the WHTF or
2 the lake from the operation of NAPS Units 1 and 2, the staff concludes that the thermal impacts
3 to striped bass in the North Anna/Pamunkey Rivers would be SMALL, and mitigation is not
4 warranted.
5
6 *Lake Anna Striped Bass Recreational Fishery*
7
8 The staff distinguishes between the striped bass fishery and the population of striped bass
9 inhabiting the reservoir: the striped bass fishery encompasses all aspects of the activity,
10 business, or practice of catching striped bass in the reservoir and includes the fish, the anglers,
11 and all related activities, such as boating, tackle sales, and guide services. In the Draft EIS
12 (NRC 2004a), the staff identified three major factors affecting the striped bass population in the
13 reservoir. These factors are (1) the VDGIF stocking strategy, (2) Unit 1 and 2 operations and
14 their effect on water temperatures, and (3) recreational fishing pressure. As described above,
15 the effects of NAPS Units 1 and 2 operations would not noticeably change with the addition of a
16 Unit 3 operation with a closed-cycle, combination wet and dry cooling system because this
17 system minimally increments the existing thermal inputs to the WHTF or the lake from the
18 operations of Units 1 and 2. Because the Lake Anna striped bass population is sustained
19 through stocking, and suitable striped bass habitat is expected to continue to exist in Lake
20 Anna, and in view of the minimal heat input to the WHTF or the lake from Unit 3, the staff
21 concludes that the heat stress impact on the striped bass and associated fishery from Unit 3
22 operations would be SMALL, and further mitigation beyond the restocking actions stated above
23 is not warranted.
24
25 **5.4.2.6 Downstream Impacts**
26
27 *This section was added to reflect the impacts to the North Anna River from changes in the water*
28 *levels for the closed-cycle, combination wet and dry cooling system and for dam operating*
29 *levels.*
30
31 In the Draft EIS, the staff indicated that the streamflow in the North Anna River downstream of
32 the Lake Anna Dam is influenced by the flow attenuation impact of the lake and policies
33 governing the release from the dam (NRC 2004a). Based on streamflow data collected
34 downstream from Doswell, Virginia (approximately 32 km [20 mi] downstream of the North Anna
35 Dam), for the period of October 1979 through September 2003, the monthly mean streamflows
36 vary from a maximum of about 23 m^3/s (804 cfs) during March to a minimum of 4 m^3/s (149 cfs)
37 during August (USGS 2006).
38
39 Because of the large surface area of the lake, small increases in lake elevations can quickly
40 accommodate significant upstream flood flows. The downstream flow from the dam is less
41 variable than it was prior to impoundment. Under normal operating conditions, release gates on
42 the dam are operated to maintain a steady lake level elevation of 76.2 m (250 ft) MSL. For lake
43 levels less than 76.2 m (250 ft) MSL, the Lake Level Contingency Plan is followed in

1   accordance with Part I.F of the VPDES permit (VDEQ 2001). This plan requires a normal
2   minimum water release rate from the Lake Anna dam of 1.1 m³/s (40 cfs). If the lake level drops
3   to 75.6 m (248 ft) MSL, releases are incrementally reduced to a minimum of 0.57 m³/s (20 cfs).
4   These minimum flow requirements are established to maintain instream flows and water quality
5   in the North Anna River downstream of the dam, and in the Pamunkey and York Rivers further
6   downstream. Steady low-flow releases from the dam can persist for several months, particularly
7   in the drier summer period, and under conditions of extreme drought (e.g., the drought occurring
8   from October 2001 to December 2002). Low-water conditions are quickly reversed when
9   normal precipitation rates resume in the region, and the water release rate of at least 1.1 m³/s
10   (40 cfs) would be reestablished.
11
12   The VDEQ identified the possibility of raising the lake level 15 to 30 cm (6 to 12 in.) to mitigate
13   the impact on North Anna River downstream flows (Dominion 2006a). The authority to raise the
14   lake level resides with the Commonwealth of Virginia.
15
16   While the staff concludes that the incremental heat load from Unit 3 would cause undetectable
17   changes in lake temperature, the evaporative losses from the cooling towers could result in
18   consumptive water losses that may noticeably impact lake levels and downstream flows. The
19   staff performed an independent review of water budget impacts of the operation of Unit 3
20   incremental declines in reservoir water levels and downstream discharges. The results in
21   Table 5.5 indicate that the fraction of time that the lake would be below 75.6 m (248 ft) would
22   increase from about 6 percent for NAPS Units 1 and 2 (baseline) to 11 percent for NAPS Units 1
23   and 2 plus Unit 3 (proposal). Because the Lake Level Contingency Plan requires the lake
24   discharges to be incrementally reduced from the normal minimum water release rate of 1.1 m³/s
25   (40 cfs) to 0.57 m³/s (20 cfs) as the lake level declines below 75.6 m (248 ft), the fraction of time
26   that the lake level is between 76.2 m (250 ft) and 75.6 m (248 ft) is essentially the same
27   between the baseline (57 percent) and the proposed (55 percent).
28
29   As set forth below, the impacts to aquatic ecology during low-flow conditions would likely be
30   greatest in the reach of the North Anna River extending from the dam to the confluence with the
31   South Anna River. To quantify the impacts to instream flows in the North Anna River, Dominion
32   calculated Indicators of Hydrologic Alteration (IHA) using a standard methodology described in
33   ER Revision 6 (Dominion 2006a). The IHA analysis concluded that the Pamunkey River flow
34   downstream of the North Anna River would be reduced by 0.5 to 5 percent for all flow regimes,
35   but that river flow during the months of April and May would be sufficient to support striped bass
36   spawning. Dominion concluded that the biological impacts of the Unit 3 closed cycle,
37   combination wet and dry cooling system to the general aquatic community of the North Anna
38   River and striped bass spawning and rearing areas in the Pamunkey would be indistinguishable
39   from the effects of operations of NAPS Units 1 and 2.
40

1 **Table 5-5.** Fraction of Time at Various Lake Level Elevation (with associated downstream
2 releases)
3

Lake level Elevation (ft)	North Anna Dam Discharge (cfs)	NAPS Existing Units 1 and 2	NAPS Existing Units 1 and 2 plus Unit 3
At or above 250 ft	Follows rating curve (>40 cfs)	37%	33%
Between 250 and 248 ft	40 cfs	57%	55%
At or below 248 ft	20 cfs	6%	11%
At or below 246 ft	20 cfs	1%	2%
Minimum elevation		245.2 ft	243.5 ft

12 The existing biological communities in the North Anna River and the waters downstream of the
13 North Anna River experience a wide variation in seasonal and even daily water temperatures,
14 and most resident species are able to tolerate potentially harmful conditions that exist because
15 of low-flow conditions, anthropogenic impacts, and other physical or chemical stressors.
16 Although the blowdown water from Unit 3 would contain biocides, antiscalants, and dispersals,
17 the concentrations of these chemicals in the blowdown water would not be expected to be high
18 enough to affect species downstream of the dam. Moreover, the use and concentrations of
19 anti-fouling chemicals would be regulated by Dominion's NPDES permit, which establishes
20 discharge criteria to protect sensitive aquatic communities. Because the incremental waste
21 heat transported to the North Anna River would be small, it is not expected to adversely affect
22 the aquatic communities of the North Anna River nor is it expected to influence the
23 temperatures of the Pamunkey or York Rivers or the Chesapeake Bay. Although flow over the
24 North Anna Dam could be reduced by up to 5 percent during some parts of the year, sufficient
25 water is expected to be available in the North Anna River to support downstream spawning of
26 striped bass in the Pamunkey River. The lowest flow regimes are likely to occur during the
27 months from June to December when striped bass are not spawning and are less likely to be
28 adversely affected by lower water levels.
29

30 A preliminary analysis of American shad, *Alosa sapidissima,* data by VDGIF suggests that a
31 correlation may exist between flow and year class strength. This could result in the need to
32 modify the water releases at the North Anna Dam during certain periods of the year.
33 Modifications to water release would be under the jurisdiction of the VDEQ.
34

35 Operation of Unit 3 could result in consumptive water use resulting from evaporative loss from
36 the cooling towers. Based on the water budget modeling conducted by the staff, the loss of
37 water from Lake Anna reservoir for the operation of Unit 3 would increase the fraction of time
38 that the reservoir would be at or below 75.6 m (248 ft) from 6 percent for the baseline
39 (i.e., operation of NAPS Units 1 and 2) to 11 percent for the operation of Units 1, 2, and 3. This
40 would extend the period of time when releases over North Anna Dam are reduced to 0.57 m^3/s
41 (20 cfs) and result in reduced downstream flow during the summer months or extended periods

1   of drought. Based on the staff's independent review of the IHA analysis conducted by
2   Dominion, sufficient water should be available during the spring and early summer to support
3   striped bass spawning and rearing requirements in the Pamunkey River downstream of NAPS.
4   Based on the foregoing, the staff concludes that the aquatic impacts to Lake Anna and the
5   downstream communities from the operation of Units 3 and 4 would be SMALL. Nevertheless,
6   the Commonwealth of Virginia may elect to implement mitigation strategies to further protect the
7   aquatic environment.
8
9   Although the staff concluded that impacts associated with operation of NAPS and proposed
10 Unit 3 on downstream aquatic communities would be SMALL, the staff also considered the
11 effects varying the North Anna Dam release rate during late summer and fall to more closely
12 emulate natural variations in stream flow. The staff determined that varying water release rates
13 might be preferable to a constant release rate in late summer and early fall because most
14 organisms in small rivers are adapted to a varying flow regime and steady flows may result in a
15 change in community structure. Modifications to the water release regime from the Lake Anna
16 Dam to mitigate impacts would be under the jurisdiction of VDEQ.
17

5.4.2.7 Shoreline Erosion and Other Physical Impacts

19
20 *Changes to this section reflect impacts to the North Anna River from changes in the water levels*
21 *for the closed-cycle, combination wet and dry cooling system and for dam operating levels.*
22
23 Because low-flow velocities in Lake Anna predominate, increased shoreline erosion, lake-bed
24 scouring, and increased turbidity levels caused by the operation of Unit 3 would not be
25 detectable or destabilizing to the aquatic resources of Lake Anna. The flow velocity in the
26 discharge channel, the connecting canals, and the main ponds of the WHTF would be slightly
27 higher than in Lake Anna because of their smaller dimensions. A closed-cycle, combination wet
28 and dry cooling system for Unit 3 would release a maximum of 351 L/s (12.4 cfs) of blowdown
29 water into the existing discharge canal. This represents an increase in the velocity of water in
30 the discharge canal and WHTF of about 0.3 percent. Because the banks of the canals are
31 currently protected by rip-rap from 73.8 to 76.2 m (242 to 250 ft) MSL, the small contribution to
32 flow and velocity from Unit 3 would not result in scouring or erosion in the canal. During the
33 operation of NAPS Units 1 and 2, the flow velocity in the WHTF is generally less than 0.3 m/s
34 (1 ft/s) and has not caused any noticeable scouring or erosion. This would not change based
35 on the small contribution to the overall flow and velocity resulting from the operation of Unit 3.
36
37 Physical impacts resulting from increased turbidity or siltation are unlikely, based on the small
38 changes in discharge flow and velocity expected from Unit 3. Siltation is expected to be
39 minimal because any medium or coarse sediment that was suspended would settle before
40 reaching the intake approach channel during normal lake conditions. Sediment entering the
41 new intake channel from floods or storm events would not present a siltation problem because
42 the channel is 3.7 m (12 ft) deeper than required by the design for the new intake, allowing

1   room for occasional sediment deposition. Fine-grained sediment entering the intakes would

2   either be removed during the water treatment process or returned to Lake Anna via the WHTF

3   at Dike 3.

4

5   There is no evidence of scouring, erosion, or excessive turbidity from the operation of NAPS

6   Units 1 and 2, and no evidence to suggest that this would change with the addition of Unit 3

7   given the proportionally small amounts of blowdown water that is or would be released during

8   normal operations. Based on the foregoing, the staff concludes that the impacts to aquatic

9   ecological resources from physical changes to Lake Anna from operation of Unit 3 closed-cycle,

10   combination wet and dry cooling would be SMALL, and mitigation is not warranted.

11

12   **5.4.2.8 Summary of Aquatic Impacts**

13

14   *Changes to this section reflect the aquatic impacts of the North Anna River as a result of*

15   *changes in the water levels for the closed-cycle, combination wet and dry cooling system and*

16   *for dam operating levels.*

17

18   The aquatic plants and animals in Lake Anna represent a balanced aquatic community. Over

19   the years, these populations have changed as the reservoir's ecosystem has matured.

20   Because of the demand for public fishing, the fish community has been changed through

21   species additions (e.g., threadfin shad) and annual stockings of striped bass and walleye.

22   Overall, the fisheries have remained healthy and balanced despite shoreline development,

23   NAPS operations, and increased fishing pressure.

24

25   Based on the information provided in ER Revision 6, the staff evaluated the impacts to aquatic

26   communities in Lake Anna, the North Anna River, and the downstream river systems to which

27   the North Anna is a tributary. Because Unit 4 would use dry cooling towers, water use would be

28   minimal and would not result in detectible impacts to the aquatic environment. The

29   closed-cycle, combination wet and dry cooling system proposed for Unit 3 would result in

30   significantly less impingement and entrainment than a once-through cooling system, and would

31   contribute minimally to the thermal load currently experienced by the lake from the operations of

32   the NAPS Units 1 and 2. Based on the impingement and entrainment modeling conducted by

33   Dominion and data from the Section 316(b) demonstration study for NAPS Units 1 and 2, as

34   discussed above, the operation of Unit 3 would increase the overall yearly impingement and

35   entrainment losses by approximately 1 percent. These losses are not expected to result in

36   noticeable changes to the fish communities in Lake Anna.

37

38   With the change in Unit 3 cooling design from the once-through system to the closed-cycle,

39   combination wet and dry cooling system, thermal impacts are expected to be reduced

40   significantly. The proposed closed-cycle, combination wet and dry cooling system for Unit 3

41   would discharge a maximum of 351 L/s (12.4 cfs) at a maximum temperature of 38°C (100°F).

42   This flow rate represents 0.3 percent of the about 120,000 L/s (4246 cfs) flow discharged by

1   NAPS Units 1 and 2, and the discharge temperature of Unit 3 blowdown water is within the
2   range currently observed in the discharge canal during July and August (37.0° to 39.1°C
3   [98.6° to 102.4°F]). Thus, the operation of Unit 3 would result in a negligible change in the
4   volume and temperature of the water entering the discharge canal, the WHTF, Lake Anna, or
5   the North Anna River.

6

7   Operation of the Unit 3 cooling towers could result in consumptive water use from evaporative
8   losses. Based on the water budget modeling conducted by the staff, Lake Anna reservoir water
9   losses resulting from the operation of Unit 3 would increase the fraction of time that the
10   reservoir would be at or below 75.6 m (248 ft) from 6 percent for the baseline (i.e., operation of
11   NAPS Units 1 and 2) to 11 percent for the operation of Units 1, 2, and 3. This would extend the
12   period of time when releases over North Anna Dam are reduced to 0.57 m^3/s (20 cfs) and result
13   in reduced downstream flow during the summer months or extended periods of drought. Based
14   on the staff's independent review of the IHA analysis conducted by Dominion, sufficient water
15   should be available during the spring and early summer to support striped bass spawning and
16   rearing requirements in the Pamunkey River downstream of NAPS. Based on the foregoing, the
17   staff concludes that the aquatic impacts to Lake Anna and the downstream communities from
18   the operation of Units 3 and 4 would be SMALL, and mitigation is not warranted.

19

20   **5.4.3 Threatened and Endangered Species**

21

22   *This section is not affected by the changes presented in ER Revision 6. However, additional*
23   *information is provided about eagles.*

24

25   No Federally listed threatened or endangered species are known to occur at or near the North
26   Anna ESP site except the bald eagle (*Haliaeetus leucocephalus*). The closest known bald
27   eagle nesting site is located more than 4 km (2.5 mi) from the North Anna ESP site. In the
28   Commonwealth of Virginia, a 0.25-mile (0.4-km) buffer zone is usually preserved to limit
29   construction activities (FWS and VDGIF 2000). Dominion stated in its May 24, 2006, RAI
30   response that it follows nesting guidelines referenced by the Virginia Department of Game and
31   Inland Fisheries and the U.S. Fish and Wildlife Service (Dominion 2006b). None of the three
32   Federally or State-listed mussel species known to exist in the region has been found in Lake
33   Anna, the North Anna River, or other local streams. The staff reviewed the potential impacts of
34   operation of Units 3 and 4 on threatened and endangered species. It is unlikely that any
35   threatened or endangered species exist on the North Anna ESP site; consequently, operation
36   associated with the cooling system changes or the increased power level would not have an
37   adverse effect on threatened or endangered species. The staff concludes that the impacts of
38   operation on threatened or endangered species would be SMALL, and further mitigation beyond
39   maintaining an adequate buffer zone is not warranted.

40

5.5 Socioeconomic Impacts

This introductory section is not affected by the changes presented in ER Revision 6 and is provided solely for context.

This section describes the socioeconomic impacts from operating Units 3 and 4 at the North Anna ESP site, and from the activities and demands of the operating workforce on the surrounding region. Socioeconomic impacts include potential impacts on individual communities, the surrounding region, and minority and low-income populations.

5.5.1 Physical Impacts

This section is not affected by the changes presented in ER Revision 6 and is provided solely for context.

This section assesses the potential physical impacts on the nearby communities that could result from the operation of new nuclear units at the North Anna ESP site. Potential impacts discussed include noise, odors, exhausts, thermal emissions, and visual intrusions. Dominion plans to manage these physical impacts to comply with applicable Federal, State, and local environmental regulations (Dominion 2006a). Dominion does not expect operation of the new units to significantly affect the North Anna ESP site and its vicinity (Dominion 2006a). The staff's evaluation is discussed in the following subsections.

5.5.1.1 Workers and the Local Public

This section is not affected by the changes presented in ER Revision 6 and is provided solely for context.

In the Draft EIS (NRC 2004a), the staff determined that residents of Mineral, Virginia, would experience minimal physical impacts from operation of the new units because of its distance from the ESP site. Onsite impacts from station operations to permanent workers could be mitigated through adequate training and use of personal protective equipment to minimize the risk of potentially harmful exposures. Accordingly, the staff concludes that the overall physical impacts of station operation to workers and the local public because of changes to the Unit 3 cooling system or the proposed power level increase did not change from that previously evaluated. The impact level category would still be SMALL, and mitigation is not warranted.

5.5.1.2 Buildings

This section is not affected by the changes presented in ER Revision 6 and is provided solely for context.

In the Draft EIS (NRC 2004a), the staff determined that operational activities are not expected to have any effect on any offsite buildings. The staff concludes that the physical impacts to offsite buildings did not change because of the change to the Unit 3 cooling system or power level increase. The impact level category would still be SMALL, and mitigation is not warranted.

5.5.1.3 Roads

This section is not affected by the changes presented in ER Revision 6 and is provided solely for context.

In the Draft EIS (NRC 2004a), the staff concluded that any needed upgrades to the regional road system would have been made in conjunction with, or as a result of, the construction of Units 3 and 4. In addition, the number of operating personnel would be significantly fewer than the number of construction personnel. The staff concludes that the physical impacts of station operation on the road system did not change. The impact level category would still be SMALL, and mitigation is not warranted.

5.5.1.4 Aesthetics

Changes to this section reflect aesthetic impacts from the operation of the cooling towers presented in ER Revision 6 and the presence of plumes.

The turbine building for Units 1 and 2 is about 30 m (100 ft) above grade and the containment buildings are about 40 m (130 ft) above grade. The tallest building for Units 3 and 4 could be 71 m (234 ft) above grade, the Unit 3 closed-cycle, combination wet and dry towers would be less than 55 m (180 ft) tall, and the Unit 4 dry towers would be about 48 m (150 ft) tall (Dominion 2006a). The ESP and cooling tower site grade would be lower than the surrounding terrain, except in the direction of Lake Anna. Given a distance of about 900 m (3000 ft) to the nearest residence, screening by trees and other vegetation on the wooded ESP site, and relative elevations, most residents near the site would not have a clear view of Units 3 and 4. However, recreational users on the Lake Anna Reservoir and some residents along the lake would be able to see the new units in addition to the other developed areas of the NAPS site already visible. The tallest building at least would rise above the treeline, but would be largely screened by trees on the south, north, and west.

The Unit 3 closed-cycle, combination wet and dry cooling towers would generate a vapor plume that could be visible above the height of the plant buildings and that could extend beyond the

1   site boundary. The plume would be most prevalent during times when the ambient temperature

2   is low and when the dry bulb and wet bulb temperatures are nearly equal. Typically this would

3   be between late autumn and early spring.

4

5   The environmental impact of the operation of the wet cooling towers was evaluated by Dominion

6   using the SACTI (Seasonal and Annual Cooling Tower Impacts) system of computer programs,

7   initially written and assembled by the Argonne National Laboratory (ANL) (ANL 1984) for the

8   Electric Power Research Institute, was used to estimate the impact of operating the cooling

9   towers. The version used by Dominion is dated November 1, 1990. The input meteorological

10   data used to estimate the impacts encompassed the period of 1998 through 2000. It included

11   data collected onsite as well as site representative data collected at the National Weather

12   Service sites in Richmond, Virginia, and Dulles Airport in northern Virginia. For purposes of the

13   analysis, the cooling towers were assumed to be operating in the EC mode, which results in the

14   greatest evaporation rates from the towers, and therefore, assumed to be the greatest level of

15   impacts (Dominion 2006a).

16

17   The results of the staff's independent analysis indicated that for all seasons, the plume would

18   extend to a maximum height of 980 m (3200 ft) and to a length of 4900 m (16,000 ft) from the

19   tower. The annual duration of plume fogging (i.e., the plume remaining at the ground level)

20   would be about 70 hr (excluding hours of natural fog), with a majority of fogging occurring at

21   about 300 m (1000 ft) to the south-southeast from the cooling towers. Fogging would, however,

22   occur as far as 1600 m (5200 ft) from the tower. Fogging is estimated to occur during all

23   seasons except summer. The analysis indicates that icing is unlikely to occur in conjunction

24   with ground-level fogging (Dominion 2006a).

25

26   Table 5-6 presents estimates, by season, of the approximate percentage of time that the plume

27   would extend above the tallest structure in the PPE (71 m [234 ft]) or would extend more than

28   800 m (about 2600 ft) from the towers.

29

30   In Dominion's plume model, the top of the tallest structure in the PPE is approximately 50 m

31   (160 ft) above the top of the traditional mechanical-draft wet cooling towers, which are only

32   22 m (74 ft) tall. The shorter traditional wet cooling tower was used in the analysis to provide an

33   upper bound for nearby plume effects such as fogging or salt drift. The Unit 3 closed-cycle,

34   combination wet and dry cooling system described in ER Revision 6 would be 55 m (180 ft) tall;

35   this would lead to a higher release point and a higher plume elevation than evaluated by

36   Dominion (Dominion 2006a). However, Dominion suggests that the combination wet and dry

37   tower could be plume-abated, which would significantly reduce the visual impact below that

38   shown in Table 5-6. The frequency results reported below are for a release point between 40 m

39   (130 ft) and 50 m (160 ft) below the height of the tallest structure on the North Anna ESP site.

40   The evaluation indicates that on an annual average basis, the plume would extend more than

41   200 m (650 ft) above the tower or extend more than 400 m (1300 ft) in length approximately

42

Table 5-6. Fraction of Time by Season That Water Vapor Plume from the Proposed Unit 3 Wet and Dry Cooling Towers Would Exceed the Height of the Tallest Structure in the PPE or Would Extend More than 800 m (2625 ft) from the Towers

Season	Plume height >40 m (131 ft) above top of towers	Plume height >50 m (164 ft) above top of towers	Plume length > 800 m (2625 ft) from towers
Winter	89	49	20
Spring	77	29	11
Summer	78	20	4
Fall	79	27	7

Source: Dominion 2006a

10 percent of the time the wet towers are operating. These results are based on the wet cooling towers operating 100 percent of the time in the EC mode, which results in the most fogging.

Deposition of salts from cooling tower drift would occur in all directions from the towers out to 1525 m (5000 ft), but would occur predominately in the areas to the north through northeast as well as to the south through southeast of the towers. The maximum estimated amount of deposition would be 12.6 kg/km^2/month at 175 m (575 ft) north-northeast of the cooling towers. The vast majority of the drift deposition would occur within 300 m (1000 ft) of the towers. Significant chemical interaction of the cooling tower plume and pollutants emitted onsite or in the vicinity of the plant is not anticipated.

In its May 24, 2006, response to the NRC's Request for Additional Information (Dominion 2006b), Dominion stated that,

> The statement concerning fogging and salt deposition to be made in a COL application (when a specific reactor design is selected), is intended to indicate that at the time of the COL (when additional design details are known), a confirmatory evaluation will be performed to show that the analysis conducted as a part of this ESP application remains bounding.

In Appendix K, the staff evaluated the impacts of Unit 3 operation on water use in relation to water levels within Lake Anna over a 35-year period; the analysis is also applicable to the aesthetics impacts as well. Lake levels would be above 75.6 m (248 ft) about 88 percent of the time with Units 1 and 2 and Unit 3 operating versus about 94 percent with only NAPS Units 1 and 2 operating. Levels below 75 m (246 ft) (just above the lowest level experienced during the 2001 to 2002 drought) could occur about 2 percent of the time with Unit 3 operating, versus 1 percent with only NAPS Units 1 and 2 operating. The minimum water level with Unit 3 operating was estimated at 74.22 m (243.5 ft), versus 74.74 m (245.2 ft) with only NAPS Units 1 and 2. Under severe drought conditions, Unit 3 could have an exacerbating effect on the drawdown of Lake

Anna, potentially adding to the duration of low water levels, which would affect the visual impact of the amount of shoreline exposed.

Because the new units would be located in the existing power station complex and the visual aspects of the site to offsite viewers would be limited by screening and topography, and based on information provided by Dominion (2006a, b) and its independent review, as discussed above, the staff concludes that the aesthetic impacts from the operation of Units 3 and 4 would be SMALL There would be an elevated steam plume from the operation of the Unit 3 cooling towers; the staff concludes that the visual impacts would be quite noticeable at times, especially during the winter (periodic MODERATE visual impact). On an annual basis, however, this impact would be limited to about 10 percent of the time and would be the least from mid-spring to early fall when outdoor recreation is most likely to occur. In addition, the staff identified that during severe drought conditions, the operation of Unit 3 would have an impact on the water levels by slightly adding to the duration and extent of shoreline mud flats that may be exposed; the staff concludes that these visual aesthetic impacts would temporarily be MODERATE. Mitigation is not warranted because of the temporary and infrequent nature of the impacts.

5.5.2 Demography

This section is not affected by the changes presented in ER Revision 6 and is provided solely for context.

In the Draft EIS (NRC 2004a), the staff evaluated the impacts of station operation on increases in population and determined that while the new operating personnel are expected to come from outside the region, their small numbers, when considering the population base of each jurisdiction, would not significantly increase the base population within each jurisdiction. Most new jobs created through the multiplier effect[a] are expected to go to workers who already reside in the region. The changes to the Unit 3 cooling system or the power level increase would not affect the size of the operations workforce. Based on the foregoing, the staff concludes that the impacts of station operation on increases in the regional population did not change. The impact level category would still be SMALL, and mitigation is not warranted.

(a)   The multiplier effect describes the situation in which each dollar spent on goods and services by a construction worker becomes income to the recipient who saves some but re-spends the rest on consumption. This re-spending becomes income to someone else, who in turn saves part and re-spends the rest. The number of times the final increase in consumption exceeds the initial dollar spent is called the "multiplier."

5.5.3 Community Characteristics

This section is not affected by the changes presented in ER Revision 6 and is provided solely for context.

In the Draft EIS (NRC 2004a), the staff evaluated the social and economic impacts to the surrounding region as a result of operation of Units 3 and 4 at the North Anna ESP site. The staff evaluated the impacts of operation and of those demands placed by the workforce on the surrounding region during a 40-year operating license period. Dominion expects to employ no more workers to (1) operate Unit 3 with a closed-cycle cooling system than it would to operate two units with a once-through cooling system or (2) operate Units 3 and 4 with a higher power level than previously evaluated.

5.5.3.1 Economy

This section is not affected by the changes presented in ER Revision 6 and is provided solely for context.

In the Draft EIS (NRC 2004a), the staff determined that most new operating personnel are expected to come from outside the region. Their employment for such an extended period of time would have economic and social impacts on the surrounding region. The new jobs, as with the construction workforce, would also create new jobs in the region through the multiplier effect. Based on the potential aspects of the operation of Units 3 and 4 on the regional economics, the staff concludes that the economic impacts did not change. The impact level category would be SMALL BENEFICIAL to MODERATE BENEFICIAL, and mitigation is not warranted.

5.5.3.2 Transportation

This section is not affected by the changes presented in ER Revision 6 and is provided solely for context.

In the Draft EIS (NRC 2004a), the staff concludes that necessary road improvements would be made during the construction phase to accommodate the much-larger construction work force and that Dominion would activate a travel management plan, as needed. Based on the infrastructure improvements that would be made for the construction of Units 3 and 4 and the expectation that the number of workers needed to operate the units would be unchanged, the staff concludes that the overall impacts of station operation on transportation did not change. The impact level category would still be SMALL, and mitigation is not warranted.

5.5.3.3 Taxes

This section is not affected by the changes presented in ER Revision 6 and is provided solely for context.

In the Draft EIS (NRC 2004a), the staff evaluated the effect of taxes from income on wages and salaries of Units 3 and 4 operational workers, and sales, use, and property taxes on these employees and on Dominion's corporate profits, most of which represent beneficial sources of income for the State and some of which would benefit the counties in the region. Property tax paid by Dominion would directly benefit Louisa County. Taxes would not change appreciably because of the proposed changes in the Unit 3 cooling system or the proposed power level increase. Based on the foregoing, the staff concludes that the impacts on income, sales, and use taxes, property taxes, and corporate profits did not change. The impact level category would still be SMALL BENEFICIAL for most of the region and LARGE BENEFICIAL for Louisa County.

5.5.3.4 Recreation

Changes to this section reflect Lake Anna water level impacts from operation of the closed-cycle, combination wet and dry cooling system.

Most of the 43,000 anglers visiting Lake Anna every year launch their boats at Lake Anna State Park and at commercial marinas. Pleasure boat traffic on the lake exceeds angler traffic by as much as 10 to 15 times. Use of stationary boat docks would be impacted when the lake level drops below 76 m (248 ft) MSL. At and below this level, many of the stationary docks become unusable. However, boat ramps would be usable for launching boats until the water level receded below the end of the ramp. During the 2001 to 2002 drought, most private boat ramps could not support launches at lake levels below 74.7 m (245.1 ft) MSL (Dominion 2004b). Even though they were adversely affected because of lower traffic and costs incurred to extend ramps, adjust docks, and move boats, most of the marina operations and fishing guides were able to adjust their operations and continue to operate profitably (Jaksch and Scott 2005). The staff does not expect adverse impacts from NAPS on marinas and guides in normal water years. Because of the adjustments already made in 2001 and 2002, the staff concludes that impacts on marina operators and fishing guides from operations of four units (even in drought years) would be SMALL to MODERATE and temporary, and mitigation is not warranted.

As discussed in the Draft EIS (NRC 2004a), visitors to the State park actually increased during 2002 above the previous years, while the number of boat launches at the park in 2002 was fewer than launches in 2001 by 13.2 percent. The number of boat launches declined by an additional 2.4 percent in 2003, which was not a drought year.[a] Thus, there appears to have been a decline in boating during the drought years, but an increase in the use of the park itself. As

---

(a)   Note that these numbers do not include boat launches from private marinas.

1   discussed in Section 5.5.1.4, operation of the Unit 3 closed-cycle, combination wet and dry
2   cooling system would slightly exacerbate conditions at the lake during times of drought.
3
4   These impacts could have economic consequences for the three counties surrounding the lake.
5   The more immediate impacts would be to the marinas and commercial businesses that earn
6   revenue on a seasonal basis from recreational users of the lake. If drought conditions persist
7   over a long enough time period, property values around the lake could be affected as well.
8   Owners of lake-front homes with preferred water views would be particularly impacted.
9   However, larger market forces such as general economic conditions, population trends, and
10   interest rates are expected to dominate property values in almost all circumstances. Minimal
11   recreational impacts to Lake Anna from operation of the units are expected to occur during
12   non-drought conditions.
13
14   As discussed in Section 5.3.2, the staff also evaluated changing the normal elevation of Lake
15   Anna and the impacts of raising normal operating lake level 15 to 30 cm (6 to 12 in.) above
16   76.2 m (250 ft) MSL. Raising the lake level could increase localized flooding potential and
17   downstream flows, and would likely affect use of some residential and marina boat ramps and
18   docks, including those at Lake Anna State Park. These facilities might need some modification
19   to avoid impacting the year-round and seasonal recreational usage of the lake. The staff
20   concludes that the recreational impacts of raising the lake level would be MODERATE.
21
22   Although the WHTF is considered by Virginia Power to be part of the nuclear facility and is
23   operated as a private industrial facility, homeowners on the shoreline of the WHTF have access
24   to it, with Virginia Power's permission, for recreational use (e.g., boating, fishing, swimming).
25   This limited access and use would remain unchanged following the addition of the cooling
26   systems for Units 3 and 4. Dominion evaluated the potential thermal and chemical impacts of
27   Unit 3 and 4 operations on the ability of homeowners to continue their recreation activities.
28   Dominion determined that there would be virtually no change in temperature of the WHTF as a
29   result of the operation of Units 3 and 4, and there would likely be no impact on thermophilic
30   organisms. While Dominion has not identified which chemicals would be added to the proposed
31   cooling towers to manage water chemistry, it also committed to evaluate potential additives in
32   accordance with "applicable EPA human health and aquatic life criteria to demonstrate that the
33   concentrations of these chemicals in the WHTF would not exceed the criteria, and thus would not
34   pose any risks to human health" (Dominion 2006a). Based on the forgoing, the staff concludes
35   that the thermal and chemical impacts from the operation of Units 3 and 4 on recreation would be
36   SMALL, and mitigation is not warranted.
37
38   Based on the individual aspects of recreational activities in the North Anna ESP site area, if the
39   normal operating level of Lake Anna remains at 76.2 m (250 ft), the staff concludes that the
40   recreational impacts of Unit 3 and 4 operations would be SMALL most of the time, but could be
41   MODERATE during the infrequent extreme droughts. Therefore, mitigation is not warranted.
42   The staff concludes that if the normal operating level of the lake is raised 15 to 30 cm

1 (6 to 12 in.), then modification of residential and marina boat ramps and docks may be
2 necessary; this action could result in a MODERATE impact.
3
4 **5.5.3.5 Housing**
5
6 *This section is not affected by the changes presented in ER Revision 6 and is provided solely for*
7 *context.*
8
9 In the Draft EIS (NRC 2004a), the staff determined that while the new operating personnel and
10 their families would be expected to come from outside the region, their small numbers, when
11 considering the population base of each jurisdiction, would not significantly increase the base
12 population within each jurisdiction even with the multiplier effect. The change to the Unit 3
13 cooling system or the power level increase would not affect the size of the operations workforce
14 previously evaluated. Therefore, the staff concludes that the impacts of station operation on
15 housing did not change. The impact level category would still be SMALL, and mitigation is not
16 warranted.
17
18 **5.5.3.6 Public Services**
19
20 *This section is not affected by the changes presented in ER Revision 6 and is provided solely for*
21 *context.*
22
23 In the Draft EIS (NRC 2004a), the staff evaluated the local and regional water and waste water
24 treatment capacities, the police, fire and medical facilities, and the demand for social and related
25 services. The changes in Unit 3 cooling and the power level increase would not change the size
26 of the operations workforce previously evaluated. Therefore, the staff concludes that the impacts
27 of station operation previously evaluated on the demand for public services did not change. The
28 impact level category would still be SMALL, and mitigation is not warranted.
29
30 **5.5.3.7 Education**
31
32 *This section is not affected by the changes presented in ER Revision 6 and is provided solely for*
33 *context.*
34
35 In the Draft EIS (NRC 2004a), the staff determined that while the new operating personnel and
36 their families are expected to come from outside the region, their small numbers, when
37 considering the population base of each jurisdiction, would not significantly increase the base
38 population within each jurisdiction even with the multiplier effect. The change to the Unit 3
39 cooling system or the power level increase would not affect the size of the operations workforce
40 previously evaluated nor the need for increased educational facilities. Therefore, the staff
41 concludes that the impacts of station operation on education did not change. The impact level
42 category would still be SMALL, and mitigation is not warranted.
43

1 # 5.6 Historic and Cultural Resource Impacts
2
3 *This section is not affected by the changes presented in ER Revision 6, and is provided solely*
4 *for context.*
5
6 In the Draft EIS (NRC 2004a), the staff reported that there would be limited potential for impacts
7 during operation of additional power units at North Anna ESP Site because all ground-disturbing
8 activities that could have an impact on historic or cultural resources would probably occur during
9 the construction phase. The change to the Unit 3 cooling system did not increase the size of the
10 land area disturbed onsite and did not change the area of potential effect. Should
11 archaeological, historic, or other cultural resources be uncovered during site excavation,
12 Dominion would implement the NAPS site-wide Excavation and Backfill Work Procedures (NAPS
13 NSS Work Procedure WP-C01), which would involve stopping work immediately and contacting
14 the appropriate organization and/or regulatory agency for proper evaluation and designation, in
15 accordance with the existing procedures (Dominion 2006a). Based on the foregoing, the staff
16 concludes that the historic and cultural impacts from operations did not change. The impact level
17 category would still be SMALL, and further mitigation beyond the existing practice is not
18 warranted.
19
20 # 5.7 Environmental Justice Impacts
21
22 *This section is not affected by the changes presented in ER Revision 6 and is provided solely for*
23 *context.*
24
25 Environmental justice refers to a Federal policy under which each Federal agency identifies and
26 addresses, as appropriate, disproportionately high and adverse human health or environmental
27 effects of its programs, policies, and activities on minority[a] low-income populations.
28
29 In the Draft EIS (NRC 2004a), the staff identified the pathways through which the environmental
30 impacts associated with the construction of Units 3 and 4 at the NAPS site could affect human
31 populations. The staff then evaluated whether minority and low-income populations could be
32 disproportionately affected by these impacts. The staff found no unusual resource dependencies
33 or practices, such as subsistence agriculture, hunting, or fishing, through which the populations
34 could be disproportionate impacted by operations of Units 3 and 4 at the NAPS site that would
35 result in those populations being adversely affected. In addition, the staff did not identify and

---

(a) The NRC Guidance for performing environmental justice reviews defines "minority" as American Indian or Alaskan Native; Asian; Native Hawaiian or other Pacific Islander; Black races, or Hispanic ethnicity ("other" may be considered a separate minority category.) The 2000 census included multi-racial data. The staff should consider multi-racial individuals in a separate minority category, in addition to the aggregate minority category (NRC 2004b).

1 location-dependent, disproportionately high and adverse impacts affecting these minority and
2 low-income populations. Because the change to the Unit 3 cooling system or the power level
3 increase would not affect such populations, the staff concludes that offsite impacts of operation
4 of Units 3 and 4 at the North Anna ESP site to minority and low-income populations did not
5 change. The impact level category would still be SMALL, and mitigation is not warranted.
6

5.8   Nonradiological Health Impacts

7
8
9 *This introductory section is not affected by changes presented in ER Revision 6 and is provided*
10 *solely for context.*
11
12 This section addresses the nonradiological health impacts of operating the proposed new units
13 at the North Anna ESP site. Health impacts to the public from the cooling system, noise
14 generated by unit operations, and electromagnetic fields are discussed. Nonradiological health
15 impacts are also evaluated for workers at the new units. Health impacts from radiological
16 sources during operations are discussed in Section 5.9.
17

5.8.1 Public Health

18
19
20 *Changes to this section reflect public health impacts from the WHTF for the closed-cycle,*
21 *combination wet and dry cooling system.*
22
23 In ER Revison 6, Dominion changed its proposed approach for the Unit 3 cooling system from
24 open-cycle cooling (i.e., cooling lake) to a closed-cycle, combination wet and dry cooling system.
25 The potential health impacts of thermophilic organisms have been investigated in the power
26 industry since the 1970s; the exiting Units 1 and 2 have open-cycle cooling systems that
27 discharge heated cooling water into the WHTF and then into the North Anna Reservoir and the
28 North Anna River (Dominion 2006a). Thermophilic microorganisms (e.g., *Naegleria fowleri*)
29 generally exist in water bodies with ambient temperatures between 25 and 80°C (77 and 176°F)
30 with maximum growth of such organisms generally occurring when ambient temperatures are
31 maintained between 50 and 60°C (122 and 140°F) (Dominion 2001b). Correspondence from the
32 Virginia Department of Health (VDH) states that *N. fowleri* begins to proliferate around 30°C
33 (86°F) and thrives at temperatures of 35 and 45°C (95 and 113°F) when compared to competing
34 organisms (VDH 2005).
35
36 Since 1975, Virginia Power has monitored water temperatures at various locations in Lake Anna,
37 the WHTF, and the discharge canal. The highest temperatures recorded in (1) the discharge
38 canal was 39°C (102.4°F) in August 2002, (2) the WHTF was 35°C (95.0°F) in July 1993, and
39 (3) Lake Anna was 34°C (92.7°F) in July 1977. These temperatures were hourly average
40 values. While ambient summer water temperatures in the sampled locations were found to be

within the range conducive to reproduction and growth of pathogenic micro-organisms, the temperatures measured were below those considered optimal for the growth of thermophilic forms.

Thermophilic microorganisms can cause primary amoebic encephalitis in humans. No cases of primary amoebic encephalitis have been documented in NAPS workers or area residents during the operating history of the plants (Dominion 2006a). The review performed by Dominion in Section 5.3.4.1 of its revised ER (Dominion 2006a) concluded that the combination wet and dry cooling system for the proposed Unit 3 would not significantly increase the temperature of the WHTF and Lake Anna and, therefore, would not create an environment conducive to the optimal growth of thermophilic organisms.

The Commonwealth of Virginia considers the WHTF a private pond and does not regulate the discharge temperature of Units 1 and 2 into the WHTF. The point of compliance for the NPDES permit is Dike 3 where the water from the WHTF is discharged into the reservoir. While the WHTF is a private pond, VEPCo has allowed private homeowners on the WHTF access for recreational purposes, including swimming. The public in the vicinity of the WHTF requested that the Virginia Department of Health (VDH) address the health effects of swimming in the WHTF because of concerns regarding the discharge temperature into the WHTF under the initial proposal for Unit 3 as a once-through cooling system, with which a substantial increase to the thermal load of WHTF was likely. In its letter dated September 15, 2005 (VDH 2005), VDH recommends that swimmers avoid recreational activities in water bodies exceeding 40°C (104°F). Subsequently, Dominion modified its proposed cooling system for Unit 3, which would result in negligible thermal impacts to the WHTF. In Enclosure 1 to Revision 6 of the ER, Dominion stated that it "in concert with VDEQ and VDH [it] is exploring options to communicate to local residents information related to existing risks" (Dominion 2006a). Based on the change to the cooling system for Unit 3, the significant reduction in Unit 3 thermal load to the WHTF and the evaluation of thermophilic organisms, the staff considers the public health impacts from the operation of Unit 3 to be SMALL, and mitigation is not warranted.

5.8.2 Occupational Health

This section is not affected by the changes presented in ER Revision 6 and is provided solely for context.

In the Draft EIS (NRC 2004a), the staff reported that human health risks for a new nuclear unit are expected to be dominated by occupational injuries (e.g., falls, electric shock, asphyxiation) to workers engaged in activities such as maintenance, testing, and plant modifications. Historically, actual injury and fatality rates at nuclear reactor facilities have been lower than the average U.S. industrial rates. The staff assumes adherence to NRC, OSHA, and State safety standards, practice, and procedures during new nuclear unit operations. The closed-cycle, combination wet and dry cooling system for Unit 3 and the increase in power level for Units 3 and 4 do not change

1   the human health risks from that previously evaluated. Based on the mitigation measures
2   identified above, the staff concludes that the overall nonradiological impacts from operations of
3   Units 3 and 4 did not change. The impact level category would still be SMALL, and mitigation is
4   not warranted.
5
6   ### 5.8.3 Noise Impacts
7
8   *Changes to this section reflect noise impacts from the closed-cycle, combination wet and dry*
9   *cooling system.*
10
11 In the Draft EIS (NRC 2004a), the staff estimated the cooling system noise levels to be
12 approximately 75 dBA at the exclusion area boundary and 55 dBA at the nearest residence. In
13 general, the effects of noise had been considered by the NRC elsewhere in NUREG-1437
14 (NRC 1996). In that generic EIS, noise levels below 60 to 65 decibels were considered to be of
15 small significance. More recently, the impacts of noise were considered by the NRC in
16 Supplement 1 of NUREG-0586 (NRC 2002a). In that generic environmental impact statement,
17 the criterion for assessing the level of significance was not expressed in terms of sound levels.
18 Rather, the level of significance was based on the effect of noise on human activities and
19 threatened and endangered species. The criterion in NUREG-0586, Supplement 1 is:
20
21   ...noise impacts ... are considered detectable if sound levels are sufficiently high
22   to disrupt normal human activities on a regular basis. ... noise impacts ... are
23   considered destabilizing if sound levels are sufficiently high that the affected area
24   is essentially unsuitable for normal human activities, or if the behavior or breeding
25   of a threatened or endangered species is affected.
26
27 Based on the information provided by Dominion and the NRC insights from the assessments in
28 NUREG-1437 and NUREG-0586 Supplement 1, the staff concluded in the Draft EIS that the
29 potential impacts of noise resulting from operation of two additional nuclear power plants with
30 cooling systems meeting the noise criteria of the PPE as defined in the ER would be small.
31
32 The changes proposed in Revision 6 of the ER describe a combination wet and dry cooling
33 system rather than the once-through cooling system for Unit 3 set forth in earlier versions of the
34 ER. The noise level for the Unit 3 cooling towers given in the PPE is 65 dBA at a distance of
35 300 m (1000 ft), which is 5 dBA higher than the noise level for the Unit 4 cooling towers. This
36 postulated noise level is consistent with the noise level previously evaluated in the Draft EIS.
37 Therefore, the staff concludes that the noise level resulting from changes in the cooling system
38 approach does not affect the conclusion on human health and impact on biota. The impact level
39 category would still be SMALL, and mitigation is not warranted.
40

5.8.4 Acute Effects of Electromagnetic Fields

Changes to this section reflect the higher power level of 4500 MW(t) for one unit (9000 MW(t) for two units).

The current transmission lines that originate from the NAPS site are capable of handling the output from two additional units. These lines consist of three 500-kV lines that were erected in the late-1970s and one 230-kV line erected in 1984. Both sets of transmission lines were designed and constructed according to National Electrical Safety Code (NESC) requirements and industry guidance that was current at that time (Dominion 2006a).

The current NESC requirements for preventing electric shock from induced current were met, and the staff concludes that the impact to the public from acute effects of electromagnetic fields did not change as a result of revisions included in ER Revision 6. The impact level category would still be SMALL, and mitigation is not warranted. The conclusion of SMALL impact by the NRC staff is predicated on the assumption made by the staff that the transmission lines carrying the additional power of the two new units would meet the NESC criteria for electric shock.

5.8.5 Chronic Effects of Electromagnetic Fields

This section is not affected by the changes presented in ER Revision 6 and is provided solely for context.

Research on the potential for chronic effects from 60-Hz electromagnetic fields from energized transmission lines was reviewed and addressed elsewhere by the NRC in NUREG-1437 (NRC 1996). At that time, research results were not conclusive. The National Institute of Environmental Health Sciences (NIEHS) directs related research through the U.S. Department of Energy. An NIEHS report (1999) contains the following conclusion:

> The NIEHS concludes that ELF-EMF (extremely low frequency-electromagnetic field) exposure cannot be recognized as entirely safe because of weak scientific evidence that exposure may pose a leukemia hazard. In our opinion, this finding is insufficient to warrant aggressive regulatory concern. However, because virtually everyone in the United States uses electricity and therefore is routinely exposed to ELF-EMF, passive regulatory action is warranted such as a continued emphasis on educating both the public and the regulated community on means aimed at reducing exposures. The NIEHS does not believe that other cancers or non-cancer health outcomes provide sufficient evidence of a risk to currently warrant concern.

This statement is not sufficient to cause the staff to consider the potential impact as significant to the public. However, because conclusive information is not available, this issue is not considered to be resolved.

5.8.6 Summary of Nonradiological Health Impacts

Changes to this section reflect nonradiological health impacts from the closed-cycle, combination wet and dry cooling system and the higher power level of 4500 MW(t) for one unit (9000 MW(t) for two units).

The staff evaluated health impacts to the public and the workers from the cooling systems, noise generated by unit operations, and acute and chronic impacts of electromagnetic fields at the higher power levels from the additional units. Based on the information provided by Dominion and its independent review, the staff concludes that the potential impacts of nonradiological effects resulting from the operation of two additional units with closed-cycle cooling systems meeting the noise criteria of the PPE would be SMALL, and mitigation is not warranted.

5.9 Radiological Health Impacts

This introductory section is not affected by the changes presented in ER Revision 6 and is provided solely for context.

This section addresses the radiological impacts of normal operations of the proposed new Units 3 and 4, including a discussion of the estimated radiation dose to a member of the public and to the biota present in the proximity of the new units. Estimated doses to workers at the proposed units are also discussed. Radiological impacts were determined using the PPE approach for liquid and gaseous radiological effluents.

5.9.1 Exposure Pathways

This section is not affected by the changes presented in ER Revision 6 and is provided solely for context.

During normal operation, small quantities of radiological materials are released to the environment through gaseous and liquid effluents from the plant. Dominion stated in ER Revision 6 that the contribution to direct radiation exposure from new reactor designs would be negligible (Dominion 2006a).

5.9.2 Radiation Doses to Members of the Public

Changes to this section reflect calculations of the radiation doses to the public from the higher power level of 4500 MW(t) for one unit (9000 MW(t) for two units).

The proposed increase in the power level from 4300 MW(t) to 4500 MW(t) per unit resulted from the increase in power level for the surrogate ESBWR reactor design from 4000 MW(t) to

1   4500 MW(t) (see Section 5.10 for a discussion of the "surrogate" label). The increase in the
2   power level to 4500 MW(t) per unit results in a small increase to the liquid and gaseous
3   radiological effluent source terms for normal operation, shown in Table 5.4-6 and Table 5.4-7,
4   respectively, of ER Revision 6 (Dominion 2006a). These tables are reproduced in Appendix H,
5   Tables H-2 and H-5 of this SDEIS. The liquid and gaseous effluent source terms result in small
6   increases to the doses to the maximally exposed individual as shown in Tables 5-7 and 5-8.
7
8   The staff performed an independent evaluation of doses to the maximally exposed individual
9   (Appendix H) and found similar results.
10
11   ### 5.9.3 Impacts to Members of the Public
12
13   *Changes to this section reflect calculations of the radiation impacts to the maximally exposed*
14   *individual and the population dose from the higher power level of 4500 MW(t) for one unit*
15   *(9000 MW(t) for two units).*
16
17   *Maximally Exposed Individual*
18
19   Table 5-9 compares the dose estimates for the maximally exposed individual exposed to liquid
20   and gaseous effluents from a single reactor unit to the 10 CFR Part 50, Appendix I, Design
21   Objectives. The liquid and gaseous radiological effluent source terms for normal operation
22   resulting from the higher power level of 4500 MW(t) per unit resulted in a small increase in doses
23   to the maximally exposed individual as shown in Table 5-9. Doses to the maximally exposed
24   individual were still within the 10 CFR Part 50, Appendix I, Design Objectives.
25
26   Table 5-10 compares the dose estimates for the maximally exposed individual exposed to liquid
27   and gaseous effluents to the 40 CFR Part 190 standards. The liquid and gaseous radiological
28   effluent source terms for normal operation resulting from the higher core thermal power level of
29   4500 MW(t) per unit resulted in a small increase in doses to the maximally exposed individual as
30   shown in Table 5-10. The sum of maximally exposed individual dose estimates for the existing
31   Units 1 and 2 and the proposed Units 3 and 4 were well within the 40 CFR Part 190 Standards.
32
33   *Population Dose*
34
35   The liquid and gaseous radiological effluent source terms for normal operation resulting from the
36   higher PPE core thermal power level of 4500 MW(t) per unit increased the population dose within
37   80 km (50 mi) of each unit from 0.32 person-Sv/yr (32 person-rem/yr) (Dominion 2005a) to 0.34
38   person-Sv/yr (34 person-rem/yr) (Dominion 2006a).
39
40   The staff performed an independent evaluation of doses to the maximally exposed individual and
41   population (Appendix H), and found similar results.
42

Table 5-7. Liquid Pathway Doses for Maximally Exposed Individuals at Lake Anna
(corresponds to Table 5-8 in the Draft EIS)

Pathway	Total Body Dose (mSv/yr)[a]	Thyroid Dose (mSv/yr)[a]	Bone Dose (mSv/yr)[a]
Fish Consumption	5.3×10^{-3}	$0^{(b)}$	2.3×10^{-2}
Invertebrate Consumption	6.9×10^{-4}	$0^{(b)}$	1.5×10^{-3}
Drinking Water	6.9×10^{-3}	1.3×10^{-2}	2.7×10^{-4}
Shoreline Recreation	3.0×10^{-4}	3.0×10^{-4}	3.0×10^{-4}
Swimming	3.2×10^{-6}	3.2×10^{-6}	3.2×10^{-6}
Boating	4.0×10^{-6}	4.0×10^{-6}	4.0×10^{-6}
Total	1.3×10^{-2}	1.3×10^{-2}	2.5×10^{-2}
Age group receiving maximum dose	Adult	Infant	Child

(a) Multiply mSv/yr times 100 to obtain mrem/yr.
(b) Thyroid dose is not applicable because infants are assumed not to consume fish and invertebrates.
Source: Dominion 2006a, Table 5.4-8. Doses were estimated for one unit.

Table 5-8. Gaseous Pathway Doses for Maximally Exposed Individual
(corresponds to Table 5-9 in the Draft EIS)

Location	Pathway	Total Body Dose (mSv/yr)[a]	Thyroid Dose (mSv/yr)[a]	Skin Dose (mSv/yr)[a]
Nearest Site Boundary (1.4 km [0.88 mi] ESE)	Plume	2.1×10^{-2}	(b)	6.2×10^{-2}
Nearest Site Boundary (1.4 km [0.88 mi] ESE)	Inhalation			
	Adult	3.0×10^{-3}	1.6×10^{-2}	(c)
	Teen	3.1×10^{-3}	2.0×10^{-2}	(c)
	Child	2.7×10^{-3}	2.3×10^{-2}	(c)
	Infant	1.6×10^{-3}	2.0×10^{-2}	(c)
Nearest Garden (1.5 km [0.94 mi] NE)	Vegetable			
	Adult	4.4×10^{-3}	4.9×10^{-2}	(c)
	Teen	5.7×10^{-3}	6.6×10^{-2}	(c)
	Child	1.1×10^{-2}	1.3×10^{-1}	(c)
Nearest Residence (1.5 km [0.96 mi] NNE)	Plume	1.4×10^{-2}	(b)	4.0×10^{-2}
Nearest Residence (1.5 km [0.96 mi] NNE)	Inhalation			
	Adult	2.0×10^{-3}	1.0×10^{-2}	(c)
	Teen	2.0×10^{-3}	1.3×10^{-2}	(c)
	Child	1.8×10^{-3}	1.5×10^{-2}	(c)
	Infant	1.0×10^{-3}	1.3×10^{-2}	(c)
Nearest Meat Cow (2.22 km [1.37 mi] SE)	Meat			
	Adult	6.7×10^{-4}	1.5×10^{-3}	(c)
	Teen	4.9×10^{-4}	1.1×10^{-3}	(c)
	Child	7.9×10^{-4}	1.7×10^{-3}	(c)

(a) Multiply mSv/yr times 100 to obtain mrem/yr.
(b) Thyroid dose is not applicable for the plume pathway.
(c) Skin dose is not applicable for the ingestion and inhalation pathways.
Source: Dominion (2006a), Table 5.4-9. Doses were estimated for one unit. There were no milk cows or goats within 8 km (5 mi) (Dominion 2006a). No infant doses were calculated for the vegetable or meat pathway because the doses that infants receive from their diet would be bounded by the dose calculated for the child.

1 **Table 5-9.** Comparison of Maximally Exposed Individual Dose Estimates from Liquid and
2 Gaseous Effluents to 10 CFR Part 50, Appendix I, Design Objectives
3 (corresponds to Table 5-10 in the Draft EIS)
4

Pathway/Type of Dose	Dominion (2006)[a,b]	Appendix I Design Objectives[c]
Liquid Effluents		
Whole body dose	0.013 mSv/yr	0.03 mSv/yr
Maximum organ dose	0.025 mSv/yr	0.1 mSv/yr
Gaseous Effluents (Noble gases only)		
Gamma air dose	0.032 mGy/yr	0.1 mGy/yr
Beta air dose	0.048 mGy/yr	0.2 mGy/yr
Whole body dose	0.024 mSv/yr	0.05 mSv/yr
Skin dose	0.062 mSv/yr	0.15 mSv/yr
Gaseous Effluents (Radioiodines and particulates)		
Organ dose	0.13 mSv/yr (thyroid)	0.15 mSv/yr

16 (a) Doses were estimated for one unit.
17 (b) Multiply mSv/yr (or mGy/yr) times 100 to obtain mrem/yr (or mrad/yr)
18 (c) Design objectives are for each light-water-cooled nuclear power reactor (10 CFR Part 50, Appendix I)
19 Source: Dominion 2006a, Table 5.4-10

20
21 **Table 5-10.** Comparison of Maximally Exposed Individual Dose Estimates from Liquid and
22 Gaseous Effluents to 40 CFR Part 190 Standards (corresponds to Table 5-11 in
23 the Draft EIS)
24

Dose	Dominion (2006) Estimate[a,b,c]	40 CFR 190 Standards
Whole body dose equivalent	0.078 mSv/yr	0.25 mSv/yr
Thyroid dose	0.28 mSv/yr	0.75 mSv/yr
Dose to another organ	0.12 mSv/yr (bone)	0.25 mSv/yr

29 (a) Doses from direct radiation were determined to be negligible (Dominion 2006a).
30 (b) Sum of dose from liquid and gaseous effluent releases for the two existing NAPS units and the proposed units
31 (Dominion 2006a)
32 (c) Multiply mSv/yr times 100 to obtain mrem/yr.
33 Source: Dominion 2006a, Table 5.4-11

34
35 Based on the small increase to the maximally exposed individual and population dose estimates
36 as a result of the higher power level of 4500 MW(t) per unit, the staff concludes that there would
37 be no observable health impacts to the public from normal operation of the proposed nuclear
38 units. The impact level category would be SMALL, and mitigation is not warranted.
39

1 ## 5.9.4 Occupational Doses to Workers
2
3 *This section is not affected by the changes presented in ER Revision 6 and is provided solely for*
4 *context.*
5
6 The increase in power level to 4500 MW(t) did not affect occupational doses to workers as the
7 estimated annual occupational dose estimate for advanced reactor designs did not consider the
8 surrogate ESBWR design. The annual occupational dose estimate of 1.5 person-Sv
9 (150 person-rem) considered the AP1000, international reactor innovative and secure design
10 (IRIS), and the gas turbine-modular helium reactor (GT-MHR) designs (Dominion and Bechtel
11 2002). Based on occupational doses being maintained within the 10 CFR Part 20 dose limits
12 and annual occupational dose estimates being within those typical of current operating
13 light-water reactor plants, the staff concludes the health impacts did not change. The impact
14 level category would still be SMALL, and mitigation is not warranted.
15
16 ## 5.9.5 Impacts to Biota
17
18 *Changes to this section reflect calculations of the radiation impacts to biota from the higher*
19 *power level of 4500 MW(t) for one unit (9000 MW(t) for two units).*
20
21 Table 5-11 compares biota dose estimates from liquid and gaseous effluents to 40 CFR Part 190
22 standards. The liquid and gaseous radiological effluent source terms for normal operation
23 resulting from the higher power level of 4500 MW(t) per unit results in a small increase in
24 calculated biota doses as shown in Table 5-11.
25
26 Table 5-12 of this section compares biota dose estimates from liquid and gaseous effluents to
27 dose guidelines published by the National Council on Radiation Protection and Measurements
28 (NRCP 1991) and International Atomic Energy Agency studies (IAEA 1992). The liquid and
29 gaseous radiological effluent source terms for normal operation resulting from the higher power
30 level of 4500 MW(t) per unit resulted in a small increase in biota doses as shown in Table 5-12.
31 Estimated biota doses were still well within the 10 mGy/day (1000 mrad/day) guideline for
32 aquatic organisms and the 1 mGy/day (100 mrad/day) guideline for terrestrial organisms.
33
34 The staff performed an independent evaluation of biota doses (Appendix H) and found similar
35 results.
36
37 The staff concludes that the increase to the biota dose estimates as a result of the higher power
38 level of 4500 MW(t) per unit did not change appreciably. The impact level category would still be
39 SMALL, and mitigation is not warranted.
40
41
42

Table 5-11. Comparison of Biota Doses from Proposed Units 3 and 4 to 40 CFR Part 190 (corresponds to Table 5-12 in the Draft EIS)

Biota	Dose from Liquid Effluent/Unit (mGy/yr)[a,b]	Dose from Gaseous Effluent/Unit (mGy/yr)[a,b]	Total Dose/Unit (mGy/yr)[a,b]	Total Dose for Two Units (mGy/yr)[a]	40 CFR 190 Total Body Dose Limit (mSv/yr)
Fish	0.099	0	0.099	0.20	0.25
Invertebrates	0.46	0	0.46	0.92	0.25
Algae	0.54	0	0.54	1.08	0.25
Muskrat	0.44	0.34	0.78	1.56	0.25
Raccoon	0.051	0.34	0.39	0.78	0.25
Heron	0.56	0.34	0.90	1.80	0.25
Duck	0.44	0.34	0.78	1.56	0.25

(a) Multiply mGy/yr or mSv/yr times 100 to obtain mrad/yr or mrem/yr.
(b) From Dominion (2006a), Table 5.4-16.

Table 5-12. Comparison of Biota Doses from Proposed Units 3 and 4 to NCRP and IAEA Studies (corresponds to Table 5-13 in the Draft EIS)

Biota	Estimated Dose for Two Units (mGy/day)[a]	Chronic Dose Rate Values from NCRP and IAEA Studies (mGy/day)[a]
Fish	5.5×10^{-4}	10
Invertebrates	2.5×10^{-3}	10
Algae	3.0×10^{-3}	10
Muskrat	4.3×10^{-3}	1
Raccoon	2.1×10^{-3}	1
Heron	4.9×10^{-3}	1
Duck	4.3×10^{-3}	1

(a) Multiply mGy/day times 100 to obtain mrad/day.

5.9.6 Radiological Monitoring

This section is not affected by changes presented in ER Revision 6 and is provided solely for context.

In the Draft EIS (NRC 2004a), the staff reported that a radiological environmental monitoring program (REMP) has been in place for the NAPS site since 1976 (NRC 1976). The REMP includes monitoring of the airborne exposure pathway, direct exposure pathway, water exposure pathway, aquatic exposure pathway from Lake Anna and the North Anna River, and the ingestion exposure pathway within a 40-km (25-mi) radius of the station. The staff reviewed the documentation for the REMP, the Offsite Dose Calculation Manual, and recent monitoring reports and determined that the current operational monitoring program is adequate to establish

1   the radiological impacts to the environment related to the construction and operation of two new
2   units at the North Anna ESP site. The staff concludes that the REMP is adequate even if the
3   units operate at the higher power level of 4500 MW(t).
4

5   ## 5.10 Environmental Impacts of Postulated Accidents
6
7   *This section addresses the impacts of postulated accidents that were not addressed in the*
8   *Draft EIS.*
9
10   In the Draft EIS (NRC 2004a), the staff considered the radiological consequences of potential
11   accidents from proposed new units at the North Anna ESP site based on the certified General
12   Electric ABWR reactor design and a surrogate Westinghouse AP1000. In Revision 6 of the ER,
13   Dominion described the consequences of potential accidents based on an ESBWR design, for
14   which General Electric submitted an application for certification in August 2005; the NRC
15   accepted the application for docketing in December 2005. This supplement considers the
16   consequences of potential accidents based on an ESBWR reactor design with source terms
17   increased by 25 percent to account for uncertainty and potential changes to the design as the
18   staff performs the design certification review (consequently, this design is referred to as the
19   "surrogate" ESBWR design). In addition, this SDEIS considers the consequences of cleanup
20   water line breaks and feedwater system pipe breaks for the ABWR, which were not considered in
21   the Draft EIS.
22
23   As used in this section, the term "accident" refers to any off-normal event not addressed in
24   Section 5.9 that results in the release of radioactive material into the environment. The focus of
25   this review is on events that could lead to releases substantially in excess of permissible limits
26   for normal operations. Normal release limits are specified in 10 CFR Part 20, Appendix B,
27   Table 2.
28
29   Numerous features combine to reduce the risk associated with accidents at nuclear power
30   plants. Safety features in the design, construction, and operation of the plants, which comprise
31   the first line of defense, are intended to prevent the release of radioactive material from the plant.
32   The design objectives and the measures for keeping levels of radioactive material in effluents to
33   unrestricted areas as low as reasonably achievable (ALARA) are specified in 10 CFR Part 50,
34   Appendix I. There are additional measures that are designed to mitigate the consequences of
35   failures in the first line of defense. These include the NRC's reactor site criteria in
36   10 CFR Part 100, which require the site to have certain characteristics that reduce the risk to the
37   public and the potential impacts of an accident, and emergency preparedness plans and
38   protective action measures for the site and environs as set forth in 10 CFR 50.47;
39   10 CFR Part 50, Appendix E; and NUREG-0654/FEMA-REP-1 (NRC 1980). All of these safety
40   features, measures, and plans make up the defense-in-depth philosophy to protect the health
41   and safety of the public and the environment.
42

1   This section discusses (1) types of radioactive material that might be released, (2) paths to the
2   environment, (3) the relationship between radiation dose and health effects, and (4) the impacts
3   of postulated reactor accidents, both design-basis accidents (DBA) and severe accidents. The
4   impacts of postulated accidents during the transportation of spent fuel are discussed in
5   Chapter 6.
6

7   The potential for dispersion of radioactive material in the environment depends on the
8   mechanical forces that physically transport the material and on the physical and chemical forms
9   of the material. Radioactive material exists in a variety of physical and chemical forms. The
10   majority of the material is in the form of nonvolatile solids. However, there is a significant amount
11   of material that is in the form of volatile solids or gases. Gaseous radioactive material includes
12   the chemically inert noble gases krypton and xenon. Radioactive forms of iodine, which are
13   created in substantial quantities in the fuel by fission, are volatile. Other radioactive material
14   formed during the routine operation of a nuclear power plant have lower volatilities and,
15   therefore, have less tendency to escape from the fuel than the noble gases and iodines.
16

17   Radiation exposure is determined by the proximity of individuals to radioactive material, the
18   duration of exposure, and factors that shield the individuals from the radiation. Pathways that
19   lead to radiation exposure include (1) external radiation from radioactive material in the air, on
20   the ground, and in the water; (2) the inhalation of radioactive material; and (3) ingestion of food
21   or water containing material initially deposited on the ground and in water.
22

23   Although radiation may cause cancers at high doses and high dose rates, currently there are no
24   data that unequivocally establish the occurrence of cancer following exposure to low doses
25   below about 100 mSv (10,000 mrem) and at low dose rates. However, radiation protection
26   experts conservatively assume that any amount of radiation exposure may pose some risk of
27   causing cancer or a severe hereditary effect and that the risk is higher for higher radiation
28   exposures. Therefore, a linear, no-threshold response model is used to describe the relationship
29   between radiation dose and detriments such as cancer induction. A recent report by the National
30   Research Council (2005), the BEIR VII report, supports the linear, no-threshold dose response
31   theory. Simply stated, any increase in dose, no matter how small, results in an incremental
32   increase in health risk. This theory is accepted by the NRC as a conservative model for
33   estimating health risks from radiation exposure, recognizing that the model probably
34   over-estimates those risks. Based on this model, the staff estimates the risk to the public from
35   radiation exposure using the nominal probability coefficient for total detriment (730 fatal cancers,
36   nonfatal cancers, and severe hereditary effects per 10,000 person-Sv [1,000,000 person-rem])
37   from ICRP Publication 60 (ICRP 1991).
38

39   Physiological effects are clinically detectable should individuals receive radiation exposure
40   resulting in a dose greater than about 0.25 Sv (25 rem) over a short period of time (hours).
41   Doses of about 2.5 to 5.0 Sv (250 to 500 rem) received over a relatively short period (hours to a
42   few days) can be expected to cause some fatalities.

5.10.1 Design-Basis Accidents

This section addresses the impacts of postulated design basis accidents for the ABWR and the surrogate ESBWR that were not addressed in the Draft EIS.

Dominion has evaluated the potential consequences of postulated accidents to demonstrate that Units 3 and 4 could be constructed and operated at the North Anna ESP site without undue risk to the health and safety of the public. These evaluations use a set of surrogate DBAs that are representative for the range of reactor designs being considered for the North Anna ESP site and site-specific meteorological data. The set of accidents covers events that range from relatively high probability of occurrence with relatively low consequences to relatively low probability of occurrence with high consequences.

This DBA review focuses on a surrogate ESBWR light-water reactor design. The bases for analyses of postulated accidents for this design are established in the application for certification. Detailed descriptions of the DBAs are presented in Chapter 15 of the ESBWR Design Control Document, Tier 2 (GE Nuclear Energy 2006). Potential consequences of DBAs are evaluated following procedures outlined in regulatory guides and standard review plans. The potential consequences of accidental releases depend on the specific radionuclides released, the activity released for each radionuclide, and meteorological conditions. A source term for the ESBWR has been provided by General Electric (2006). Dominion increased the source term by 25 percent for the surrogate ESBWR to account for uncertainty because the design review of the ESBWR has not been completed. Methods for evaluating potential accidents are based on guidance in Regulatory Guide 1.183 (NRC 2000b). The staff will determine the significance of any change in the source term if an applicant for a CP or COL references any North Anna ESP.

In its review of the information in Revision 6 of the ER, the staff identified several inconsistencies between the DBA source terms and the resulting doses. On May 10, 2006, the staff requested that Dominion resolve the inconsistencies (NRC 2006). Dominion resolved the inconsistencies in a May 24, 2006, response (Dominion 2006b) to the staff request for additional information.

For environmental reviews, consequences are evaluated assuming realistic meteorological conditions. Meteorological conditions are represented in these consequence analyses by an atmospheric dispersion factor, which is also referred to as χ/Q. Acceptable methods of calculating χ/Q for DBAs from meteorological data are set forth in Regulatory Guide 1.145 (NRC 1983).

Dominion had provided the staff with meteorological data for 1996, 1997, and 1998 for the North Anna ESP site. These data have been reviewed by the staff and found to be representative of the meteorological conditions at the site (see Section 2.3 of the Draft EIS). The meteorological instrumentation and its maintenance are consistent with staff guidance (i.e., Regulatory Guide 1.23 [AEC 1972]), and the data quality is consistent with standards set forth in that guidance.

1 Therefore, the data are considered acceptable for use in evaluation of the consequences
2 of DBAs.
3
4 Table 3.1-9 of the PPE, referenced in Section 3.1.3 of the ER, lists χ/Q values. These values are
5 not appropriate for environmental reviews. Realistic (50th percentile) χ/Q values for use in the
6 environmental review of DBAs are provided in ER section 7.1.4. However, Dominion only
7 provided only one χ/Q for the low population zone (LPZ). NRC guidance (e.g., Regulatory
8 Guides 1.145 and 1.183) indicates that LPZ χ/Qs should be calculated for each of the four time
9 periods that comprise the "course of the accident" (i.e., 30 days [720 hours]). Therefore, the staff
10 calculated χ/Q values for the four time periods using data provided by Dominion.
11
12 Table 5-13 presents the atmospheric dispersion factors that the staff used to calculate doses for
13 DBAs. The first column lists the time periods and boundaries for which χ/Q values and dose
14 estimates are needed. For the exclusion area boundary (EAB), the postulated DBA dose and
15 the short-term χ/Q (i.e., 2 hours) value is calculated, and for the LPZ, they are calculated for the
16 "course of the accident" (i.e., 30 days [720 hours]). For the LPZ, there is a χ/Q value for each of
17 the four time periods. The second column lists the χ/Q values. The χ/Q values for the EAB and
18 the first LPZ time period are the same as those calculated by Dominion. The χ/Q values for the
19 last three LPZ time periods are those estimated by the staff.
20
21 The staff concludes that the atmospheric dispersion characteristics of the North Anna ESP site
22 are acceptable for estimating the potential consequences of postulated DBAs for reactor designs
23 with postulated design χ/Q values falling within the site χ/Q values. The staff intends to verify
24 that the χ/Q values used in analyzing the reactor design proposed at the CP/COL stage are
25 equal to or greater than the χ/Q values specified in the ESP if an applicant for a CP or COL
26 references any North Anna ESP.
27
28 Table 5-14 lists the set of surrogate DBAs considered by Dominion and presents the staff's
29 analysis of the consequences of each postulated DBA in terms of total effective dose equivalent
30 (TEDE). TEDE is the sum of the committed effective dose equivalent (CEDE) from inhalation
31 and the deep dose equivalent from external exposure. Dose conversion factors from Federal
32 Guidance Report 11 (Eckerman et al. 1988) were used by the staff to calculate the CEDE.
33 Similarly, dose conversion factors from Federal Guidance Report 12 (Eckerman and
34 Ryman 1993) were used by Dominion to calculate the deep dose equivalent. The review criteria
35 used in the staff's safety review of DBA doses are included in Table 5-14 to illustrate how small
36 the calculated environmental consequences (TEDE doses) are. In all cases, the calculated
37 TEDEs are a small fraction of the review criterion. Considering the magnitude of the doses in
38 Table 5-14 and the relationship between low doses and health effects such as fatal and non-fatal
39 cancers and severe hereditary effects described in 5.10, the staff concludes that the potential
40 environmental impacts of design basis accidents are small.
41
42 Table 5-15 lists the doses for the Cleanup Water Line Break for the ABWR reactor. The doses
43 calculated for this accident are small fractions of the dose limits. According to Revision 6 of the

Table 5-13 Atmospheric Dispersion Factors (χ/Q, s/m^3) for the North Anna ESP Site Design-Basis Accident Calculations

Time Period[a] and Boundary	Site
0 to 2 hr, Exclusion Area Boundary	3.34×10^{-5}
0 to 8 hr, Low Population Zone	2.17×10^{-6}
8 to 24 hr, Low Population Zone	1.5×10^{-6}
1 to 4 day, Low Population Zone	1.2×10^{-6}
4 to 30 day, Low Population Zone	9.0×10^{-7}

(a) Times are relative to the beginning of the release to the environment.

Table 5-14. Design-Basis Accident Doses for the Surrogate ESBWR Reactor

Accident	Standard Review Plan Section[b]	TEDE in Sv[a] EAB	TEDE in Sv[a] LPZ	Review Criterion
Main Steam Line Break	15.6.4			
Pre-Existing Iodine Spike		3.13×10^{-3}	2.03×10^{-4}	2.5×10^{-1}[c]
Accident-Initiated Iodine Spike		1.57×10^{-4}	1.02×10^{-5}	2.5×10^{-2}[d]
Loss-of-Coolant Accident	15.6.5	2.08×10^{-3}	1.33×10^{-3}	2.5×10^{-1}[c]
Feedwater System Pipe Break	15.2.8	6.85×10^{-3}	4.45×10^{-9}	2.5×10^{-2}[d]
Cleanup Water Line Break		2.59×10^{-4}	1.68×10^{-5}	2.5×10^{-2}[d]
Failure of Small Lines Carrying Primary Coolant Outside Containment	15.6.2	4.49×10^{-5}	6.82×10^{-6}	2.5×10^{-2}[d]
Fuel Handling	15.7.4	1.84×10^{-3}	1.16×10^{-4}	6.3×10^{-2}[d]

(a) To convert Sv to rem, multiply Sv by 100.
(b) NUREG-0800 (NRC 1987).
(c) 10 CFR 50.34(a)(1) and 10 CFR 100.21 criteria.
(d) Standard Review Plan criterion.

Table 5-15. Additional Design-Basis Accident Doses for the ABWR Reactor

Accident	Standard Review Plan Section[b]	TEDE in Sv[a] EAB	TEDE in Sv[a] LPZ	Review Criterion
Cleanup Water Line Break		4.68×10^{-6}	3.04×10^{-7}	2.5×10^{-2}[c]

(a) To convert Sv to rem, multiply Sv by 100.
(b) NUREG-0800 (NRC 1987).
(c) Standard Review Plan criterion.

1   ER and the Dominion May 24, 2006, response to the staff's RAI, the impacts of the Feedwater
2   System Pipe Break DBA are bounded by the impacts of the Cleanup Water Line Break
3   (Dominion 2006b).
4

5   Although Dominion chose the PPE approach in the overall ESP application, it based its initial
6   evaluation of the impacts of DBAs on characteristics of the ABWR and the AP1000 reactor
7   designs with the explicit assumption that these impacts would bound the impacts of other
8   ALWRs designs (Dominion 2006a). In Revision 6 to the ESP application and its May 24, 2006,
9   response to the staff RAIs of May 10, 2006, Dominion provided an evaluation of the impacts of
10   postulated DBAs for the ESBWR reactor design (Dominion 2006a, b). The staff has reviewed
11   these analyses. The results of the Dominion and staff analyses indicate that the impacts of
12   postulated DBAs, if the surrogate ESBWR were to be located at the North Anna ESP site, would
13   be small compared to the TEDE doses used as safety review criteria, which are included in
14   Table 5-14 to illustrate how small the calculated environmental consequences (TEDE doses) are.
15   In all cases, the calculated TEDEs are a small fraction of the review criterion. Considering the
16   magnitude of the doses in Table 5-14 and the relationship between low doses and health effects
17   such as fatal and non-fatal cancers and severe hereditary effects described in Section 5.10, the
18   staff concludes that the potential environmental impacts of design basis accidents would be
19   SMALL, and mitigation is not warranted.
20

21   ## 5.10.2 Severe Accidents

22

23   *This section addresses the impacts of postulated severe accidents for an ESBWR that were not*
24   *addressed in the Draft EIS.*
25

26   For the ABWR and AP1000 reactor designs, Dominion based its evaluation of the potential
27   consequences of postulated severe accidents on the evaluation for current generation reactors
28   previously evaluated by the staff in NUREG-1437 (NRC 1996). Three pathways were
29   considered: (1) the atmospheric pathway in which radioactive material is released to the air,
30   (2) the surface water pathway in which airborne radioactive material falls out on open bodies of
31   water, and (3) the groundwater pathway in which groundwater is contaminated by a basemat
32   melt-through with subsequent contamination of surface water by the groundwater.
33

34   In addition, in Revision 6 of the ER, Dominion performed a site-specific analysis of the potential
35   consequences of postulated severe accidents for an ESBWR at the North Anna ESP site.
36   Dominion used the MACCS2 computer code (Chanin et al. 1990; Jow et al. 1990) for the
37   analysis. Along with Revision 6 to the North Anna ESP application, Dominion also provided the
38   MACCS2 input and output files to the NRC (Dominion 2006a).
39

40   The MACCS computer code was developed to evaluate the potential offsite consequences of
41   severe accidents for the sites covered by NUREG-1150 (NRC 1990). MACCS2 (Chanin and
42   Young 1997) is the current version of MACCS. The MACCS and MACCS2 codes evaluate the

1 consequences of atmospheric releases of radioactive material following a postulated severe
2 accident. The pathways analyzed include external exposure to the passing plume, external
3 exposure to material deposited on the ground and skin, inhalation of material in the passing
4 plume and resuspended from the ground, and ingestion of contaminated food and surface water.
5 The primary enhancements in MACCS2 are that MACCS2 has (1) a more flexible emergency
6 response model, (2) an expanded library of radionuclides, and (3) an improved food-chain model
7 (Chanin and Young 1997).
8
9 Three types of severe accident consequences were assessed: (1) human health, (2) economic
10 cost, and (3) land area affected by contamination. Human-health effects are expressed in terms
11 of the number of cancers that might be expected if a severe accident were to occur. These
12 effects are directly related to the cumulative radiation dose received by the general population.
13 MACCS2 estimates both early cancer fatalities and latent fatalities. Early fatalities are related to
14 high doses or dose rates and can be expected to occur within a year of exposure (Jow et al.
15 1990). Latent fatalities are related to exposure of a large number of people to low doses and
16 dose rates and could occur after a latent period of several (2 to 15) years. Population health risk
17 estimates are based on the population distribution within an 80-km (50-mi) radius of the plant,
18 whereas average individual health risks are based on the distribution of population close to the
19 plant. Economic costs of a severe accident include the costs associated with short-term
20 relocation of people, decontamination of property and equipment, interdiction of food supplies,
21 land, and equipment use, and condemnation of property. The affected land area is a measure of
22 the areal extent of the residual contamination following a severe accident.
23
24 Risk is the product of the frequency of an accident, also called the core damage frequency,
25 and the consequences of the accident. For example, a severe accident with containment
26 leakage at the Technical Specification rate for the surrogate ESBWR is estimated to have a core
27 damage frequency of 2.80×10^{-8} per reactor year (Ryr^{-1}). The cumulative population dose
28 associated with this accident at the North Anna ESP site is calculated to be 2.43×10^{2} person-Sv
29 (2.43×10^{4} person-rem). The population dose risk for this class of accidents is the product
30 of 2.80×10^{-8} Ryr^{-1} and 2.43×10^{2} person-Sv, or 6.80×10^{-6} person-Sv Ryr^{-1}
31 (6.80×10^{-4} person-rem Ryr^{-1}).
32
33 Core damage frequency estimates are made using well developed methods that have been
34 updated based on investigation of the accident at Three Mile Island, Unit 2 and research
35 following the accident. Core damage frequency estimation methods used to generate the
36 estimates presented in this EIS are described in NUREG-1150, *Severe Accident Risks: An*
37 *Assessment for Five U.S. Nuclear Power Plants* (NRC 1990). These methods explicitly consider
38 both pre-accident and post-accident human errors. The core damage frequencies listed in this
39 EIS are core damage frequencies estimated for the ESBWR reactor design as set forth in the
40 ESBWR certification application. The following sections discuss the estimated risks associated
41 with the air, surface water, and groundwater pathways.
42

1   *Air Pathway*
2
3   The MACCS2 code directly estimates consequences associated with releases to the air
4   pathway. The results of the MACCS2 runs are presented in Table 5-16. The core damage
5   frequencies given in this table are for internally initiated accident sequences while the plant is at
6   power. Internally initiated accident sequences include those sequences initiated by human error,
7   equipment failures, and loss of offsite power. The core damage frequencies for externally
8   initiated events and during shutdown would be comparable to or lower than those for internally
9   initiated events.
10
11   Table 5-16 shows that the probability weighted consequences (i.e., the risks) of severe accidents
12   for the surrogate ESBWR located on the North Anna ESP site are small for all risk categories
13   considered. In Table 5-17, the health risks estimated for the surrogate ESBWR at the North
14   Anna ESP site are compared with health risk estimates for the five reactors considered in
15   NUREG-1150 (NRC 1990). Although risks associated with both internally and externally-initiated
16   events were considered for the Peach Bottom and Surry reactors in NUREG-1150, only risks
17   associated with internally initiated events are presented in Table 5-17. The health risks shown
18   for the surrogate ESBWR at the North Anna ESP site are significantly lower than the risks
19   associated with the current generation of operating reactors presented in NUREG-1150. For
20   perspective, Table 5-18 compares the health risks from severe accidents for the surrogate
21   ESBWR at the North Anna ESP site with the risks for the current generation of operating reactors
22   at various sites.
23
24   The Commission has set safety goals for average individual early fatality and latent cancer
25   fatality risks from internally initiated events (NRC 1986). These goals are that average individual
26   early fatality risk be less than 5×10^{-7} Ryr^{-1} and that average individual latent cancer fatality risk be
27   less than 2×10^{-6} Ryr^{-1}. The average individual early fatality risk is calculated using the population
28   distribution within 1.6 km (1 mi) of the plant boundary. The average individual latent cancer
29   fatality risk is calculated using the population distribution within 16 km (10 mi) of the plant. For
30   the plants considered in NUREG-1150, these risks were well below the Commission's safety
31   goals. Risks calculated for the surrogate ESBWR at the North Anna ESP site are lower than the
32   risks associated with the current generation reactors considered in NUREG-1150, and are well
33   below the NRC safety goals.
34
35   The staff compared the core damage frequencies and population dose risk estimates for the
36   surrogate ESBWR at the North Anna ESP site with comparable statistics summarizing the results
37   of contemporary severe accident analyses performed for 28 current generation operating
38   reactors at 23 sites. The results of these analyses are included in the final site-specific
39   Supplements 1 through 20 to the *Generic Environmental Impact Statement for License Renewal*,
40   NUREG-1437, and in the ERs included with license renewal applications for those plants for
41   which supplements have not been published. All of the analyses were completed after
42   publication of NUREG-1150, and the 23 analyses used MACCS2, which was released in 1997.

Table 5-16. Mean Environmental Risks from Surrogate ESBWR Severe Accidents at the North Anna ESP Site

Release Category Description (Accident Class)	Core Damage Frequency (Ryr^{-1})	Population Dose (person-Sv Ryr^{-1})[a]	Environmental Risk					
			Fatalities (Ryr^{-1})		Cost[d] ($ Ryr^{-1})	Land Requiring Decontamination[e] (ha Ryr^{-1})	Population Dose from Water Ingestion (person-Sv Ryr^{-1})[a]	
			Early[b]	Latent[c]				
TSL Containment leakage at Technical Specification Limit	2.8 x 10^{-8}	6.80 x 10^{-6}	(f)	3.05 x 10^{-7}	0.05	1.30 x 10^{-6}	1.06 x 10^{-8}	
CCW Containment failures due to core concrete interaction lower drywell debris bed covered	2.9 x 10^{-10}	3.60 x 10^{-6}	(f)	1.61 x 10^{-7}	0.24	3.48 x 10^{-6}	1.48 x 10^{-8}	
EVE Ex-vessel steam explosion fails containment	(f)	1.93 x 10^{-5}	2.68 x 10^{-10}	9.75 x 10^{-7}	3.98	2.03 x 10^{-5}	5.85 x 10^{-7}	
FR Release through controlled (filtered) venting from suppression chamber	(f)	(f)	(f)	(f)	(f)	(f)	(f)	
CCD Containment failures due to core concrete interaction lower drywell debris bed uncovered	(f)	2.08 x 10^{-6}	2.41 x 10^{-11}	9.31 x 10^{-8}	0.44	2.32 x 10^{-6}	5.97 x 10^{-8}	
OPW2 Containment failures due to late (>24 hours) loss of containment heat removal	(f)	4.02 x 10^{-7}	(f)	1.81 x 10^{-8}	0.06	5.47 x 10^{-7}	(f)	

1
2
3
4
5
6
7
8
9
10
11

Table 5-16. (contd)

Release Category Description (Accident Class)	Core Damage Frequency (Ryr⁻¹)	Population Dose (person-Sv Ryr⁻¹)(a)	Fatalities (Ryr⁻¹) Early(b)	Fatalities (Ryr⁻¹) Latent(c)	Cost(d) ($ Ryr⁻¹)	Land Requiring Decontamination(e) (ha Ryr⁻¹)	Population Dose from Water Ingestion (person-Sv Ryr⁻¹)(a)
BOC Break outs de of conta nment	(f)	3.73×10^{-7}	2.22×10^{-10}	2.26×10^{-8}	0 05	(f)	1.04×10^{-8}
BYP Conta nment bypassed because of C S fa ure w th arge (>12" ho e) open ng Lower drywe debr s bed covered	(f)	(f)	4.65×10^{-11}	(f)	(f)	(f)	(f)
DCH D rect conta nment heat ng (h gh pressure RPV fa ure) event damages conta nment	(f)	(f)	(f)	(f)	(f)	(f)	(f)
OPVB Conta nment fa s due to fa ure of vapor suppress on system (vacuum breaker) funct on	(f)	(f)	(f)	(f)	(f)	(f)	(f)
OPW1 Conta nment fa s due to ear y (<24 hours) oss of conta nment heat remova	(f)	(f)	(f)	(f)	(f)	(f)	(f)
Tota	2.88×10^{-8}	3.29×10^{-5}	5.62×10^{-10}	1.59×10^{-6}	4 85	2.84×10^{-5}	6.89×10^{-7}

(a) To convert Sv to rem mu t p y Sv by 100
(b) Ear y fata t es are fata t es re ated to h gh doses or dose rates that genera y can be expected to occur w th n a year of the exposure (Jow et a 1990)
(c) Latent fata t es are fata t es re ated to ow doses or dose rates that cou d occur after a atent per od of severa (2 to 15) years
(d) Cost r sk nc udes costs assoc ated w th short-term re ocat on of peop e decontam nat on nterd ct on and condemnat on t does not nc ude costs assoc ated w th hea th effects (Jow et a 1990)
(e) Land r sk s farm and requ r ng decontam nat on pr or to resumpt on of agr cu tura usage To convert hectares to acres mu t p y by 2 47
(f) Less than 1 percent of the tota

Table 5-17. Comparison of Environmental Risk for New Surrogate ESBWR Units at the North Anna ESP Site with Risks for Five Sites Evaluated in NUREG-1150

Reactor Site	Core Damage Frequency (Ryr⁻¹)	50-mi (80-km) Population Dose Risk (person-Sv Ryr⁻¹)[a]	Fatalities Ryr⁻¹		Average Individual Fatality Ryr⁻¹	
			Early	Latent	Early	Latent Cancer
Grand Gulf[b]	4.0×10^{-6}	5×10^{-1}	8×10^{-9}	9×10^{-4}	3×10^{-11}	3×10^{-10}
Peach Bottom[b]	4.5×10^{-6}	7×10^{-0}	2×10^{-8}	5×10^{-3}	5×10^{-11}	4×10^{-10}
Sequoyah[b]	5.7×10^{-5}	$1 \times 10^{+1}$	3×10^{-5}	1×10^{-2}	1×10^{-8}	1×10^{-8}
Surry[b]	4.0×10^{-5}	$5 \times 10^{+0}$	2×10^{-6}	5×10^{-3}	2×10^{-8}	2×10^{-9}
Zion[b]	3.4×10^{-4}	$5 \times 10^{+1}$	1×10^{-4}	2×10^{-2}	9×10^{-9}	8×10^{-9}
Surrogate ESBWR at North Anna ESP Site[c]	2.9×10^{-8}	3.3×10^{-5}	5.6×10^{-10}	1.6×10^{-6}	5.5×10^{-13}	3.4×10^{-12}

(a) To convert Sv to rem, multiply Sv by 100.
(b) Risks were calculated using the MACCS code and presented in NUREG-1150 (NRC 1990).
(c) Calculated with MACCS2 code using North Anna site-specific input.

1 **Table 5-18.** Comparison of Environmental Risks from Severe Accidents Initiated by Internal
2 Events for the Surrogate ESBWR at the North Anna ESP Site with Risks Initiated by
3 Internal Events for 28 Current Operating Plants Undergoing License Renewal
4

Reactor Site	Core Damage Frequency (yr^{-1})	50-mi (80-km) Population Dose Risk (person-Sv Ryr^{-1})[a]
Current Reactor Maximum[b]	2.4×10^{-4}	6.9×10^{-1}
Current Reactor Mean[b]	3.6×10^{-5}	1.5×10^{-1}
Current Reactor Median[b]	2.8×10^{-5}	1.4×10^{-1}
Current Reactor Minimum[b]	1.9×10^{-6}	5.5×10^{-3}
ESBWR at North Anna ESP Site[c]	2.9×10^{-8}	3.3×10^{-5}

5
6
7
8
9
10

11 (a) To convert Sv to rem, multiply Sv by 100.
12 (b) Based on MACCS and MACCS2 calculations for current plants undergoing operating license renewal.
13 (c) Calculated with MACCS2 code using North Anna site-specific input.
14

15 Table 5-18 shows that the core damage frequencies estimated for an ESBWR are significantly
16 lower than those of current generation reactors. Similarly, the population doses estimated for an
17 ESBWR at the North Anna ESP site are well below the mean and median values for current
18 generation reactors undergoing license renewal.
19

20 Population dose and risk estimates of the North Anna site in Tables 5-16 through 5-18 are for the
21 year 2030 and are based on data from the 1990 census. Population growth estimates presented
22 in Chapter 2 of the ER (Dominion 2006a) indicate that between 2030 and 2065 the population in
23 the region is expected to increase by factors of 1.2 to 1.6 depending on distance and direction
24 from the site. Even if the projected growth from 2030 to 2065 were as large as a factor of 2, the
25 risks would still be well below the Commission's safety goals.
26

27 The staff has considered the risk estimates given in Tables 5-16, the comparisons of atmospheric
28 pathway risks in Tables 5-17 and 5-18, and the comparison of average individual early fatality and
29 average individual latent cancer fatality risks in Table 5-17 with the Commission's safety goals.
30 Based on the foregoing, the staff concludes that the impacts for the proposed North Anna ESP
31 site for the air pathway releases for severe accidents would be small for operation of the
32 surrogate ESBWR.
33

34 *Surface Water Pathways*
35

36 Surface water pathways are an extension of the air pathway. These pathways cover the effects
37 of radioactive material deposited on open bodies of water. The surface water pathways of
38 interest include exposure to external radiation from submersion in water and activities near the

1   water, ingestion of water, and ingestion of fish and other aquatic creatures. Of these pathways,

2   the MACCS2 code only evaluates the ingestion of contaminated water. The risks associated with

3   this surface water pathway calculated for the North Anna ESP site are included in the last column

4   of Table 5-16. For each accident class, the population dose risk from ingestion of water is a small

5   fraction of the dose risk from the air pathway.

6

7   Lake Anna is used for recreational activities including swimming and fishing. Doses from these

8   surface water pathways are not modeled in MACCS or MACCS2. NUREG-1437 (NRC 1996)

9   provides an estimate of typical population exposure risk for the aquatic food pathway for plants

10   located on small rivers. The North Anna ESP site is classified as being on a small river. For

11   these plants, the risk associated with the aquatic food pathway is about 4×10^{-3} person-Sv Ryr^{-1}

12   (4×10^{-1} person-rem Ryr^{-1}). The total risk for the existing NAPS Units 1 and 2 is about 2.5×10^{-1}

13   person-Sv Ryr^{-1} (2.5×10^{1} person-rem Ryr^{-1}) (NRC 2002a). Thus, the generic aquatic pathway risk

14   is less than 2 percent of the total risk. Analysis of water-related exposure pathways at the Fermi

15   reactor (NRC 1981) suggests that population exposures from swimming are significantly lower

16   than exposures from the aquatic ingestion pathway.

17

18   Virginia Power controls the land to the high water mark of Lake Anna (Dominion 2006a). In the

19   event of a large release of radioactive material, Virginia Power and the Commonwealth of Virginia

20   could control access to the lake, which is the major surface water body in the vicinity of the North

21   Anna ESP site. By exercising that control, Virginia Power could reduce exposures through the

22   surface water pathways.

23

24   After considering the water ingestion dose estimates, the NUREG-1437 evaluations, and Virginia

25   Power's control over access to Lake Anna, as set forth above, the staff concludes that the

26   impacts for the proposed North Anna ESP site from surface water pathway releases for severe

27   accidents are small for operation of the surrogate ESBWR.

28

29   *Groundwater Pathway*

30

31   Neither MACCS nor MACCS2 evaluates the environmental risks associated with severe accident

32   releases of radioactive material to groundwater. However, this pathway has been addressed in

33   NUREG-1437 in the context of renewal of licenses for the current generation reactors.

34   NUREG-1437 assumes a 1×10^{-4} Ryr^{-1} probability of occurrence of a severe accident with a

35   basemat melt-through leading to potential groundwater contamination, and the staff concluded

36   that groundwater generally contributed a small fraction of the risk attributable to the atmospheric

37   pathway. Although the staff assumed that the probability of occurrence of a release via the

38   groundwater pathway is significantly larger than the probability of a release via the atmospheric

39   pathway, the groundwater pathway is more tortuous and affords a greater time for implementing

40   protective actions and therefore results in a lower risk to the public. As a result, the staff

41   concludes that the impacts for the proposed North Anna ESP site from groundwater pathway

42   releases for severe accidents are small for operation of the surrogate ESBWR.

1   *Summary of Severe Accident Impacts*
2
3   Although Dominion chose the PPE approach in the overall ESP application, it based its initial
4   evaluation of the environmental impacts of severe accidents on characteristics of the ABWR and
5   the surrogate AP1000 reactor designs with the explicit assumption that these impacts would
6   bound the impacts of other ALWRs designs (Dominion 2005). In Revision 6 to the ESP
7   application, Dominion provided an evaluation of the environmental impacts of severe accidents for
8   the surrogate ESBWR design (Dominion 2006a). The NRC staff reviewed the analysis in the
9   ER and conducted its own confirmatory analysis using the MACCS2 code. The results of both the
10   Dominion and NRC analyses indicate that the environmental risks associated with severe
11   accidents if an ESBWR were to be located at the North Anna ESP site would be small compared
12   to risks associated with operation of current generation reactors at the North Anna site and other
13   sites. These risks are well below the NRC safety goals. On these bases, the staff concludes that
14   the probability weighted consequences of severe accidents at the North Anna ESP site would be
15   SMALL. However, alternatives to mitigate severe accidents are not resolved. If an ESP is issued
16   and an applicant for a CP or COL references it, the applicant needs to address the site-specific
17   and design-specific severe accident mitigation alternatives.
18
19   The environmental impacts for severe accidents that have been considered include potential
20   radiation exposures to individuals and to the population as a whole, the risk of near- and
21   long-term adverse health effects that such exposures could entail, and the potential economic and
22   societal consequences of accidental contamination of the environment. Although severe
23   accidents have a potential for early fatalities and economic costs that cannot arise from normal
24   operations, the overall assessment of environmental risk of accident, assuming protective action,
25   shows that it is roughly comparable with the risk from normal operation. The risks of an early
26   fatality from potential accidents at the site are small in comparison with the risks of an early
27   fatality from other human activities in a comparably sized population. Consequently, the staff
28   concludes that the potential environmental impacts of severe accidents are small.
29
30   ## 5.10.3 Summary of Postulated Accident Impacts
31
32   *This section summarizes the potential environmental impacts of postulated accidents for a*
33   *ESBWR at the North Anna ESP site. These accidents were not addressed in earlier versions of*
34   *the ER or in the Draft EIS.*
35
36   The staff evaluated the environmental impacts from postulated design basis accidents and
37   postulated severe accidents using an additional reactor design, a surrogate ESBWR, to
38   characterize the impacts from ALWRs. Based on the information provided by Dominion, and its
39   independent review as discussed above, the staff concludes that the potential environmental
40   impacts of postulated accidents would be SMALL. However, alternatives to mitigate severe

1   accidents are not resolved. If an ESP is issued and an applicant for a CP or COL references it,
2   the applicant needs to address the site-specific and design-specific severe accident mitigation
3   alternatives.
4

5.11 Measures and Controls to Limit Adverse Impacts During Operation

7

*Changes to this section reflect the higher power level and the closed-cycle, combination wet and
dry cooling system.*

10

The following general measures and controls on which the staff relied in its evaluation of
environmental impacts during operation of the proposed new units at the North Anna ESP site
include those which Dominion would be required to implement by applicable permits and
authorizations (Federal, State, and local requirements contained in Table 1.2-1 of the ER) as well
as feasible measures and controls contained in Table 5.10-1 of the ER:

- solid waste management, erosion and sediment control, air emission control, noise
  control, storm water management, spill response and cleanup, hazardous material
  management

- imposed on water discharges from the proposed new units (ER Sections 5.1.1, 5.3.2,
  5.5.1)

- installing and operating air emission sources (ER Section 5.8.1)

- Virginia Power procedures applicable to environmental control and management

Dominion specifically identified the following general plans or specific mitigation measures in its
ER (Dominion 2006a) on which the staff relied in its evaluation:

- Current transmission line maintenance practices would continue if two new units were built
  at the ESP site (ER Section 5.6.1.1).

- Locations of rare or sensitive plant species within transmission line corridors would be
  identified so modified treatment practices can be used in these areas to avoid adverse
  impacts (ER Section 5.6.1.1).

- The intake structure for the proposed new units at the ESP site would meet Section 316(b)
  of the Clean Water Act and the implementing regulations, as applicable (ER
  Section 5.3.1.2).

1 • A fish return system based on the latest technology available during detailed engineering
2   would be considered for incorporation into the intake system (ER Section 5.3.1.2).
3
4 • Vegetative shielding would block a clear view of the new units from nearby residences,
5   and a visual impact study would be performed and the results would be described in the
6   COL application (ER Section 5.8.1.5).
7
8 • Noise levels would be controlled in accordance with applicable local county regulations
9   (ER Section 5.8.1.2).
10
11 • Potential increases in traffic would be mitigated through effective traffic management
12   (ER Section 5.1.1).
13
14 • Any ground disturbing activities would be conducted in coordination with VDHR and
15   professional archaeological practices (ER Section 4.1.3).
16

17 Dominion evaluated the measures and controls shown in Table 5.10-1 of the ER (Dominion 2005)
18 and considered them feasible from both a technical and economic standpoint. In addition,
19 Dominion expects these measures and controls to be adequate for avoiding or mitigating potential
20 adverse impacts associated with operation of the proposed new units. The staff considered these
21 measures and controls in its evaluation of station operation impacts.
22

5.12 Summary of Operational Impacts

24

25 *Changes to this section reflect the summary of operations impacts from the proposed project*
26 *presented in ER revision 6*

27

28 Impact level categories denoted in Table 5-19 as SMALL, MODERATE, or LARGE were assigned
29 to each resource area based on the staff's evaluations and conclusions regarding expected
30 adverse environmental impacts, if any. A brief statement explains the basis for the impact level.
31 Some impacts, such as the addition of tax revenue from Dominion for the local economies, are
32 likely to be beneficial impacts to the community, and are noted as such.
33

Table 5-19. Characterization of Operational Impacts at the North Anna ESP Site

Category	Comments	Impact Level
Land use impacts		--
The site and vicinity	Operation of new units within existing site. Possible new housing and retail space added in vicinity due to potential growth.	SMALL
Transmission line rights-of-way	No new transmission line rights-of-way would be needed.	SMALL
Air quality impacts	Meteorological impacts are expected to be negligible. Pollutants emitted during operations considered insignificant and limits could be incorporated under existing Exclusionary Permit.	SMALL
Water-related impacts		--
Hydrological alterations	Changes in the quantity and distribution of heat in the lake are expected to be negligible.	SMALL
Water use		
Normal years	During normal water years, the impact would be small.	SMALL
Drought years	During critical low-water years, the impacts could be temporarily moderate.	MODERATE
Water quality	Water effluents would be regulated by the VPDES permit, but their exact composition would depend on information not yet available.	Unresolved
Ecological impacts		--
Terrestrial ecosystems	No detectable impacts expected; no important species in area.	SMALL
Aquatic ecosystems	The fish community is balanced. Proportion of resources subject to impingement and entrainment would be small. The impact on striped bass would be minimal even during drought years or late summer. Refugia would be available.	SMALL
Threatened and Endangered Species		--
Terrestrial Species	No threatened or endangered species known to inhabit area.	SMALL
Aquatic Species	No threatened or endangered species known to inhabit area.	SMALL

Table 5-19. (contd)

Category	Comments	Impact Level
Socioeconomic impacts		--
Physical Impacts		--
Workers/Public	Workers would use protective equipment and receive training to mitigate any possible impact. North Anna location is relatively remote, so the public would not be impacted.	SMALL
Buildings	No impact to onsite or offsite buildings.	SMALL
Roads	Upgrades and other mitigation before or during construction would cover the lesser impact of operational work forces.	SMALL
Aesthetics	Visual impact would be minimal due to remote location. Lower water levels, and their effect on shoreline exposure during severe drought could temporarily impact area. Elevated steam plume for the Unit 3 cooling towers, especially in the winter. These impacts are expected to be temporarily at the moderate level.	SMALL TO MODERATE
Demography	Number of new employees small in proportion to population base.	SMALL
Community Characteristics		
Economy	Increased jobs would benefit the area economically, up to a moderate beneficial impact (Louisa and Orange Counties) is possible.	SMALL BENEFICIAL TO MODERATE BENEFICIAL
Transportation	Improvements made for construction would be sufficient to cover any adverse impact from small number of additional operational workers.	SMALL
Taxes	Depends on residence location; generally impacts are beneficial, especially for property taxes and employment. Beneficial impacts of additional taxes would be large for Louisa County.	SMALL BENEFICIAL TO LARGE BENEFICIAL
Recreation	Overall impacts to recreation minimal due to remote location. Traffic around and use of lake could increase. Lower water levels, and their effect on shoreline exposure and recreational usage during severe drought could temporarily impact area.	SMALL TO MODERATE

Table 5-19. (contd)

	Category	Comments	Impact Level
1	Housing	Adequate housing is available in Henrico and Spotsylvania Counties and in Richmond to handle operational workers. Orange and Louisa Counties could experience a temporary shortage of upscale housing, possibly at the moderate impact level.	SMALL
2	Public Services	Adequate in all counties for any population increase due to operation workforce.	SMALL
3	Education	Current schools and planned additions would handle additional students.	SMALL
4	**Historic and cultural resources**	A cultural resource program is in place for minimizing impacts from routine land disturbances.	SMALL
5	**Environmental justice**	No unusual resource dependence in the area.	SMALL
6	**Nonradiological health impacts**	Health impacts monitored and controlled in accordance with Occupational Safety and Health Administration regulations.	SMALL
7	**Radiological health impacts**	Doses to public and occupational workers are monitored and controlled in accordance with NRC limits.	SMALL
8	**Impacts of postulated accidents**		--
9	Design basis accidents	Doses for advanced light water reactors are expected to be a small fraction of the regulatory dose limits. Staff would verify that source terms for postulated DBAs on chosen reactor designs are within those considered in the ESP EIS.	SMALL
10	Severe accidents	Risks for ALWRs would be small. If gas-cooled reactor is selected at the CP/COL stage then the staff will evaluate the severe accident impacts for gas-cooled reactors. Severe accident mitigation alternatives are unresolved.	SMALL

11

12 # 5.13 References

13

14 *This section was changed to include new references to support the analysis of changes to the*
15 *cooling system and power output as presented in ER Revision 6.*

16

17 10 CFR Part 20. Code of Federal Regulations, Title 10, *Energy*, Part 20, "Standards for
18 Protection Against Radiation."

19

10 CFR Part 50. Code of Federal Regulations, Title 10, *Energy*, Part 50, "Domestic Licensing of Production and Utilization Facilities."

10 CFR Part 51. Code of Federal Regulations, Title 10, *Energy*, Part 51, "Environmental Protection Regulations for Domestic Licensing and Related Regulatory Functions."

10 CFR Part 52. Code of Federal Regulations, Title 10, *Energy*, Part 52, "Early Site Permits; Standard Design Certifications; and Combined Licenses for Nuclear Power Plants."

10 CFR Part 100. Code of Federal Regulations, Title 10, *Energy*, Part 100, "Reactor Site Criteria."

40 CFR Part 190. Code of Federal Regulations, Title 40, *Protection of Environment*, Part 190, "Environmental Radiation Protection Standards for Nuclear Power Operations."

Argonne National Laboratory (ANL). 1984. *A User's Manual: Cooling-Tower-Plume Prediction Code.* Published as EPRI CS-3403-CCM by Electric Power Research Institute, Palo Alto, California.

Chanin D.I., J.L. Sprung, L.T. Ritchie, and H.N. Jow. 1990. *MELCOR Accident Consequence Code System (MACCS), User's Guide.* NUREG/CR-4691, Volume 1. Washington, D.C.

Chanin D.I. and M.L. Young. 1997. *Code Manual for MACCS2: Volume1*, User's Guide. SAND97-0594, Sandia National Laboratories, Albuquerque, New Mexico.

Cheek T.E., J.M. Van Den Avyle, and C.C. Coutant. 1985. "Influence of Water Quality on Distribution of Striped Bass in a Tennessee River Impoundment." *Transactions of the American Fisheries Society* 114:67-76.

Chesapeake Bay Program. 2002. Chesapeake Bay Dissolved Oxygen Criteria, Draft No. 2. Accessed at http://www.chesapeakebay.net/pubs/ subcommittee/wqsc/doctg/doc-docritdoc-01-03-2002.PDF on August 23, 2005.

Clean Water Act (also called the Federal Water Pollution Control Act). 33 USC 1251,et seq..

Dominion Energy Inc. and Bechtel Power Corp. (Dominion and Bechtel). 2002. *Study of Potential Sites for the Deployment of New Nuclear Plants in the United States.* U.S. Department of Energy Cooperative Agreement No. DE-FC07-02ID14313. Available at www.ne.doe.gov/NucPwr2010/ESP_Study/ESP_Study_Dominion1.pdf

Dominion Nuclear North Anna, LLC (Dominion). 2001a. "Migratory Birds." *Water/Waste Environmental Protection Manual.* Chapter 13, Glen Allen, Virginia.

Operational Impacts at the Proposed Site

1   Dominion Nuclear North Anna, LLC (Dominion). 2001b. *Appendix E – Applicant's Environmental*
2   *Report – Operating License Renewal Stage, North Anna Power Station Units 1 and 2, License*
3   *Nos. NPF-4 and NPF-7,* Virginia Electric & Power Company, May 2001.

5   Dominion Nuclear North Anna, LLC (Dominion). 2004a. *North Anna Early Site Permit Application*
6   *- Part 3 - Environmental Report.* Revision 3, Glen Allen, Virginia.

8   Dominion Nuclear North Anna, LLC (Dominion). 2004b. Letter dated June 28, 2004 from
9   E. S. Grecheck (DNNA) to the NRC responding to the Virginia Department of Environmental
10   Quality comment letter Glen Allen, Virginia.

12   Dominion Nuclear North Anna, LLC (Dominion). 2004c. Letter dated May 17, 2004 from
13   E. S. Grecheck (DNNA) to the NRC submitting additional information in response to an NRC
14   request dated March 12, 2004. Glen Allen, Virginia.

16   Dominion Nuclear North Anna, LLC (Dominion). 2006a. *North Anna Early Site Permit Application*
17   *– Part 3 – Environmental Report.* Revision 6, Glen Allen, Virginia.

19   Dominion Nuclear North Anna, LLC (Dominion). 2006b. Letter from E. Grecheck (DNNA) to NRC
20   dated May 24, 2006, submitting additional information in response to an NRC request dated May
21   10, 2006, (ML061510131).

23   Eckerman K.F., A.B. Wolbarst, and A.C.B. Richardson. 1988. *Limiting Values of Radionuclide*
24   *Intake and Air Concentrations and Dose Conversion Factors for Inhalation, Submersion, and*
25   *Ingestion.* Federal Guidance Report No. 11, -520/1-88-202, U.S. Environmental Protection
26   Agency, Washington, D.C.

28   Eckerman K.F. and J.C. Ryman. 1993. *External Exposure to Radionuclides in Air, Water, and*
29   *Soil.* Federal Guidance Report No. 12, -402-R-93-081, U.S. Environmental Protection Agency,
30   Washington, D.C.

32   Golden J., R.P. Ouellette, S. Saari, and P.N. Cheremisinoff. 1980. *Environmental Impact Data*
33   *Book.* Ann Arbor Science Publishers, Inc., Ann Arbor, Michigan.

35   Goodyear C.P. 1978. *Entrainment Impact Estimates Using the Equivalent Adult Approach,*
36   FWS/OBS-78/65, U.S. Department of Interior, Fish and Wildlife Service, National Power Plant
37   Team, Ann Arbor, Michigan.

39   Hanover County Department of Public Utilities. 2004. Letter to Chief, Rules and Directives
40   Branch, NRC, regarding Dominion's ESP application, January 7, 2004.

Health Physics Society (HPS). 2004. *Radiation Risk in Perspective.* The Health Physics Society, McLean, Virginia.

Ho E. And E. E. Adams. 1984. Final Calibration of the Cooling Lake Model, North Anna Power Station. Report No. 295, Ralph M. Parsons Laboratory, Massachusetts Institute of Technology, Cambridge, Massachusetts.

International Atomic Energy Agency (IAEA). 1992. *Effects of Ionizing Radiation on Plants and Animals at Levels Implied by Current Radiation Protection Standards.* Technical Report Series No. 332, Vienna, Austria.

International Commission on Radiological Protection (ICRP). 1991. *1990 Recommendations of the International Commission on Radiological Protection.* ICRP Publication 60, Pergamon Press, Oxford, United Kingdom.

Jaksch J. and M. Scott. 2005. North Anna ESP Site Audit Trip Report – Socioeconomics. 12-8-2003 through 12-12-2003 with Additional Telephone Interviews 2-26-2003 through 12-12-2003 with Additional Telephone Interviews 2-26-2004 through 9-29-2004 and 7-11-2005 through 7-15-2005. Available at http://www.nrc.gov/reading-rm/adams.html, Accession No. ML052170374.

Jenkins R.E. and N.M. Burkhead. 1994. *Freshwater Fishes of Virginia.* American Fisheries Society, Bethesda, Maryland.

Jow H.N., J.L Sprung, J.A. Rollstin, L.T. Ritchie, and D.I. Chanin. 1990. *MELCOR Accident Consequence Code System (MACCS), Model Description.* NUREG/CR-4691, Volume 2, U.S. Nuclear Regulatory Commission, Washington, D.C.

National Council on Radiation Protection and Measurements (NCRP). 1991. *Effects of Ionizing Radiation on Aquatic Organisms.* NCRP Report No. 109, Bethesda, Maryland.

National Environmental Policy Act of 1969 (NEPA). 42 USC 4321, et seq.

National Institute of Environmental Health Sciences (NIEHS). 1999. *NEISH Report on Health Effects from Exposure to Power Line Frequency and Electric and Magnetic Fields.* Publication No. 99-4493, Research Triangle Park, North Carolina.

National Research Council. 2005. *Health Risks from Exposure to Low Levels of Ionizing Radiation: BIER VII - Phase 2.* Committee to Assess Health Risks from Exposure to Low Levels of Ionizing Radiation, National Research Council, National Academy Press, Washington, D.C.

1 Olmsted L.L. and J.P. Clugston. 1986. "Fishery Management in Cooling Impoundments,"
2 Pages 227-237 in *Reservoir Fisheries Management*, (G.E. Hall and M.J. Van Den Ayvle, editors),
3 Reservoir Committee, Southern Division American Fisheries Society, Bethesda Maryland.
4
5 U.S. Atomic Energy Commission (AEC). 1972. Onsite Meteorological Programs. Regulatory
6 Guide1.23 (or Safety Guide 23), Washington, D.C.
7
8 U.S. Geological Survey (USGS). 2006. USGS Surface Water Data for USA: USGS Surface -
9 Water Monthly Statistics for the Nation. Accessed at http://waterdata.usgs.gov/nwis/monthly/
10 ?referred_module=sw&site_no=01671020&por_01671020_1=189374,00060,1,1979-10,2004-
11 09&start_dt=1979-10&end_dt=2003-09&format=html_ table&date_format=YYYY-MM-
12 DD&rdb_compression=file&submitted_form=parameter_selection _list on June 22, 2006.
13
14 U.S. Nuclear Regulatory Commission (NRC). 1976. *Addendum to the Final Environmental*
15 *Statement Related to Operation of North Anna Power Station Units 1 and 2.* Virginia Electric and
16 Power Company. NUREG-0134, Docket Nos. 50-338 and 50-339, Washington, D.C.
17
18 U.S. Nuclear Regulatory Commission (NRC). 1980. *Criteria for Preparation and Evaluation of*
19 *Radiological Emergency Response Plans and Preparedness in Support of Nuclear Power Plants.*
20 NUREG-0654/FEMA-REP-1, Rev. 1, U.S. Nuclear Regulatory Commission, Washington, D.C.
21
22 U.S. Nuclear Regulatory Commission (NRC). 1981. *Final Environmental Impact Statement*
23 *Related to the Operation of Enrico Fermi Atomic Power Plant, Addendum No.1.* NUREG-0769
24 Addendum 1. Washington, D.C.
25
26 U.S. Nuclear Regulatory Commission (NRC). 1983. *Atmospheric Dispersion Models for*
27 *Evaluating Design Basis Accidents at Nuclear Power Plants.* Regulatory Guide 1.145 Rev. 1,
28 Washington, D.C.
29
30 U.S. Nuclear Regulatory Commission (NRC). 1986. "Safety Goals for the Operation of Nuclear
31 Power Plants; Policy Statement." *Federal Register*, Volume 51, p. 30028, August 21, 1986.
32
33 U.S. Nuclear Regulatory Commission (NRC). 1987. Standard Review Plan for the Review of
34 Safety Analysis Reports for Nuclear Power Plants. NUREG-0800 (formerly issued as
35 NUREG-75/087), Volumes 1 and 2. NRC, Washington, D.C.
36
37 U.S. Nuclear Regulatory Commission (NRC). 1990. *Severe Accident Risks: An Assessment for*
38 *Five U.S. Nuclear Power Plants.* NUREG-1150, Volume 1. Washington, D.C.
39
40 U.S. Nuclear Regulatory Commission (NRC). 1996. *Generic Environmental Impact Statement for*
41 *License Renewal of Nuclear Plants.* NUREG-1437, Volumes 1 and 2, Washington, D.C.
42

1   U.S. Nuclear Regulatory Commission (NRC). 2000a. Environmental Standard Review Plan.
2   NUREG-1555, Volumes 1 and 2, Washington, D.C.

4   U.S. Nuclear Regulatory Commission (NRC). 2000b. *Alternative Radiological Source Terms for*
5   *Evaluating Design Basis Accidents at Nuclear Power Plants.* Regulatory Guide 1.183,
6   Washington, D.C.

8   U.S. Nuclear Regulatory Commission (NRC). 2002a. *Generic Environmental Impact Statement*
9   *on Decommissioning of Nuclear Facilities Supplement 1 Regarding the Decommissioning of*
10   *Nuclear Power Reactors.* NUREG-0586, Supplement 1, Volumes 1 and 2, Washington, D.C.

12   U.S. Nuclear Regulatory Commission (NRC). 2002b. *Generic Environmental Impact Statement*
13   *for License Renewal of Nuclear Plants; Supplement 7 Regarding North Anna Power Station, Units*
14   *1 and 2.* NUREG-1437 Supplement 7. November 2002. Available at
15   http://www.nrc.gov/reading-rm/doc-collections/nuregs/staff/sr1437/supplement7

17   U.S. Nuclear Regulatory Commission (NRC). 2004a. *Draft Environmental Impact Statement*
18   *(EIS) for an Early Site Permit (ESP) at the North Anna ESP Site.* NUREG-1811,
19   Washington, D.C.

21   U.S. Nuclear Regulatory Commission (NRC). 2004b. NRR Office Instruction Change Notice
22   LIC-203, Revision 1. Procedural Guidance for Preparing Environmental Assessments and
23   Considering Environmental Issues. May 24, 2004. Appendix D, Environmental Justice Guidance
24   and Flow Chart. Washington, D.C.

26   U.S. Nuclear Regulatory Commission (NRC). 2004c. Letter dated March 12, 2004 from Andrew
27   J. Kugler (NRC) to David A. Christian (Dominion) requesting additional information.

29   U.S. Nuclear Regulatory Commission (NRC). 2006 Letter from Nitin Patel (NRC) to David A.
30   Christian (Dominion) requesting additional information. May 10, 2006

32   Virginia Department of Conservation and Recreation (VDCR). 2004. Natural Heritage Resources
33   Factsheet. *Migratory Songbird Habitat in Virginia's Coastal Plain.* Accessed at
34   http://www.dcr.state.va/dnh/songfact.htm on March 4, 2004.

36   Virginia Department of Environmental Quality (VDEQ). 2001. *Virginia Pollutant Discharge*
37   *Elimination System Permit No. VA0052451,* Authorization to Discharge Under the Virginia
38   Pollutant Discharge Elimination System and the Virginia State Water Control Act, Commonwealth
39   of Virginia, Department of Environmental Quality, permit's effective date, January 11, 2001;
40   expiration date, January 11, 2006.

Operational Impacts at the Proposed Site

1   Virginia Department of Health (VDH). 2005. Letter from R. B. Stroube (VDH) to R. Burnley
2   (VDEQ) regarding heat effects of immersion in ambient waters. September 15, 2005.
3
4   Virginia Electric and Power Company (VEPCo). 1985. *Impingement and Entrainment Studies for*
5   *North Anna Power Station,*1978-1983, Prepared by Water Quality Department, Richmond,
6   Virginia.
7
8   Virginia Electric and Power Company (VEPCo). 2002. *Environmental Study of Lake Anna and*
9   *the Lower North Anna River:* Annual Report for 2001 including Summary for 1998-2000.
10   Richmond, Virginia.
11
12

6.0 Fuel Cycle, Transportation, and Decommissioning

This chapter of the Supplement to the Draft Environmental Impact Statement was changed to reflect new information and analysis of fuel cycle, transportation, and decommissioning impacts related to an increase in the maximum power level from 4300 MW(t) to 4500 MW(t) proposed in Revision 6 of the Environmental Report. Section summaries are provided for context.

This chapter addresses the environmental impacts from (1) the uranium fuel cycle and solid waste management, (2) transportation of radioactive material, and (3) decommissioning following operation of two new nuclear power units. The environmental impacts were assessed for the North Anna early site permit (ESP) site and the alternative sites. Distinctions between the impacts of advanced light-water reactor (LWR) designs and gas-cooled reactor designs are discussed. Elements of the analysis presented in the Draft Environmental Impact Statement (Draft EIS)(NRC 2004) that are relevant in this Supplement to the Draft Environmental Impact Statement (SDEIS) are related to changes in the power level of proposed Units 3 and 4 (referred to hereafter as Units 3 and 4).

The power level change is an increase from a maximum of 4300 MW(t) to 4500 MW(t) per unit. This change is intended to align with the new power level designated for the surrogate economic simplified boiling water reactor (surrogate ESBWR).

In its evaluation of uranium fuel cycle impacts for the North Anna ESP site, Dominion used the plant parameter envelope (PPE) approach for the advanced LWR designs but not for the two gas-cooled reactors. In its evaluation of the impacts from transportation of radioactive materials, Dominion did not use the PPE approach but rather evaluated each reactor design individually. Therefore, an applicant referencing any North Anna ESP would have to perform a new evaluation if a different design is proposed at the construction permit (CP) or combined license (COL) stage.

The staff reviewed the potential impacts of the higher output level on the fuel cycle, transportation, and decommissioning. Throughout Chapter 6 of the Draft EIS, there are references to the previous maximum power level of 4300 MW(t) (now 4500 MW(t)) and 1500 MW(e) (now 1520 MW(e)). Each place in the text of the Draft EIS that previously addressed the analyses for 4300 MW(t) or 1500 MW(e) will be changed in the Final EIS to reflect the analyses for the 4500 MW(t) or 1520 MW(e) values and power levels. Changes in this chapter are identified and discussed only for those sections in which the higher power level of 4500 MW(t) affects the actual impact analysis.

6.1 Fuel Cycle Impacts and Solid Waste Management

This introductory section is not affected by the changes presented in ER Revision 6 and is provided solely for context.

This section of the Draft EIS discussed the environmental impacts from the uranium fuel cycle and solid waste management for both the advanced LWR designs and gas-cooled reactor designs. The impacts of the two types of design are presented separately because Title 10 of the Code of Federal Regulations (10 CFR), Section 51.51 provides specific criteria for evaluating the environmental impacts only for LWR designs.

6.1.1 Light-Water Reactors

Changes to this section reflect the revised PPE value of 4500 MW(t) for the power level for one unit (9000 MW(t) for two units) provided in ER Revision 6. The corresponding net electrical output is 1520 MW(e) for one unit (3040 MW(e) for two units).

In the Draft EIS, the staff evaluated impacts at a net electric output of 3200 MW(e); this level is still higher than the 3040 MW(e) power level proposed in ER Revision 6. This section was changed to reflect the higher power level as presented in ER Revision 6.

The regulations in 10 CFR 51.51(a) state that

> Every environmental report prepared for the construction permit stage of a light-water-cooled nuclear power reactor, and submitted on or after September 4, 1979, shall take Table S–3, *Table of Uranium Fuel Cycle Environmental Data*, as the basis for evaluating the contribution of the environmental effects of uranium mining and milling, the production of uranium hexafluoride, isotopic enrichment, fuel fabrication, reprocessing of irradiated fuel, transportation of radioactive materials, and management of low level wastes and high level wastes related to uranium fuel cycle activities to the environmental costs of licensing the nuclear power plant. Table S–3 shall be included in the environmental report and may be supplemented by a discussion of the environmental significance of the data set forth in the table as weighed in the analysis for the proposed facility.

The PPE for the North Anna ESP site used input parameters from the following LWR designs:

- Advanced Canada Deuterium Uranium Reactor (ACR-700) – This reactor, developed by Atomic Energy Canada Limited, is an evolutionary extension of the CANDU 6 plant using very slightly enriched uranium fuel and light-water cooling.

1   • Advanced Boiling Water Reactor (ABWR) – This reactor, developed by General Electric
2   Company (GE), is a standardized plant that has been certified under the NRC
3   requirements in 10 CFR Part 52, Appendix A. The ABWR is fueled with slightly enriched
4   uranium and has light-water cooling.
5
6   • Advanced Pressurized Water Reactor (AP1000) – This earlier version of the AP1000, a
7   reactor design developed by Westinghouse Electric Company, using slightly enriched
8   uranium and light-water cooling. This design is not the standard AP1000 design that
9   was certified by the NRC in 10 CFR Part 52, Appendix D; therefore, this design is
10   referred to as the "surrogate AP1000."
11
12   • Surrogate Economic Simplified Boiling Water Reactor (ESBWR) – This surrogate reactor
13   design is based on a design developed by GE using slightly enriched uranium fuel and
14   has light-water cooling. Dominion revised its application to reflect a higher power level
15   volume of 4500 MW(t) (Dominion 2006a). The ESBWR design certification application is
16   currently under review by the NRC.
17
18   • International Reactor Innovative and Secure (IRIS) Next-Generation Pressurized Water
19   Reactor (PWR) – This reactor, under development by a consortium led by Westinghouse
20   Electric Company, is a modular light-water reactor.
21
22 These light-water designs all use uranium dioxide fuel; therefore, the values in Table S–3 can
23 be used to assess environmental impacts. Table S–3 values are normalized for a reference
24 1000-MW(e) LWR at an 80 percent capacity factor. The PPE power level for the North Anna
25 ESP site is 9000 MW(t), assuming two ESBWR units would be located on the ESP site or at any
26 of the alternative sites, with a PPE capacity factor of 96 percent (Dominion 2006a). This
27 corresponds to 3040 MW(e). The 10 CFR 51.51(a) Table S–3 values are reproduced in
28 Table 6-1, which follows.
29
30 In the Draft EIS (NRC 2004), the staff evaluated a net electrical output of 3200 MW(e) for the
31 site; therefore, the higher power level (i.e., 3040 MW(e)) in ER Revision 6 (Dominion 2006a)
32 remains bounded by the staff's evaluation. The staff's evaluation assumed that fuel cycle
33 impacts for the ESP site would be approximately four times the impact values in Table S–3
34 (see Table 6-1). This is referred to as the 1000-MW(e) LWR scaled model.
35
36 This section was revised to reflect the revised PPE value of 4500 MW(t) for the power level for
37 one unit (9000 MW(t) for two units) provided in Dominion (2006a). The corresponding net
38 electrical output is 1520 MW(e) for one unit (3040 MW(e) for two units).
39
40 Summaries of Section 6.1.1 subsections follow.
41

6.1.1.1 Land Use

This section was not affected by the changes presented in ER Revision 6 and is provided solely for context.

In the Draft EIS (NRC 2004), the staff reported that the total annual land needed for the fuel cycle supporting the 1000-MW(e) scaled model is about 184 ha (452 ac). Approximately 20 ha (52 ac) are permanently committed land, and 164 ha (400 ac) are temporarily committed. In comparison, a coal-fired power plant with the same MW(e) output as the LWR scaled model and that uses strip-mined coal results in the disturbance of about 324 ha (800 ac) per year for fuel alone. Because the land use needs remain bounded by the staff's previous evaluation, the staff concludes that the impact on land use did not change because of the higher power level. The impact level category would still be SMALL, and mitigation is not warranted.

6.1.1.2 Water Use

This section was not affected by the changes presented in ER Revision 6 and is provided solely for context.

In the Draft EIS (NRC 2004), the staff reported that principal water-use for the fuel cycle supporting a 1000-MW(e) scaled model is that needed to remove waste heat from the power stations supplying electrical energy to the enrichment step of this cycle. The maximum consumptive water use (assuming that all plants supplying electrical energy to the nuclear fuel cycle using cooling towers) would be about 6 percent of the 1000-MW(e) scaled model using cooling towers. Because the water use needs remain bounded by the staff's previous evaluation, the staff concludes that impacts on water use did not change because of the higher power level. The impact level category would still be SMALL, and mitigation is not warranted.

6.1.1.3 Fossil Fuel Impacts

This section was not affected by the changes presented in ER Revision 6 and is provided solely for context.

In the Draft EIS (NRC 2004), the staff reported that electrical energy and process heat are used during various phases of the fuel cycle process. Electrical energy associated with the fuel cycle represents about 5 percent of the annual electrical power production of the reference 1000-MW(e) LWR. Process heat is primarily generated by the combustion of natural gas. This gas consumption, if used to generate electricity, would be less than 0.4 percent of the electrical output from the model plant. Because the fossil fuel needs remain bounded by the staff's previous evaluation, the staff concludes that fossil fuel impacts did not change because of the higher power level. The impact level category would still be SMALL, and mitigation is not warranted.

1   ### 6.1.1.4 Chemical Effluents
2
3   *This section was not affected by the changes presented in ER Revision 6 and is provided solely*
4   *for context.*
5
6   In the Draft EIS (NRC 2004), the staff reported that the quantities of chemical effluents released
7   from operation of fuel cycle facilities would be approximately four times less than for the
8   reference 1000-MW(e) LWR. The principal effluents are SO_x, NO_x, and particulates. Because
9   the chemical effluents likely to be released remain bounded by the staff's previous evaluation,
10   the staff concludes that impacts from chemical effluents did not change because of the higher
11   power level. The impact level category would still be SMALL, and mitigation is not warranted.
12
13   ### 6.1.1.5 Radioactive Effluents
14
15   *This section was not affected by the changes presented in ER Revision 6 and is provided solely*
16   *for context.*
17
18   In the Draft EIS (NRC 2004), the staff reported that radioactive effluents estimated to be
19   released to the environment from waste management activities and certain other phases of the
20   fuel cycle process are set forth in Table S–3 of 10 CFR 51.51(a) (see Table 6-1). The
21   estimated 100-yr environmental dose commitment to the U.S. population from radioactive
22   gaseous and liquid releases resulting from the fuel cycle is approximately 66 person-Sv
23   (6600 person-rem) (whole body) per reference reactor-year, which includes doses from
24   radon-222 and technetium-99. Using the risk estimation method from International Commission
25   on Radiological Protection (ICRP) Publication 60 (ICRP 1991), this dose commitment equates
26   to an estimated total of approximately 4.8 fatal cancers, nonfatal cancers, and severe hereditary
27   effects to the U.S. population annually. This risk is quite small compared to the number of fatal
28   cancers, nonfatal cancers, and severe hereditary effects that would be estimated to the
29   U.S. population annually from exposure to natural sources of radiation using the same risk
30   estimation method. Because the radioactive effluents likely to be released remain bounded by
31   the staff's previous evaluation, the staff concludes that the impact from radioactive effluents did
32   not change because of the higher power level. The impact level category would still be SMALL,
33   and mitigation is not warranted.
34
35   ### 6.1.1.6 Radioactive Wastes
36
37   *This section was not affected by the changes presented in ER Revision 6 and is provided solely*
38   *for context.*
39
40   In the Draft EIS (NRC 2004), the staff reported that the quantities of buried radioactive waste
41   material (low-level, high-level, and transuranic wastes) are specified in Table S–3 (see

1 **Table 6-1**. Table of Uranium Fuel Cycle Environmental Data[a] – Taken From Table S–3
2 (Normalized to Model LWR Annual Fuel Requirement [AEC 1972; AEC 1974]
3 or Reference Reactor Year [NRC 1976])(corresponds to Table 6-1 in the Draft EIS)

Environmental Considerations	Total	Maximum Effect per Annual Fuel Requirement or Reference Reactor Year of Model 1000 MW(e) LWR
Natural Resource Use		
Land (acres):		
Temporarily committed[b]	100	
Undisturbed area	79	
Disturbed area	22	Equivalent to a 100 MW(e) coal-fired power plant.
Permanently committed	13	
Overburden moved (millions of MT)	2.8	Equivalent to 95 MW(e) coal-fired power plant.
Water (millions of gallons):		
Discharged to air	160	= 2 percent of model 1000 MW(e) LWR with cooling tower.
Discharged to water bodies	11,090	
Discharged to ground	127	
Total	11,377	<4 percent of model 1000 MW(e) LWR with once-through
Fossil fuel:		
Electrical energy (thousands of MW-hr)	323	<5 percent of model 1000 MW(e) LWR output.
Equivalent coal (thousands of MT)	118	Equivalent to the consumption of a 45 MW(e) coal-fired power plant.
Natural gas (millions of standard cubic feet)	135	<0.4 percent of model 1000 MW(e) energy output.
Effluents--Chemical (MT)		
Gases (including entrainment):[c]		
SO_x	4400	
NO_x[d]	1190	Equivalent to emissions from 45 MW(e) coal-fired plant for a year.
Hydrocarbons	14	
CO	29.6	
Particulates	1154	
Other gases:		
F	0.67	Principally from uranium hexafluoride (UF_6) production, enrichment, and reprocessing. The concentration is within the range of state standards–below level that has effects on human health.
HCl	0.014	

Table 6-1. (contd)

Environmental Considerations	Total	Maximum Effect per Annual Fuel Requirement or Reference Reactor Year of Model 1000 MW(e) LWR
Liquids:		
SO_4	9.9	From enrichment, fuel fabrication, and reprocessing
NO_3	25.8	steps. Components that constitute a potential for
Fluoride	12.9	adverse environmental effect are present in dilute
Ca^{++}	5.4	concentrations and receive additional dilution by
Cl^-	8.5	receiving bodies of water to levels below permissible
Na^+	12.1	standards. The constituents that require dilution and
NH_3	10	the flow of dilution water are: NH_3-600 cfs, NO_3-20 cfs,
Fe	0.4	Fluoride-70 cfs.
Tailings solutions (thousands of MT)	240	From mills only–no significant effluents to environment.
Solids	91,000	Principally from mills–no significant effluents to environment.
Effluents--Radiological (curies)		
Gases (including entrainment):		
Rn-222	......	Presently under reconsideration by the Commission.
Ra-226	0.02	
Th-230	0.02	
Uranium	0.034	
Tritium (thousands)	18.1	
C-14	24	
Kr-85 (thousands)	400	
Ru-106	0.14	Principally from fuel reprocessing plants.
I-129	1.3	
I-131	0.83	
Tc-99	......	Presently under consideration by the Commission.
Fission products and transuranics	0.203	
Liquids:		
Uranium and daughters	2.1	Principally from milling–included tailings liquor and returned to ground–no effluents; therefore, no effect on environment.
Ra-226	0.0034	From UF_6 production.
Th-230	0.0015	
Th-234	0.01	From fuel fabrication plants–concentration 10 percent of 10 CFR Part 20 for total processing 26 annual fuel requirements for model LWR.
Fission and activation products	5.9×10^{-6}	
Solids (buried on site):		
Other than high level (shallow)	11,300	9100 Ci comes from low level reactor wastes and 1500 Ci comes from reactor decontamination and decommissioning—buried at land burial facilities. 600 Ci comes from mills—included in tailings returned to ground. Approximately 60 Ci comes from conversion and spent fuel storage. No significant effluent to the environment.

Table 6-1. (contd)

Environmental Considerations	Total	Maximum Effect per Annual Fuel Requirement or Reference Reactor Year of Model 1000 MW(e) LWR
TRU and HLW (deep)	1.1 x 10^7	Buried at Federal Repository.
Effluents—thermal (billions of British thermal units)	4063	<5 percent of model 1000 MW(e) LWR
Transportation (person-rem):		
Exposure of workers and general public	2.5	
Occupational exposure (person-rem)	22.6	From reprocessing and waste management.

(a) In some cases where no entry appears it is clear from the background documents that the matter was
     addressed and that, in effect, the table should be read as if a specific zero entry had been made. However,
     there are other areas that are not addressed at all in the table. Table S–3 does not include health effects from
     the effluents described in the table, or estimates of releases of radon-222 from the uranium fuel cycle or
     estimates of technetium-99 released from waste management or reprocessing activities. These issues may be
     the subject of litigation in the individual licensing proceedings.

     Data supporting this table are given in the "Environmental Survey of the Uranium Fuel Cycle," WASH-1248,
     April 1974; the "Environmental Survey of the Reprocessing and Waste Management Portion of the LWR Fuel
     Cycle," NUREG-0116 (Supp.1 to WASH-1248, NRC 1976); the "Public Comments and Task Force Responses
     Regarding the Environmental Survey of the Reprocessing and Waste Management Portions of the LWR Fuel
     Cycle," NUREG-0216 (Supp. 2 to WASH-1248) (NRC 1977b); and in the record of the final rulemaking
     pertaining to Uranium Fuel Cycle Impacts from Spent Fuel Reprocessing and Radioactive Waste Management,
     Docket RM-50-3. The contributions from reprocessing, waste management, and transportation of wastes are
     maximized for either of the two fuel cycles (uranium only and no recycle). The contribution from transportation
     excludes transportation of cold fuel to a reactor and of irradiated fuel and radioactive wastes from a reactor
     which are considered in Table S–4 of Sec. 51.20(g). The contributions from the other steps of the fuel cycle
     are given in columns A-E of Table S–3A of WASH-1248.

(b) The contributions to temporarily committed land from reprocessing are not prorated over 30 years, because the
     complete temporary impact accrues regardless of whether the plant services one reactor for one year or
     57 reactors for 30 years.

(c) Estimated effluents based upon combustion of equivalent coal for power generation.

(d) 1.2 percent from natural gas use and process.

Table 6-1). For low-level waste disposal at land burial facilities, the Commission notes in
Table S–3 that there will be no significant radioactive releases to the environment. For
high-level and transuranic waste, the Commission notes that these are to be buried at a
repository, such as the candidate repository at Yucca Mountain, Nevada, and that no release to
the environment is expected to be associated with such disposal because it has been assumed
that all of the gaseous and volatile radionuclides contained in the spent fuel are released to the
atmosphere before the disposal of the waste. Because the quantities of radioactive wastes
remain bounded by the staff's previous evaluation, the staff concludes that the impact from
radioactive wastes did not change because of the higher power level. The impact level
category would still be SMALL, and mitigation is not warranted.

6.1.1.7 Occupational Dose

This section was not affected by the changes presented in ER Revision 6 and is provided solely for context.

In the Draft EIS (NRC 2004), the staff reported that the annual occupational dose attributable to all phases of the fuel cycle for the 1000-MW(e) LWR scaled model is about 24 person-Sv (2400 person-rem). This is based on the 6 person-Sv (600 person-rem) occupational dose estimate attributed to all phases of the fuel cycle for the model 1000-MW(e) LWR (NRC 1996). Occupational doses would be maintained to meet the dose limits in 10 CFR Part 20. Because the occupational dose estimates remain bounded by the staff's previous evaluation, the staff concludes that the health impacts from occupational dose did not change because of the higher power level. The impact level category would still be SMALL, and mitigation is not warranted.

6.1.1.8 Transportation

This section was not affected by the changes presented in ER Revision 6 and is provided solely for context.

In the Draft EIS (NRC 2004), the staff reported that the transportation dose to workers and the public was about 0.025 person-Sv (2.5 person-rem) annually for the reference 1000-MW(e) LWR per Table S–3 (see Table 6-1). This corresponds to a dose of 0.1 person-Sv (10 person-rem) for the 1000-MW(e) LWR scaled model. For comparative purposes, the estimated collective dose from natural background radiation to the population within 80 km (50 mi) of the ESP site is 9200 person-Sv/yr (920,000 person-rem/yr) (Dominion 2006a). Because the transportation dose estimates remain bounded by the staff's previous evaluation, the staff concludes that the impact from transportation did not change because of the higher power level. The impact level category would still be SMALL, and mitigation is not warranted.

6.1.1.9 Conclusion

Changes to this section reflect the higher power level proposed in ER Revision 6.

In the Draft EIS (NRC 2004), the staff concluded that the impacts on land use to support the 1000-MW(e) LWR scaled model would be SMALL and that the impacts on water use for these combinations of thermal loadings and water consumption would be SMALL relative to the water use and thermal discharges of the proposed project. The staff also concluded that the fossil fuel impacts from the direct and indirect consumption of electrical energy for fuel cycle operations would be SMALL relative to the net power production of the proposed project, and that the impacts of chemical and radioactive effluents and transportation would be SMALL.

1   Because the PPE combined power level increase to 3040 MW(e) for two units at the North Anna
2   ESP site or at any of the alternative sites remains bounded by the staff's previous evaluation at
3   3200 MW(e) for both units, the staff concludes that the LWR impacts from fuel cycle activities
4   did not change. The impact level category would still be SMALL, and mitigation is not
5   warranted.
6
7   ### 6.1.2   Gas-Cooled Reactors
8
9   *Changes to this section reflect the revised PPE value of 4500 MW(t) for the power level for one*
10   *unit (9000 MW(t) for two units) provided in ER Revision 6. The corresponding net electrical*
11   *output is 1520 MW(e) for one unit (3040 MW(e) for two units). The staff reassessed the*
12   *methods used to calculate the maximum number of Gas Turbine-Modular Helium Reactor units*
13   *and PBMR units that could be sited on the North Anna ESP site or at the alternative sites. In*
14   *the Draft EIS, the staff evaluated impacts at a net electric output of 3200 MW(e); this is higher*
15   *than the 3040 MW(e) proposed in ER Revision 6.*
16
17   Table S–3 from 10 CFR 51.51(a) can be used as a basis for bounding the environmental
18   impacts from the uranium fuel cycle only for LWRs. Dominion performed an assessment of the
19   environmental impacts of the fuel cycle for gas-cooled reactor designs by comparing key
20   parameters for these reactor designs to those used to generate the impacts in Table S–3
21   (Dominion 2006a). Key parameters are energy usage, material involved, and number of
22   shipments for each major fuel cycle activity (i.e., mining, milling, conversion, enrichment, fuel
23   fabrication, and radioactive waste disposal). Dominion sought to demonstrate in the ER that the
24   impacts for the gas-cooled reactor designs were comparable to the environmental impacts
25   identified for LWRs in the technical basis document, WASH-1248, "Environmental Summary of
26   the Uranium Fuel Cycle," and its Supplement 1 (NUREG-0116) for Table S–3 (NRC 1976).
27
28   As discussed in Section 6.1, the fuel cycle impacts in Table S–3 were based on a reference
29   1000 MW(e) LWR operating at an annual capacity factor of 80 percent for a net electric output
30   of 800 MW(e). This is termed the "reference reactor year." For the purposes of evaluating fuel
31   cycle impacts for the North Anna ESP site or for any of the alternative sites, it was assumed that
32   the additional LWR's site-wide fuel impacts would be based on a total net electric output of
33   3200 MW(e). This was termed the 1000 MW(e) LWR scaled model and resulted in a factor
34   approximately four times (i.e., 3200/800) the impacts in Table S–3.
35
36   One of the other-than-LWRs considered by Dominion, the Gas Turbine-Modular Helium Reactor
37   (GT-MHR), is a four module 2400-MW(t), nominal 1140-MW(e) unit assumed to operate at an
38   annual capacity factor of 88 percent for a net electric output of 1003 MW(e). Therefore, the
39   maximum number of GT-MHR units that could be sited at the North Anna ESP site or at any of
40   the alternative sites and remain below the 3200 MW(e) total net electric output for the site is
41   three (i.e., 3 x 1003).

1 The second other-than-LWR reactor considered by Dominion, the pebble bed modular reactor
2 (PBMR), is an eight module, 3200 MW(t), nominal 1320 MW(e) unit assumed to operate at an
3 annual capacity factor of 95 percent for a net electric output of 1253 MW(e). Therefore, the
4 comparable number of PBMR units to remain below the 3200 MW(e) total net electric output for
5 the site is two (i.e., 2 x 1253).
6
7 Dominion (2006a) compared the impacts between the Table S–3 LWR to the gas-cooled reactor
8 designs. The comparison used an annual fuel loading as a starting point and then proceeded in
9 reverse direction through the fuel cycle (i.e., fuel fabrication, enrichment, conversion, milling,
10 mining, radioactive waste). Table 6-2 provides an estimate of the impacts for each phase of the
11 uranium fuel cycle assuming that the North Anna ESP site or any of the alternative sites would
12 host three four-module GT-MHR units or two eight-module PBMR units.
13

14 **6.1.2.1 Fuel Fabrication**
15
16 *This section was not affected by the changes presented in ER Revision 6 and is provided solely*
17 *for context.*
18
19 In the Draft EIS (NRC 2004), the staff reported that the quantity of UO_2 needed for reactor fuel is
20 a key parameter. The more UO_2 needed, the greater the environmental impacts (i.e., more
21 energy, greater emissions, and increased water usage). The 1000-MW(e) LWR scaled model
22 described in Section 6.1.1 would use the equivalent of 160 MT of enriched UO_2 annually. This
23 compares to 18 to 19 MT of enriched UO_2 annually for the gas-cooled reactor technologies.
24
25 GT-MHR fuel consists of microspheres of uranium oxycarbide (UCO) coated with multiple layers
26 of pyrocarbon and silicon carbide referred to as TRISO coating. Two types of microspheres are
27 used in the GT-MHR fuel, one enriched to 19.8 percent uranium-235 and one with natural
28 uranium. The microspheres and graphite shims are bound together into a rod-shaped compact
29 that is stacked into graphite blocks referred to as fuel elements. A reactor core consists of
30 1020 fuel elements.
31
32 PBMR fuel consists of UO_2 kernels (enriched to 12.9 percent uranium-235) that are TRISO
33 coated, similar to the GT-MHR fuel. The TRISO-coated particles are imbedded into a graphite
34 matrix to form a fuel sphere that is 60 mm (2.4 in.) in diameter. Each fuel sphere contains
35 approximately 15,000 TRISO-coated particles. Approximately 260,000 fuel spheres make up a
36 core of a single reactor module.
37
38 The fuel described above for gas-cooled reactors are fabricated differently than fuel for LWRs.
39 There are no currently operating large-scale fuel fabrication facilities producing gas-cooled
40 reactor fuels in the United States, so a direct comparison of environmental impacts is not
41 possible. Based on some environmental impacts from a small-scale fuel fabrication facility

1 **Table 6-2.** Fuel Cycle Environmental Impacts from Gas-Cooled Reactor Designs for the
2 North Anna ESP Site[a] (corresponds to Table 6-3 in the Draft EIS)
3

Reactor Technology Facility/Activity	GT-MHR (4 Modules) (2400 MW(t) total) ($\approx$1140 MW(e) total) 88 percent Capacity: Multiplier=3	PBMR (8 Modules) (3200 MW(t) total) (~1320 MW(e) total) 95 percent Capacity: Multiplier=2
Mining Operations		
Annual ore supply (Million MT)	1.01	0.67
Milling Operations		
Annual yellowcake (MT)	909	606
UF$_6$ Production		
Annual UF$_6$ (MT)	1137	758
Enrichment Operations		
Enriched UF$_6$ (MT)	24	25
Annual separative work unit (SWU) (MT)	612	388
Fuel Fabrication Plant Operations		
Enriched UO$_2$ (MT)	18	19
Annual fuel loading (MTU)	16	17
Solid Radioactive Waste		
Annual LLW from reactor operations (Ci)	3300 Ci;[b] 400m^3	131 Ci;[b] 2400 drums
LLW from reactor decontamination and decommissioning Ci per reference reactor-year	Data not available	Data not available

4
5
6
7
8
9
10
11
12
13
14
15
16
17
18
19
20
21

22 (a) Values calculated by multiplying annual values from Table 5.7-1 of the ER (Dominion 2006a) by a multiplier
23   of 3 for the GT-MHR and a multiplier of 2 for the PBMR.
24 (b) To convert from curies to becquerels, multiply by 3.7 x 10^{10} Bq/Ci.

25 References: 10 CFR 51.51(a), Table S–3 Table of Uranium Fuel Cycle Environmental Data
26 Notes:
27 1. The enrichment SWU calculation was performed using the United States Enrichment Corporation, Inc.
28   (USEC) SWU calculator and assumes a 0.30 percent tails assay.
29 2. The information on the reference reactor (mining, milling, UF$_6$, enrichment, fuel fabrication values) taken from
30   NUREG-0116, Table 3.2, no recycling (NRC 1976).
31 3. The information on the reference reactor (solid radioactive waste) taken from 10 CFR 51.51, Table S–3.
32 4. The calculated information on the reference reactor uses the same methodology as for the reactor
33   technologies.
34 5. The normalized information is based on 1000 MW(e) and the reactor vendor-supplied unit capacity factor.
35 6. For the new reactor technologies, the annual fuel loading was provided by the reactor vendor.
36 7. The USEC SWU calculator also calculated the kilograms of uranium feed. This number was multiplied by
37   1.48 to get the necessary amount of UF$_6$.
38 8. The annual yellowcake number was generated using the relationship 2.61285 lb. of U$_3$O$_8$ to 1 kg U of UF$_6$;
39   1.185 kg of U$_3$O$_8$ to 1.48 kg.
40 9. The annual ore supply was generated assuming an 0.1% ore body and a 90% recovery efficiency.
41 10. Cobalt-60 with a 5.26 year half-life and iron-55 with a 2.73 year half-life are the main nuclides listed for the
42   PBMR decontamination and decommissioning waste.

43

producing gas-cooled reactor fuel, Dominion concluded that the environmental impacts from producing gas-cooled reactor fuel would be "not inconsistent" with those of LWRs (Dominion 2006a). By comparison with the fuel fabrication impacts for LWR technologies, the staff concludes that the environmental impacts from producing gas-cooled reactor fuel did not change. The impact level category would still likely be SMALL. For a gas-cooled reactor these impacts will need to be assessed at the CP or COL stage, when the staff will consider the environmental data that is available on a large-scale, fuel fabrication facility for gas-cooled reactors.

6.1.2.2 Enrichment

This section was not affected by the changes presented in ER Revision 6 and is provided solely for context.

In the Draft EIS (NRC 2004), the staff reported that there are two quantities of interest for enrichment (Dominion 2006a). These were (1) the amount of energy required to enrich the fuel measured in separative work units (SWUs) and (2) the amount of UF_6 needed. An SWU is a measure of energy needed to enrich the fuel. The major environmental impacts for the entire uranium fuel cycle are from the emissions of the fossil fuel plants used to supply energy for the gaseous diffusion plants that enrich the uranium. An enrichment technology developed since the impacts in Table S–3 (see Table 6-1) were developed and evaluated includes the gas centrifuge process, which uses 90 percent less energy than the gaseous diffusion process (NRC 1996).

To produce 160 MT of enriched UO_2 for the 1000-MW(e) LWR scaled model, the enrichment plant needs to produce about 208 MT of UF_6, which requires over 500 MT of SWUs (Dominion 2006a). For gas-cooled reactor technologies, the needed enriched UF_6 ranges from 24 to 25 MT of UF_6. The amount of energy to produce these quantities of enriched UF_6 for the gas-cooled reactor designs ranges from 388 to 612 MT of SWU. The upper range is approximately 20 percent higher than the energy needed for the reference LWR. Dominion concluded that the large reduction in energy associated with using an alternate enrichment technology (e.g., centrifuge) and its associated environmental impacts would more than offset the increase in SWUs (Dominion 2006a). Based on the foregoing, the staff concludes that the environmental impacts of enriching gas-cooled fuels by comparison with the impacts of enriching LWR fuel did not change. The impact level category would still likely be SMALL. For a gas-cooled reactor, these impacts will need to be assessed at the CP or COL stage, when the staff will consider impacts from the enrichment technology in use at that time.

6.1.2.3 Uranium Hexafluoride Production – Conversion

This section was not affected by the changes presented in ER Revision 6 and is provided solely for context.

In the Draft EIS (NRC 2004), the staff reported that there are two uranium conversion processes: a wet process and a dry process. In an earlier evaluation, NUREG-1437 (NRC 1996), the NRC stated that environmental releases are small from the conversion facilities compared to the overall fuel cycle impacts and that changing from 100 percent use of one process to 100 percent use of the other would make no significant difference in the overall impacts. Conversion technologies that would be used today to produce UF_6 are similar to those considered when determining the environmental impacts that were part of Table S–3 of 10 CFR 51.51(a) (see Table 6-1).

The conversion facility would need to produce 1440 MT of UF_6 annually for the reference 1000-MW(e) LWR scaled model compared to 758 to 1137 MT of UF_6 for the gas-cooled reactors based on the USEC SWU calculator (Dominion 2006a). Because the other-than-LWR values are still less than the amount of UF_6 needed for the LWR and the associated impacts are expected to be less, the staff concludes that the environmental impacts from producing UF_6 for gas-cooled reactors did not change. The impact level categories would still be SMALL.

6.1.2.4 Uranium Milling

This section was not affected by the changes presented in ER Revision 6 and is provided solely for context.

In the Draft EIS (NRC 2004), the staff reported that annual yellowcake production is the metric of interest for uranium milling. Plants needing to produce less yellowcake than the reference plant would consequently need less energy, have fewer emissions, and use less water.

The uranium mill for the 1000-MW(e) LWR scaled model would produce about 1200 MT of yellowcake. Because the uranium mill for the gas-cooled reactor technologies would need to produce 606 to 909 MT of yellowcake, which is still less than the amount of yellowcake needed for the scaled LWR (Dominion 2006a), the staff concludes that the environmental impacts from uranium milling for gas-cooled reactors did not change. The impact level category would still be SMALL.

6.1.2.5 Uranium Mining

This section was not affected by the changes presented in ER Revision 6 and is provided solely for context.

In the Draft EIS (NRC 2004), the staff reported that annual ore supply is the metric of interest for uranium mining. The less ore mined, the smaller the environmental impacts (i.e., less energy used, fewer emissions, less water usage). For the 1000-MW(e) LWR scaled model, 1.09 million MT of raw ore would be needed to produce 1200 MT of yellowcake. Because 0.67 to 1.01 million MT of ore would be needed for the gas-cooled reactor technologies, a range that is less than the amount of ore needed for the reference 1000-MW(e) scaled-model LWR, the staff concludes that the environmental impacts from uranium mining for the gas-cooled reactors did not change. The impact level category would still be SMALL.

6.1.2.6 Solid Low-Level Radioactive Waste – Operations

This section was not affected by the changes presented in ER Revision 6 and is provided solely for context.

In the Draft EIS (NRC 2004), the staff cited Table S–3 of 10 CFR 51.51(a), which is reproduced as Table 6-1. The table indicates that there are 3.4×10^{14} Bq (9100 Ci) of low-level waste generated annually from operations of the reference LWR; the 1000-MW(e) LWR scaled model would result in 1.3×10^{15} Bq (36,400 Ci) of low-level waste annually. Gas-cooled reactor technologies are projected to generate 4.8×10^{12} Bq to 1.2×10^{14} Bq (131 to 3300 Ci) of low-level waste scaled annually, far below the amounts generated by the reference LWR. Because these amounts are still well below the amounts generated by the reference LWR, the staff concludes that the environmental impacts from low-level radioactive waste operations for gas-cooled reactors did not change. The impact level category would still be SMALL.

6.1.2.7 Solid Low-Level Radioactive Waste – Decontamination and Decommissioning

This section was not affected by the changes presented in ER Revision 6 and is provided solely for context.

In the Draft EIS (NRC 2004), the staff cited Table S–3 of 10 CFR 51.51(a), which is reproduced as Table 6-1. The table indicates that 5.6×10^7 MBq (1500 Ci) per Reference-Reactor Year "...comes from reactor decontamination and decommissioning - buried at land burial facilities." Dominion noted that gas-cooled reactor technologies would (1) operate much cleaner than the reference 1000-MW(e) LWR, as evidenced by lower estimates of low-level waste generated and (2) produce less heavy metal radioactive waste because of the higher thermal efficiency and higher fuel burnup (Dominion 2006). The gas-cooled reactor designs are also more compact than the reference LWR design, which would be expected to result in less decontamination and decommissioning waste; additionally, low-level waste impacts from decontamination and decommissioning of a gas-cooled reactor are expected to be comparable to or less than that of the reference LWR. Based on the foregoing, the staff concludes that the environmental impacts from solid low-level radioactive waste generated during decontamination and decommissioning for gas-cooled reactors would likely be SMALL, but these impacts will need to be assessed again at the CP or COL stage if an applicant selects a gas-cooled design.

6.1.2.8 Conclusions

This section was not affected by the changes presented in ER Revision 6 and is provided solely for context.

The staff expects that the environmental impacts from the uranium fuel cycle activities and solid waste management activities for the proposed gas-cooled reactors would be SMALL. However, because of the uncertainty in the final design of the gas cooled reactors and the change in technology that could be applied to uranium fuel cycle activities, this issue is unresolved. Should an applicant reference one of these designs, additional reviews would be needed at the CP or COL stage in the following areas: fuel fabrication, enrichment, and solid low-level waste operation during decontamination and decommissioning.

6.2 Transportation of Radioactive Materials

This introductory section was not affected by the changes presented in ER Revision 6 and is provided solely for context.

This section addresses both the radiological and nonradiological environmental impacts from normal operating and accident conditions resulting from (1) shipment of unirradiated fuel to the North Anna ESP site, (2) shipment of spent fuel to a monitored retrievable storage facility or a permanent repository, and (3) shipment of low-level radioactive waste and mixed waste to offsite disposal facilities. Distinctions between transportation impacts of advanced LWR designs and gas-cooled reactor designs are discussed.

The NRC evaluated the environmental effects of transportation of fuel and waste for light-water-cooled nuclear power reactors in WASH-1238 (AEC 1972) and NUREG-75/038 (NRC 1975) and found the impact to be small. These documents provided the basis for Table S–4 in 10 CFR 51.52, which summarizes the environmental impacts of transportation of fuel and waste to and from one LWR of 3000 to 5000 MW(t) (1000 to 1500 MW(e)). Impacts are provided for normal transportation conditions and accidents in transport for a reference 1100-MW(e) LWR. Dose to transportation workers during normal transportation operations was estimated to result in a collective dose of 0.04 person-Sv (4 person-rem) per reference reactor-year. The

1 combined dose to the public along the route and dose to onlookers was estimated at
2 0.03 person-Sv (3 person-rem) per reference reactor-year. Environmental risks (radiological)
3 during accident conditions were determined to be SMALL. Nonradiological impacts during
4 accident conditions were estimated as one fatal injury in 100 reactor years and one nonfatal
5 injury in 10 reference reactor-years. Subsequent reviews of transportation impacts in
6 NUREG-0170 (NRC 1977a) and Sprung et al. (2000) concluded that impacts were bounded by
7 Table S–4 in 10 CFR 51.52.
8
9 In accordance with 10 CFR 51.52(a), a full description and detailed analysis of transportation
10 impacts is not required when licensing an LWR (i.e., impacts are assumed to be bounded by
11 Table S–4) if the reactor meets the following criteria:
12
13 • The reactor has a core thermal power level not exceeding 3800 MW(t).
14
15 • Fuel is in the form of sintered UO_2 pellets having a uranium-235 enrichment not
16   exceeding 4 percent by weight, and pellets are encapsulated in zirconium-clad fuel rods.
17
18 • Average level of irradiation of the fuel from the reactor does not exceed
19   33,000 MWd/MTU, and no irradiated fuel assembly is shipped until at least 90 days
20   after it is discharged from the reactor.
21
22 • With the exception of irradiated fuel, all radioactive waste shipped from the reactor is
23   packaged and in solid form.
24
25 • Unirradiated fuel is shipped to the reactor by truck; irradiated fuel is shipped from the
26   reactor by truck, rail, or barge; and radioactive waste other than irradiated fuel is shipped
27   from the reactor by truck or rail.
28
29 None of the proposed advanced reactors, including the surrogate ESBWR (i.e., ESBWR with
30 increased power level), meet all the conditions in 10 CFR 51.52(a), so a full description and
31 detailed analysis of transportation is required.
32
33 ## 6.2.1 Transportation of Unirradiated Fuel
34
35 *This introductory section is not affected by changes presented in ER Revision 6 and is provided*
36 *solely for context.*
37
38 The staff performed an independent review of the environmental impacts of transporting
39 unirradiated (fresh) fuel to the proposed North Anna ESP site. Environmental impacts of
40 transportation accidents during normal operating conditions are discussed in this section.
41 Appendix G of the Draft EIS provides the details of the analysis.

1 **6.2.1.1 Normal Conditions**

2

3 *Changes to this section reflect health impacts from the transportation of unirradiated fuel to the*
4 *North Anna ESP site or any alternative site for two units with a higher power level of*
5 *9000 MW(t).*

6

7 In the Draft EIS (NRC 2004), the staff reported that normal conditions, sometimes referred to as
8 "incident-free" transportation, are transportation activities in which shipments reach their
9 destination without releasing any radioactive cargo to the environment. Impacts from these
10 shipments would be from the low levels of radiation that penetrate the unirradiated fuel shipping
11 casks.

12

13 This section presents estimates of the numbers of shipments of unirradiated fuel needed to load
14 the initial reactor core, annual reloads, and totals over 40 years for each type of advanced
15 reactor. The shipment estimates were then normalized to the reference reactor electrical
16 generating capacity given in WASH-1238 that forms the basis for Table S–4 in 10 CFR 51.52.
17 ER Revision 6 did not result in a change of the reactor-type-specific numbers of unirradiated
18 fuel shipments are presented in Table 6-3. This table updates Table 6-4 in the Draft EIS and
19 includes information on the surrogate ESBWR. The impacts of the power level increase for
20 surrogate ESBWR unirradiated fuel shipments are described below.

21

22 Dominion indicated in the ER Revision 6 that this change would not significantly affect the
23 quantity of unirradiated fuel required for the initial core and annual refueling requirements.
24 Typically, a higher power output would require more fuel and therefore more fuel shipments.
25 However, the surrogate ESBWR has a higher unit capacity than was assumed in the Draft EIS
26 (96 percent versus 95 percent) and the fuel will have a higher average burnup. In addition, In
27 Enclosure 1 to Revision 6 of the ER, Dominion states that the ESBWR fuel assemblies are
28 about 28 percent lighter than ABWR fuel assemblies (Dominion 2006a). This lower weight is
29 offset by the requirements for 30 percent more ESBWR fuel assemblies than ABWR fuel
30 assemblies. The result is that the total number of fuel shipments for the surrogate ESBWR
31 would increase by 1 to 2 percent, well within the uncertainty of the estimates. Therefore, the
32 number of shipments of unirradiated fuel is essentially unchanged from the estimates presented
33 in the Draft EIS. The power level increase would also not affect the radiation dose rates from
34 the unirradiated fuel shipments and other parameters used to define the shipping routes and
35 receptors. Therefore, the ESBWR power level increase does not affect the per-shipment dose
36 or annual dose impacts presented in the Draft EIS for unirradiated fuel. However, it does have
37 a small impact on the number of shipments normalized to the reference LWR net electric
38 generating capacity. The normalized unirradiated fuel shipments and impacts for the
39 ABWR/ESBWR in the Draft EIS are larger than the surrogate ESBWR. This is because the
40 estimates are normalized to net electric output (i.e., impacts per MW(e)). Because the
41 surrogate ESBWR has a higher net electric output than the ABWR/ESBWR previously

1   **Table 6-3.**   Numbers of Truck Shipments of Unirradiated Fuel for Each Advanced Reactor
2   Type (corresponds to Table 6-4 in the Draft EIS)
3

| Reactor Type | Number of Shipments per Reactor Unit | | | Unit Electric Generation, MW(e)[c] | Capacity Factor[c] | Normalized, Shipments per 1100 MW(e)[d,e] |
	Initial Core[a]	Annual Reload	Total[b]			
Reference LWR (WASH-1238)	18	6	252	1100	0.8	252
ABWR	30	6.1	267	1500	0.95	165
Surrogate ESBWR[e]	30	6.1	267	1520[f]	0.96[f]	162
Surrogate AP1000	14	3.8	161	1150	0.95	130
ACR-700	30	15.4	628	1462[g]	0.9	420
IRIS	34	4.3	201	1005[h]	0.96	184
GT-MHR	51	20	831	1140[i]	0.88	729
PBMR	44	20	824	1320[j]	0.95	579

14   NOTE: The reference LWR shipment values have all been normalized to 880 MW(e) net electrical
15   generation.
16   (a)   Shipments of the initial core have been rounded up to the next highest whole number.
17   (b)   Total shipments of unirradiated fuel over a 40-year plant lifetime (i.e., initial core load plus 39 years
18   of average annual reload quantities).
19   (c)   Unit capacities and capacity factors were taken from INEEL (2003).
20   (d)   Normalized to net electric output for WASH-1238 reference LWR (i.e., 1100 MW(e) plant at
21   80 percent or net electrical output of 880 MW(e)).
22   (e)   Ranges of capacities are given in INEEL (2003) for these reactor unirradiated fuel shipments. The
23   unirradiated fuel shipment data for these reactors were derived using the upper limit of the ranges.
24   (f)   Values taken from ER Revision 6 (Dominion 2006a).
25   (g)   The ACR-700 unit includes two reactors at 731 MW(e) per reactor.
26   (h)   The IRIS unit includes three reactors at 335 MW(e) per reactor.
27   (i)   The GT-MHR unit includes four reactors (modules) at 285 MW(e) per reactor.
28   (j)   The PBMR unit includes eight reactors (modules) at 165 MW(e) per reactor.

29
30   evaluated in the Draft EIS, and the un-normalized impacts are essentially equal, the staff
31   concludes that the impacts per MW(e) are slightly smaller. For the surrogate ESBWR, the
32   normalized shipment value is 162 shipments per 1100 MW(e) rather than 165 shipments per
33   1100 MW(e) previously reported.
34

1   The analysis of maximally-exposed individuals under normal transport conditions is not affected
2   by the changes in ER Revision 6.
3
4   In addition, the normal radiological impacts of transporting unirradiated fuel to advanced reactor
5   sites before normalization to the reference LWR electric generation remain unchanged.
6   However, because the net electric output capacity for the surrogate ESBWR was changed to
7   1520 MW(e), the normalized impacts were updated in Table 6-4; this table corresponds to
8   Table 6-5 in the Draft EIS.
9
10   Based on the foregoing, the staff still concludes that all of the total detriment estimates would be
11   less than 1×10^{-4} fatal cancers, nonfatal cancers, and severe hereditary effects per reference
12   reactor year. These risks are very small compared to the fatal cancers, nonfatal cancers, and
13   severe hereditary effects that would be expected to occur annually in the same population from
14   exposure to natural sources of radiation. However, these impacts are not considered to be
15   resolved for other-than-LWR designs and would need to be assessed at the CP or COL stage
16   when specific information is available regarding other-than-LWR fuel performance and shipping
17   containers, if the applicant references such designs.
18
19   **6.2.1.2 Accidents**
20
21   *This section is not affected by the changes presented in ER Revision 6 and is provided solely*
22   *for context.*
23
24   In the Draft EIS (NRC 2004), the staff reported that accident risks are a combination of accident
25   frequency and consequence. Accident frequencies for transportation of fuel to and from future
26   reactors are expected to be lower that those used in the analysis in WASH-1238 (AEC 1972),
27   which forms the basis for Table S–4 of 10 CFR 51.52, because of improvements in highway
28   safety and security, and an expected decrease in traffic accident, injury, and fatality rates.
29   There is no significant difference in consequences of accidents severe enough to result in a
30   release of unirradiated fuel particles to the environment between advanced LWRs and current-
31   generation LWRs because the fuel form, cladding, and packaging are similar to those analyzed
32   in WASH-1238. Consequently, the impacts of accidents during transport of unirradiated fuel for
33   advanced LWRs to the North Anna ESP site or any alternative site are expected to be smaller
34   than the impacts listed in Table S–4 for current generation LWRs.
35
36   With respect to advanced gas-cooled reactors, accident rates (accidents per unit distance) and
37   associated accident frequencies (accidents per year) would be expected to follow the same
38   trends as for LWRs (i.e., overall reduction relative to the accident rates used in the WASH-1238
39   analysis). The consequences of accidents involving gas-cooled reactor unirradiated fuel,
40   however, are more uncertain. The staff assumed that gas-cooled reactor unirradiated fuel
41   would have the same abilities as LWR unirradiated fuel to maintain functional integrity following

1 **Table 6-4.** Radiological Impacts of Transporting Unirradiated Fuel to Advanced Reactor Sites
2 (corresponds to Table 6-5 in the Draft EIS)
3

Plant Type	Normalized Average Annual Shipments	Cumulative Annual Dose; person-Sv/yr per 1100 MW(e)[a]		
		Workers	Public - Onlookers	Public - Along Route
Reference LWR (WASH-1238)	6.1	1.1×10^{-4}	4.2×10^{-4}	1.0×10^{-5}
ABWR	4.2	7.1×10^{-5}	2.7×10^{-4}	6.6×10^{-6}
Surrogate ESBWR	4.1	6.9×10^{-5}	2.7×10^{-4}	6.5×10^{-6}
Surrogate AP1000	3.3	5.6×10^{-5}	2.2×10^{-4}	5.2×10^{-6}
ACR-700	10.5	1.8×10^{-4}	7.0×10^{-4}	1.7×10^{-5}
IRIS	4.6	7.9×10^{-5}	3.1×10^{-4}	7.4×10^{-6}
GT-MHR	18.2	3.1×10^{-4}	1.2×10^{-3}	2.9×10^{-5}
PBMR	14.5	2.5×10^{-4}	9.6×10^{-4}	2.3×10^{-5}
10 CFR 51.52, Table S–4 Condition	<1 per day	4.0×10^{-2}	3.0×10^{-2}	3.0×10^{-2}

4
5
6
7
8
9
10
11
12
13
14

15 (a) Multiply person-Sv/yr times 100 to obtain doses in person-rem/yr.
16
17 a traffic accident. This assumption is considered to be conservative because gas-cooled
18 reactor fuel operates at significantly higher temperatures, and thus maintains integrity under
19 more severe thermal conditions than LWR fuel. Detailed information about the behavior of the
20 gas-cooled reactor fuel under impact conditions was not available. However, packaging
21 systems for unirradiated gas-cooled reactor fuel will meet the same performance requirements
22 as unirradiated LWR fuel packages including fissile material controls to prevent criticality during
23 normal and accident conditions. Consequently, it is expected that packaging systems for
24 unirradiated gas-cooled reactor fuels would provide equivalent protection as those packages
25 designed for unirradiated LWR fuels. In addition, the fuel forms for the gas-cooled reactors are
26 similar to LWRs (i.e., UO_2 for the PBMR and uranium oxycarbide for the GT-MHR versus UO_2
27 for LWRs); therefore, the inherent failure resistance provided by unirradiated gas-cooled reactor
28 fuels should be similar to that provided by LWR fuel. Because unirradiated gas-cooled and
29 LWR fuels and associated packaging systems would still provide similar resistance to various
30 environmental conditions, the staff concludes that the impacts of accidents involving
31 unirradiated gas-cooled reactor fuel would not be significantly different than for unirradiated
32 LWR fuel and will be within the impacts listed in Table S–4 for current generation LWRs and did
33 not change. However, these impacts are not considered to be resolved, and would need to be

1　assessed at the CP or COL stage when specific information is available regarding other-than-
2　LWR fuel performance, if the applicant references such designs.
3
4　**6.2.2　Transportation of Spent Fuel**
5
6　*This section is not affected by the changes presented in ER Revision 6 and is provided solely*
7　*for context.*
8
9　In the Draft EIS (2004), the staff performed an independent review of the environmental impacts
10　of transporting spent fuel from the proposed North Anna ESP site or any alternative site to a
11　spent fuel disposal repository. The Yucca Mountain, Nevada, site has been identified as a
12　possible location for a geologic repository. The staff considers an estimate of the impacts of the
13　transportation of spent fuel to a possible repository in Nevada to be a reasonable bounding
14　estimate of the transportation impacts to a storage or disposal facility because of the distances
15　involved and the representativeness of the distribution of members of the public in urban,
16　suburban, and rural areas (i.e., population distributions) along the shipping routes.
17　Environmental impacts of normal operating conditions and transportation accidents are
18　discussed in this section.
19
20　This analysis is based on shipment of spent fuel by legal-weight trucks in casks with
21　characteristics similar to casks currently available (i.e., massive, heavily shielded, cylindrical
22　metal pressure vessels). Each shipment is assumed to consist of a single shipping cask loaded
23　on a modified trailer. These assumptions are consistent with assumptions made by the NRC
24　elsewhere in the evaluation of the environmental impacts of transportation of spent fuel in
25　Addendum 1 to NUREG-1437 (NRC 1999). These assumptions are conservative because the
26　alternative assumptions involve rail transportation or heavy-haul trucks, which would reduce the
27　overall number of spent fuel shipments (NRC 1999), thus reducing impacts.
28
29　Environmental impacts of transportation of spent fuel were calculated using the RADTRAN 5
30　computer code (Neuhauser et al. 2003). Routing and population data used in RADTRAN 5 for
31　truck shipments were obtained from the TRAGIS routing code (Johnson and
32　Michelhaugh 2000). The population data in the TRAGIS code are based on the 2000 census.
33
34　**6.2.2.1 Normal Conditions**
35
36　*Changes to this section reflect the higher power level.*
37
38　This section presents estimates of the numbers of shipments of spent fuel to be transported to a
39　spent fuel repository from each type of advanced reactor. The shipment estimates were then
40　normalized to the reference reactor electrical output capacity given in WASH-1238 that forms
41　the basis for Table S–4 in 10 CFR 51.52. The power level increase also does not affect the

1 radiation dose rates from spent fuel shipping casks because the regulatory maximum dose
2 rates were assumed in the Draft EIS and it does not affect the route characteristics or additional
3 parameters important to estimating normal transportation impacts. Therefore, the staff
4 concludes that the normal condition per-shipment impacts for the surrogate ESBWR are not
5 significantly different than those presented in Table 6-6 of the Draft EIS for the North Anna ESP
6 site and the alternative sites.
7
8 The power level increase does have a small impact on the annual number of shipments
9 normalized to the reference LWR net electric output capacity. The normalized spent fuel
10 shipments and impacts were updated in Table 6-5; this table updates Table 6-7 in the Draft EIS
11 and includes information on the surrogate ESBWR. The APWR/ESBWR estimates in the Draft
12 EIS were larger than the surrogate ESBWR because they are normalized to net electric output
13 (i.e., impacts per MW(e)). Because the surrogate ESBWR has a higher net electric output than
14 the ABWR/ESBWR in the Draft EIS, and the un-normalized (i.e., per-shipment) impacts are
15 essentially equal, the impacts per MW(e) are slightly smaller. Because of an increase in
16 burnup, the normalized shipment value is 40 shipments/yr/1100 MW(e) for the surrogate
17 ESBWR rather than 41 shipments/yr/1100 MW(e) as stated in the Draft EIS.
18
19 The updated dose estimates do not change the conclusion that the total detriment estimates
20 associated with the population doses would be less than 1×10^{-1} fatal cancers, nonfatal cancers,
21 and severe hereditary effects per reference reactor year. These risks are very small compared
22 to the fatal cancers, nonfatal cancers, and severe hereditary effects that would be expected to
23 occur annually in the same population from exposure to natural sources of radiation.
24
25 **6.2.2.2 Accident Conditions**
26
27 *Changes to this section reflect the higher power level.*
28
29 The staff used the RADTRAN 5 computer code to estimate the impacts of transportation
30 accidents involving spent fuel shipments. RADTRAN 5 considers a spectrum of potential
31 transportation accidents, ranging from those with high frequencies and low consequences
32 (e.g., "fender benders") to those with low frequencies and high consequences (i.e., accidents in
33 which the shipping container is exposed to severe mechanical and thermal conditions). Details
34 of the analysis are discussed in Appendix G of the Draft EIS.
35
36 Radionuclide inventories are important parameters in the calculation of accident risks. The
37 radionuclide inventories used in the Draft EIS were based on the information provided by the
38 Idaho National Environmental and Engineering Laboratory (INEEL) in *Early Site Permit*
39 *Environmental Report Sections and Supporting Documentation* (INEEL 2003). This report
40 included hundreds of radionuclides for each advanced reactor type. A screening analysis was
41 conducted to select the dominant contributors to accident risks to simplify the RADTRAN 5
42 calculations. The screening identified the radionuclides that would contribute more than

Table 6-5. Routine (Incident-Free) Population Doses from Spent Fuel Transportation, Normalized to Reference Light-Water Reactors (corresponds to Table 6-7 in the Draft EIS)

Reactor Type	Reference LWR (WASH-1238)			ABWR			Surrogate AP1000			ACR-700		
Shipments per Year	60			41			40			90		
Environmental Effects, person-Sv per reactor-year[a]												
Reactor Site	Crew	Onlookers	Along Route	Crew	Onlookers	Along Route	Crew	Onlookers	Along Route	Crew	Onlookers	Along Route
North Anna	0.06	0.21	0.01	0.04	0.14	<0.01	0.04	0.14	<0.01	0.09	0.32	0.01
Portsmouth	0.06	0.19	<0.01	0.04	0.13	<0.01	0.04	0.12	<0.01	0.08	0.28	0.01
Savannah River Site	0.06	0.21	0.01	0.04	0.14	<0.01	0.04	0.14	<0.01	0.08	0.32	0.01
Surry	0.06	0.21	0.01	0.04	0.14	<0.01	0.04	0.14	<0.01	0.10	0.32	0.01

Reactor Type	Surrogate ESBWR			IRIS			GT MHR			PBMR		
Shipments per Year	40			35			34			12		
Environmental Effects, person-Sv per reactor-year[a]												
Reactor Site	Crew	Onlookers	Along Route	Crew	Onlookers	Along Route	Crew	Onlookers	Along Route	Crew	Onlookers	Along Route
North Anna	0.04	0.14	0.36	0.036	0.12	0.0032	0.034	0.12	0.0031	0.012	0.04	0.001
Portsmouth	0.04	0.13	0.29	0.031	0.11	0.0025	0.03	0.11	0.0024	0.01	0.036	0.00082
Savannah River Site	0.04	0.14	0.40	0.034	0.12	0.0035	0.033	0.12	0.0033	0.011	0.039	0.0011
Surry	0.04	0.14	0.38	0.037	0.12	0.0033	0.035	0.12	0.0032	0.12	0.04	0.011

(a) Multiply person Sv/yr times 100 to obtain doses in person rem/yr.

99.999 percent of the dose from inhalation of radionuclides released following a transportation accident. No radionuclide inventory data were provided by INEEL (2003) for the ACR-700 and IRIS advanced reactors; therefore, transportation accident risks were not quantified for these reactor types and would need to be assessed at the CP or COL stage if the applicant referenced either of these designs.

ER Revision 6 proposed a higher power level for the surrogate ESBWR, which affects the radionuclide inventories in spent fuel. Because the revised radionuclide inventory for the ESBWR was not addressed in *Early Site Permit Environmental Report Sections and Supporting Documentation* (INEEL 2003), a preliminary analysis was conducted to estimate the impacts of the revised radionuclide inventories on the spent fuel transportation accident impacts given in the Draft EIS. The radionuclide inventories for each reactor type were updated for the surrogate ESBWR radionuclide inventories provided in Dominion's response to the NRC staff's Request for Additional Information dated May 24, 2006 (Dominion 2006b), in Table 6-5; this table corresponds to Table 6-8 in the Draft EIS.

1 **Table 6-6.** Radionuclide Inventories Used in Transportation Accident Risk Calculations for
2 Each Advanced Reactor Type, Bq/MTU[a] (corresponds to Table 6-8 in the
3 Draft EIS)

4

Radionuclide	ABWR Inventory, Bq/MTU	Surrogate ESBWR Inventory, Bq/MTU	Surrogate AP1000 Inventory, Bq/MTU	GT-MHR Inventory, Bq/MTU	PBMR Inventory, Bq/MTU
Am-241	4.96×10^{13}	4.81×10^{13}	2.69×10^{13}	8.18×10^{13}	7.55×10^{13}
Am-242m	1.24×10^{12}	1.02×10^{12}	4.85×10^{11}	5.03×10^{11}	8.51×10^{11}
Am-243	1.20×10^{12}	1.21×10^{12}	1.24×10^{12}	5.14×10^{11}	4.77×10^{12}
Ce-144	4.22×10^{14}	5.00×10^{14}	3.28×10^{14}	2.15×10^{15}	1.19×10^{15}
Cm-242	2.04×10^{12}	1.80×10^{12}	1.05×10^{12}	1.51×10^{12}	2.78×10^{12}
Cm-243	1.37×10^{12}	1.28×10^{12}	1.14×10^{12}	2.02×10^{11}	1.96×10^{12}
Cm-244	1.80×10^{14}	1.84×10^{14}	2.87×10^{14}	2.83×10^{13}	5.48×10^{14}
Cm-245	2.43×10^{10}	2.50×10^{10}	4.48×10^{10}	1.65×10^{8}	5.29×10^{10}
Co-60[a]	1.01×10^{14}	1.06×10^{14}	0	0	0
Cs-134	1.78×10^{15}	1.92×10^{15}	1.78×10^{15}	2.21×10^{15}	4.03×10^{15}
Cs-137	4.59×10^{15}	4.70×10^{15}	3.44×10^{15}	1.08×10^{16}	1.41×10^{16}
Eu-154	3.81×10^{14}	3.90×10^{14}	3.38×10^{14}	3.23×10^{14}	3.74×10^{14}
Eu-155	1.93×10^{14}	2.00×10^{14}	1.71×10^{14}	8.77×10^{13}	1.08×10^{14}
Pm-147	1.25×10^{15}	1.31×10^{15}	6.51×10^{14}	6.92×10^{15}	5.07×10^{15}
Pu-238	2.27×10^{14}	2.28×10^{14}	2.25×10^{14}	1.17×10^{14}	4.55×10^{14}
Pu-239	1.43×10^{13}	1.43×10^{13}	9.44×10^{12}	2.25×10^{13}	1.11×10^{13}
Pu-240	2.28×10^{13}	2.30×10^{13}	2.01×10^{13}	3.96×10^{13}	3.32×10^{13}
Pu-241	4.51×10^{15}	4.51×10^{15}	2.58×10^{15}	8.33×10^{15}	7.18×10^{15}
Pu-242	8.29×10^{10}	8.29×10^{10}	6.73×10^{10}	1.56×10^{11}	4.51×10^{11}
Ru-106	6.07×10^{14}	6.88×10^{14}	5.74×10^{14}	1.48×10^{15}	1.68×10^{15}
Sb-125	1.99×10^{14}	2.14×10^{14}	1.42×10^{14}	2.21×10^{14}	2.51×10^{14}
Sr-90	3.27×10^{15}	3.36×10^{15}	2.29×10^{15}	8.95×10^{15}	1.08×10^{16}
Y-90	3.27×10^{15}	3.36×10^{15}	2.29×10^{15}	8.95×10^{15}	1.08×10^{16}

29 (a) Cobalt-60 is an activation product. Only the ABWR/ESBWR submittal in INEEL (2003) provided inventory
30 data for activation products; it was scaled up for the surrogate ESBWR.

31

32 The total inventory for the surrogate ESBWR radionuclides that contribute more than
33 99.999 percent of the dose from inhalation are increased by 3 percent relative to the Draft EIS
34 for the ABWR/ESBWR. The dose from many of the actinides (e.g., americium-241,
35 americium-242m, curium-242, curium-243, and plutonium-239) would be lower for the surrogate
36 ESBWR than was shown in the Draft EIS; however, fission product inventories (and doses)
37 would be somewhat higher. The range of differences is from -17 percent (americium-242m) to
38 +18 percent (cerium-144).

1   The likely impacts on the spent fuel transportation accident risks were estimated using
2   radionuclide-specific risk estimates from the Draft EIS RADTRAN 5 outputs. The risk estimates
3   were adjusted linearly with the radionuclide inventories discussed above. For example, the
4   calculations indicated that americium-241 inventories for the surrogate ESBWR decreased by
5   3 percent relative to the previous power level ESBWR. Thus, the americium-241 contribution to
6   the total risk was decreased by 3 percent. This was repeated for the rest of radionuclides in the
7   RADTRAN calculations. The result was that the accident risk for the higher power level case
8   was about 5 percent higher; the updated results are provided in Table 6-7; this table
9   corresponds to Table 6-9 of the Draft EIS. This increase is approximately the same as the
10 percentage increase in power levels (4300 MW(t) for the ABWR/ESBWR in the Draft EIS to
11 4500 MW(t) for the ESBWR in the ER Revision 6).
12
13 The increase in power level did not change the conclusion that all of the total detriment
14 estimates associated with the doses in the Table 6-7 would be less than 1×10^{-6} fatal cancers,
15 nonfatal cancers, and severe hereditary effects per year for the North Anna ESP and any
16 alternative site.
17
18 ### 6.2.2.3 Conclusion
19
20 *Changes to this section reflect the higher power level.*
21
22 The values determined by this analysis represent the contribution of such effects to the
23 environmental costs of licensing the reactor. Because of the conservative approaches and data
24 used to calculate doses, actual environmental effects are not likely to exceed those calculated
25 in the EIS. Thus, the staff concludes that the overall transportation accident risks associated
26 with advanced LWR reactor (including surrogate ESBWR) spent fuel shipments from the North
27 Anna ESP site or any alternative site are SMALL and are consistent with the risks associated
28 with transportation of spent fuel from current generation reactors presented in Table S–4 of
29 10 CFR 51.52. The fuel performance characteristics, shipping casks, and accident risks for
30 other-than-LWR designs are not resolved and would need to be assessed at the CP or COL
31 stage if the applicnat references such designs.
32
33 ### 6.2.3   Transportation of Radioactive Waste
34
35 *Changes to this section reflect the higher power level.*
36
37 In the Draft EIS (NRC 2004), the staff discussed the environmental effects of transporting waste
38 from proposed ESP site and alternative sites. The following results apply to all advanced
39 reactors, including the surrogate ESBWR described in ER Revision 6:
40
41

Table 6-7. Annual Spent Fuel Transportation Accident Impacts for Advanced Reactors, Normalized to Reference 1000 – MW(e) LWR Net Electrical Generation (corresponds to Table 6-9 in the Draft EIS)

	Advanced Reactor Type				
MTU/yr	**ABWR** 20.3	**Surrogate ESBWR** 20.3	**Surrogate AP1000** 19.7	**GT-MHR** 6	**PBMR** 5.8
Population Dose, person-Sv/yr[a]					
North Anna	4.7×10^{-6}	5.0×10^{-6}	4.2×10^{-7}	1.9×10^{-7}	3.1×10^{-7}
Portsmouth	5.2×10^{-6}	5.5×10^{-6}	4.0×10^{-7}	1.8×10^{-7}	3.0×10^{-7}
Savannah River Site	5.3×10^{-6}	5.6×10^{-6}	4.7×10^{-7}	2.1×10^{-7}	3.5×10^{-7}
Surry	4.8×10^{-6}	5.1×10^{-6}	4.3×10^{-7}	2.0×10^{-7}	3.2×10^{-7}
Latent Cancer Fatalities per Year					
North Anna	2.8×10^{-7}	3.0×10^{-7}	2.5×10^{-8}	1.1×10^{-8}	1.9×10^{-8}
Portsmouth	3.1×10^{-7}	3.3×10^{-7}	2.4×10^{-8}	1.1×10^{-8}	1.8×10^{-8}
Savannah River Site	3.2×10^{-7}	3.4×10^{-7}	2.8×10^{-8}	1.3×10^{-8}	2.1×10^{-8}
Surry	2.9×10^{-7}	3.1×10^{-7}	2.6×10^{-8}	1.2×10^{-8}	1.9×10^{-8}
Total Detrimental Health Effects per Year					
North Anna	4.0×10^{-7}	4.2×10^{-7}	3.6×10^{-8}	1.6×10^{-8}	2.7×10^{-8}
Portsmouth	4.4×10^{-7}	4.6×10^{-7}	3.4×10^{-8}	1.6×10^{-8}	2.6×10^{-8}
Savannah River Site	4.5×10^{-7}	4.8×10^{-7}	4.0×10^{-8}	1.8×10^{-8}	3.0×10^{-8}
Surry	4.1×10^{-7}	4.3×10^{-7}	3.7×10^{-8}	1.7×10^{-8}	2.7×10^{-8}

(a) Multiply person-Sv/yr times 100 to obtain person-rem/yr.

- Radioactive waste (except spent fuel) would be packaged in solid form.

- Radioactive waste (except spent fuel) would be shipped from the reactor by truck or rail.

- The weight limitation of 33,100 kg (73,000 lb) per truck and 90,700 kg (100 tons) per cask per railcar would be met.

1   • The traffic density would be less than the one truck shipment per day or three railcars
2   per month limitation.
3
4 Table 6-8 presents updated estimates of annual waste volumes and annual waste shipment
5 numbers for the advanced reactor types, including the surrogate ESBWR, normalized to the
6 reference 1100-MW(e) LWR defined in WASH-1238 (AEC 1972), this table corresponds to
7 Table 6-10 of the Draft EIS. Dominion indicated in its May 24, 2006 response to RAIs that no
8 change is anticipated in the volume of radioactive waste produced (Dominion 2006b). The
9 quantity of radioactive waste generated is more closely related to Dominion's operational
10 practices than the reactor's power output. The staff concludes that increases in solid
11 radioactive waste generation estimates for the surrogate ESBWR, if any, would be small and
12 that any increase would be within the range of uncertainty of the waste generation estimates.
13
14 The normalized annual waste generation rate and waste shipments for the surrogate ESBWR
15 would be slightly smaller than previously evaluated because the estimates are normalized to net
16 electric output (i.e., impacts per MW(e)). Because the surrogate ESBWR has a higher net
17 electric output than the ABWR/ESBWR in the Draft EIS, and the waste volumes, shipments, and
18 impacts are essentially equal, the impacts per MW(e) are slightly smaller. The normalized
19 amount of solid waste generated annually would decrease from 62 to 60 m^3/1100 MW(e) plant
20 and the normalized annual shipments would decrease from 27 to 26 shipments/1100 MW(e).
21 The waste generation and shipment estimates were adjusted to reflect a slightly higher capacity
22 factor of 0.96 versus the capacity factor used in the Draft EIS (i.e., 0.95).
23
24 As shown in the table, only the PBMR would be expected to generate a larger volume of
25 radioactive waste than the reference LWR in WASH-1238 (AEC 1972). However, the GT-MHR
26 and PBMR information in INEEL (2003) assumed that the applicant would ship wastes using
27 two different packaging systems: one that hauls 28 m^3 per shipment (1000 ft^3 per shipment) and
28 one that hauls 5.7 m^3 per shipment (200 ft^3 per shipment). Under those conditions, the number
29 of shipments of radioactive waste per year, normalized to 1100 MW(e) electric generation
30 capacity, would be about six shipments per year per 1100 MW(e) (880 net MW(e)) for the
31 GT-MHR and seven shipments per year per 1100 MW(e) for the PBMR. These estimates are
32 well below the reference LWR (46 shipments per year per 1100 MW(e)). However, impacts
33 from other-than-LWR designs are not resolved because of the lack of verifiable information.
34
35 The higher power level does not change the staff conclusion that the sum of the daily shipments
36 of unirradiated fuel, spent fuel, and radioactive waste is well below the one truck shipment per
37 day condition given in 10 CFR 51.52, Table S–4 for all reactor types. Doubling the shipment
38 estimates to account for empty return shipments of fuel and waste is still well below the
39 one-truck-shipment-per-day condition.

1 **Table 6-8.** Summary of Radioactive Waste Shipments for Advanced Reactors (corresponds to
2 Table 6-10 of the Draft EIS)
3

Reactor Type	INEEL (2003) Waste Generation Information	Annual Waste Volume, m³/yr per Unit	Electrical Output, MW(e) per Unit	Normalized Rate, m³/1100 MW(e) Unit (880 MW(e) Net)[a]	Shipments/ 1100 MW(e) (880 MW(e) Net) Electrical Output[b]
5 Reference 6 LWR 7 (WASH-1238)	100 m³/yr per unit	108	1100	108	46
8 ABWR	100 m³/yr per unit	100	1500	62	27
9 Surrogate 10 ESBWR	100 m³/yr per unit	100	1520[c]	60	26
11 Surrogate 12 AP1000	55 m³/yr per unit	56	1150	45	20
13 ACR-700	47.5 m³/yr per unit	47.5	731	64	28
14 IRIS	25 m³/yr per modules	74 (3 modules)	1,005 (3 modules)	67	29
15 GT-MHR	98 m³/yr (4 module Plant)	98 (4 modules)	1,320 (8 modules)	86	37[d]
16 PBMR	100 drums/yr per modules	168 (8 modules)	1,320 (8 modules)	118	51[d]

17 Conversions: 1 m³ = 35.31 ft³. Drum volume = 210 liters (0.21 m³).
18 (a) Capacity factors used to normalize the waste generation rates to an equivalent electrical generation
19 output are given in Table 6-3 for each reactor type. All are normalized to 880 MW(e) net electrical
20 output (1100-MW(e) unit with an 80 percent capacity factor).
21 (b) The number of shipments per 1100 MW(e) was calculated assuming the WASH-1238 average
22 waste shipment capacity of 2.34 m³ per shipment (108 m³/yr divided by 46 shipments/yr).
23 (c) This value was taken from the ER, Revision 6 (Domion 2006a)
24 (d) The applicant states in INEEL (2003) that 90 percent of the waste could be shipped on trucks
25 carrying 28 m³ (1000 ft³) of waste and the remaining 10 percent in shipments carrying 5.7 m³
26 (200 ft³) of radioactive waste. This would result in six to seven shipments per year after
27 normalization to the reference LWR electrical output.
28

1   ### 6.2.4   Conclusions

2

3   *Changes to this section reflect the higher power level.*

4

5   An analysis was conducted of the impacts under normal operating and accident conditions of
6   transporting unirradiated fuel to advanced reactor sites and spent fuel and wastes from
7   advanced reactor sites to disposal facilities. To make comparisons to Table S–4, the
8   environmental impacts are normalized to a reference reactor year. The reference reactor is an
9   1100-MW(e) reactor that has an 80-percent capacity factor, for a total electrical output of
10   880 MW(e) per year. The environmental impacts can be adjusted to calculate impacts per site
11   by multiplying the normalized impacts by the ratio of the total electric output for the advanced
12   reactor sites to the electric output of the reference reactor.

13

14   Because of the conservative approaches and data used to calculate doses, actual
15   environmental effects are not likely to exceed those calculated in the EIS. Thus, the staff
16   concludes that the environmental impacts of transportation of fuel and radioactive wastes to and
17   from advanced LWR designs would be SMALL, and would be consistent with the risks
18   associated with transportation of fuel and radioactive wastes from current-generation reactors
19   presented in Table S–4 of 10 CFR 51.52. For gas-cooled designs, the impacts are likely to be
20   small, but this issue is not resolved because of the lack of verifiable information on these
21   designs. At the CP or COL stage, an applicant referencing these designs would need to
22   provide the necessary data and the staff would need to validate the assumptions used in this
23   transportation analysis.

24

25   Assumptions that will need validation if a gas-cooled design is selected include:

26

27   • Verifying that unirradiated and spent fuel from gas-cooled reactors have the same
28   abilities as LWR unirradiated and spent fuel to maintain fuel and cladding integrity
29   following a traffic accident.

30

31   • Verifying that shipping cask design assumptions (for example, cask capacities) are
32   equal to or bounded by the assumptions in this analysis.

33

34   • Verifying that unirradiated fuel initial core/refueling requirement, spent fuel generation
35   rates, and radioactive waste generation rate assumptions are equal to or bounded by
36   the assumptions in this analysis.

37

38   • Verifying that shipping cask capacities and accident source terms, including spent fuel
39   inventories, severity fractions, and release fractions, are equal to or bounded by the
40   assumptions in this analysis.

41

1 Should the ACR-700 or IRIS reactors be chosen for the ESP site, a transportation accident
2 analysis will be performed as spent fuel inventories were not available for this analysis.
3

6.3 Decommissioning Impacts

5
6 *This section is not affected by the changes presented in ER Revision 6 and is provided solely*
7 *for context.*
8
9 At the end of the operating life of a power reactor, the NRC regulations require that the facility
10 undergo decommissioning. Decommissioning is the removal of a facility safely from service and
11 the reduction of residual radioactivity to a level that permits termination of the NRC license. The
12 regulations governing decommissioning of power reactors are found in 10 CFR 50.75
13 and 50.82.
14
15 Environmental impacts from the activities associated with the decommissioning of any LWR
16 before or at the end of an initial or renewed license are evaluated in the *Generic Environmental*
17 *Impact Statement on Decommissioning of Nuclear Facilities, Supplement 1, Regarding the*
18 *Decommissioning of Nuclear Power Reactors*, NUREG-0586, (NRC 2002). If an applicant for a
19 CP or COL referencing the North Anna ESP applies for a license to operate one or more
20 additional units at the North Anna ESP site, there is a requirement to provide a report containing
21 a certification that financial assurance for radiological decommissioning will be provided. At the
22 time an application is submitted, the requirements in 10 CFR 50.33, 50.75, and 52.77 (and any
23 other applicable requirements) would have to be met.
24
25 At the ESP stage, applicants are not required to submit information regarding the process of
26 decommissioning, such as the method chosen for decommissioning, the schedule, or any other
27 aspect of planning for decommissioning. Dominion did not provide this information in its
28 application. For the new nuclear unit or units, if LWR designs are chosen or if other-than-LWRs
29 that were considered in NUREG-0586, Supplement 1 are chosen, the impacts from
30 decommissioning are expected to be within the bounds described in NUREG-0586,
31 Supplement 1. In such cases, the staff expects the impact from decommissioning are likely to
32 be small. However, for whatever design that is selected, the impacts from decommissioning are
33 not resolved and would have to be assessed at the CP or COL stage.
34

6.4 References

36
37 10 CFR Part 20. Code of Federal Regulations. Title 10, *Energy*, Part 20, "Standards for
38 Protection Against Radiation."
39

1   10 CFR Part 50. Code of Federal Regulations. Title 10, *Energy*, Part 50, "Domestic Licensing
2   of Production and Utilization Facilities."
3
4   10 CFR Part 51. Code of Federal Regulations. Title 10, *Energy*, Part 51, "Environmental
5   Protection Regulations for Domestic Licensing and Related Regulatory Functions."
6
7   10 CFR Part 52. Code of Federal Regulations. Title 10, *Energy*, Part 52, "Early Site Permits,
8   Standard Design Certifications, and Combined Licenses for Nuclear Power Plants."
9
10   Dominion Nuclear North Anna, LLC (Dominion). 2006a. *North Anna Early Site Permit
11   Application – Part 3 – Environmental Report.* Revision 6, Glen Allen, Virginia.
12
13   Dominion Nuclear North Anna, LLC (Dominion). 2006b. Letter from E. Grecheck (DNNA) to
14   NRC dated May 24, 2006, submitting additional information in response to an NRC request
15   dated May 10, 2006, (ML061510131).
16
17   Idaho National Engineering and Environmental Laboratory (INEEL). 2003. *Early Site Permit
18   Environmental Report Sections and Supporting Documentation.* Engineering Design File
19   Number 3747. Idaho Falls, Idaho.
20
21   International Commission on Radiological Protection (ICRP). 1991. ICRP Publication 60: 1990
22   Recommendations of the International Commission on Radiological Protection. Annals of the
23   ICRP, Vol 21/1-3, Elsevier, The Netherlands
24
25   Johnson P.E. and R.D. Michelhaugh. 2000. *Transportation Routing Analysis Geographic
26   Information System (WebTRAGIS) User's Manual.* ORNL/TM-2000/86, Oak Ridge National
27   Laboratory, Oak Ridge, Tennessee. Available at
28   http://www.ornl.gov/~webworks/cpr/v823/rpt/106749.pdf
29
30   Neuhauser K.S., F.L. Kanipe, and R.F. Weiner. 2003. *RADTRAN 5 User Guide.*
31   SAND2003-2354, Sandia National Laboratories, Albuquerque, New Mexico. Available on the
32   Internet at http://infoserve.sandia.gov/sand_doc/2003/032354.pdf
33
34   Sprung J.L., D.J. Ammerman, N.L. Breivik, R.J. Dukart, F.L. Kanipe, J.A. Koski, G.S. Mills,
35   K.S. Neuhauser, H.D. Radloff, R.F. Weiner, and H.R. Yoshimura. 2000. *Reexamination of
36   Spent Fuel Shipment Risk Estimates.* NUREG/CR-6672, U.S. Nuclear Regulatory Commission,
37   Washington, D.C.
38
39   U.S. Atomic Energy Commission (AEC). 1972. *Environmental Survey of Transportation of
40   Radioactive Materials to and from Nuclear Power Plants.* WASH-1238, Washington, D.C.
41

1 U.S. Atomic Energy Commission (AEC). 1974. *Environmental Survey of the Uranium Fuel*
2 *Cycle.* WASH-1248, Washington, D.C.
3
4 U.S. Nuclear Regulatory Commission (NRC). 1975. *Environmental Survey of Transportation of*
5 *Radioactive Materials to and from Nuclear Power Plants, Supplement 1.* NUREG-75/038,
6 Washington, D.C.
7
8 U.S. Nuclear Regulatory Commission (NRC). 1976. *Environmental Survey of the Reprocessing*
9 *and Waste Management Portions of the LWR Fuel Cycle.* NUREG-0116, Supplement 1 to
10 WASH-1248, Washington, D.C.
11
12 U.S. Nuclear Regulatory Commission (NRC). 1977a. *Final Environmental Statement on*
13 *Transportation of Radioactive Material by Air and Other Modes.* NUREG-0170, Vol.1,
14 Washington, D.C.
15
16 U.S. Nuclear Regulatory Commission (NRC). 1977b. *Public Comments and Task Force*
17 *Responses Regarding the Environmental Survey of the Reprocessing and Waste Management*
18 *Portions of the LWR Fuel Cycle,* NUREG-0216 (Supplement 2 to WASH-1248),
19 Washington, D.C.
20
21 U.S. Nuclear Regulatory Commission (NRC). 1996. *Generic Environmental Impact Statement*
22 *for License Renewal of Nuclear Plants.* NUREG-1437, Volumes 1 and 2, Washington, D.C.
23
24 U.S. Nuclear Regulatory Commission (NRC). 1999. *Generic Environmental Impact Statement*
25 *for License Renewal of Nuclear Plants, Main Report,* "Section 6.3 – Transportation, Table 9.1,
26 Summary of findings on NEPA issues for license renewal of nuclear power plants, Final Report."
27 NUREG-1437, Volume 1, Addendum 1, Washington, D.C.
28
29 U.S. Nuclear Regulatory Commission (NRC). 2002. *Generic Environmental Impact Statement*
30 *on Decommissioning of Nuclear Facilities, Supplement 7 Regarding the Decommissioning of*
31 *Nuclear Power Reactors.* NUREG-0586, Vol. 1, Washington, D.C.
32
33 U.S. Nuclear Regulatory Commission (NRC). 2004. *Draft Environmental Impact Statement*
34 *(EIS) for an Early Site Permit (ESP) at the North Anna ESP Site.* NUREG-8511 Draft, Office of
35 Nuclear Reactor Regulations, Division of Regulatory Improvement Programs, Washington, D.C.

7.0 Cumulative Impacts

This chapter of the Supplement to the Draft Environmental Impact Statement is presented in its entirety and was changed to reflect the cumulative impact of the higher power output level, and additional cooling system design alternative (i.e., once-through cooling) as a result of cooling system changes proposed for Unit 3 in Revision 6 of the Environmental Report.

The U.S. Nuclear Regulatory Commission (NRC) staff considered potential cumulative impacts during its evaluation of information applicable to each of the potential impacts of constructing and operating reactors at the proposed North Anna Power Station (NAPS) early site permit (ESP) site for reactor designs that fall within the plant parameter envelope (PPE) presented in Revision to Dominion Nuclear North Anna, LLC's (Dominion) Environmental Report (ER) (Dominion 2006). For the purpose of this analysis, past actions are those occurring after Lake Anna was created, but prior to operation of the existing NAPS Units 1 and 2. Present actions are those from the start of operation of existing NAPS Units 1 and 2 until the start of construction of the proposed ESP Units 3 and 4 (hereafter referred to as Units 3 and 4). Future actions are those that are reasonably foreseeable through construction and operation of Units 3 and 4, including decommissioning. The geographical area over which past, present, and future actions could contribute to cumulative impacts depends on the type of impact evaluated.

The impacts of the proposed action, as described in Chapters 4 and 5, are combined with other past, present, and reasonably foreseeable future actions in the vicinity of the NAPS site that would affect the same resources impacted by NAPS Units 1 and 2 regardless of what entity (Federal or non-Federal) or person undertakes such other actions. These combined impacts are defined as "cumulative" in Title 40 of the Code of Federal Regulations (CFR) Part 1508.7 and include individually minor but collectively significant actions taking place over a period of time. It is possible that an impact that may be SMALL by itself could result in a MODERATE or LARGE impact when considered in combination with the impacts of other actions on the affected resource. Likewise, if a resource is regionally declining or imperiled, even a SMALL individual impact could be important it if contributes to or accelerates the overall resource decline.

7.1 Land Use

For purposes of this analysis, the geographic area considered for cumulative impacts resulting from construction and operation of Units 3 and 4 includes the three-county area of Louisa, Orange, and Spotsylvania Counties, Virginia, because the impacts to land use are insignificant outside the three county area. The staff reviewed the available information on the land-use impacts of constructing two additional nuclear units at the North Anna ESP site. Accordingly, the staff concludes that while lower tax rates or better services could encourage development,

1   the comprehensive land-use plans would control development. As a result, cumulative land-use
2   impacts did not change. The impact level category would still be SMALL, and mitigation is not
3   warranted.
4

7.2   Air Quality

6

7   The NAPS site is located in an area that is in attainment for criteria pollutants. In Section 5.2 of
8   the *Draft Environmental Impact Statement (EIS) for an Early Site Permit (ESP) at the North*
9   *Anna ESP Site* (Draft EIS) the staff evaluated the impacts of the discharge of warm moist air
10   from the wet cooling tower portion of the wet and dry cooling system (NRC 2004). The existing
11   units use a once-through cooling system and Unit 4 would use a dry cooling system, neither of
12   which discharges warm moist air. Therefore, the cumulative impacts of the change from the
13   cooling system is the same as the impact analyzed in Section 5.2. The increase in power for
14   Units 3 and 4 had no affect on air quality. The changes to the cooling system and the power
15   increase would not result in additional releases from vehicles, auxiliary boilers, emergency
16   generators, and energized transmission lines. The discharge from Unit 3 wet cooling tower
17   portion of the wet and dry cooling system would have a SMALL impact. In addition, the
18   Commonwealth of Virginia regulates emissions to the atmosphere. The air quality impacts of
19   construction and operations are estimated to be small. No other significant impacts from other
20   actions were identified. Accordingly, the staff concludes that the cumulative impacts of air
21   quality did not change. The impact level category would still be SMALL, and mitigation is not
22   warranted.
23

7.3   Water Use and Quality

25

26   There would be two primary surface water resource parameters affected by the operation of
27   Unit 3 on the Lake Anna reservoir: (1) the lake level and (2) the downstream flow. In turn,
28   changes in these parameters impact water use, aquatic ecosystems, and socioeconomics. The
29   cumulative effects on those parameters are discussed in each category in this section. The
30   thermal effects on the lake, as evaluated in the Draft EIS (NRC 2004) for once-through cooling
31   are not an issue in light of the currently proposed combination wet and dry cooling system.
32

33   The staff, while preparing this assessment, did not identify any other currently planned
34   industrial, commercial, or public installations that would consume water within the general
35   vicinity of the North Anna ESP site. The intake of water from, and the discharge of water to,
36   Lake Anna from the new units would be regulated by the Virginia Department of Environmental
37   Quality (VDEQ) just as the existing NAPS Units 1 and 2 are currently regulated by the VDEQ.
38   The intake and discharge limits for each installation are established considering the overall or
39   cumulative impact of all of the other regulated activities in the area. The staff expects that
40   compliance with Clean Water Act and regulations of the Commonwealth of Virginia are

1   adequate to minimize the cumulative effects on water resources. Operation of Units 3 and 4
2   would require National Pollutant Discharge Elimination System (NPDES) permits from the
3   Commonwealth. NPDES permits must be renewed every 5 years, which will ensure that the
4   Commonwealth of Virginia addresses changes in water quality over time. The Commonwealth
5   has the authority to designate the North Anna drainage as a surface water management area,
6   which will ensure that water supply changes over time are addressed. Unit 4 would use dry
7   cooling towers and not water from Lake Anna for cooling; therefore, its operation would have
8   little operational impact on Lake Anna.
9
10   In Chapter 5, the staff evaluated the effects of the existing Units 1 and 2 and the effects of
11   adding Units 3 and 4 on Lake Anna. A cumulative evaluation of the effects of Units 3 and 4 on
12   Lake Anna starts with the existing lake conditions and adds the effects of construction and
13   operation of Units 3 and 4 to reach a cumulative impact assessment. Based on the fact that the
14   Lake Anna drainage is largely rural and the shoreline is largely residential, the staff concludes
15   that it is unlikely that future development will appreciably alter the hydrology. In non-drought
16   years, the projected incremental decline of the lake level attributable to Unit 3 using the
17   combination wet and dry cooling tower system is relatively minor and less than the effect of
18   using once-through cooling. The lowest pool elevation and greatest incremental decline would
19   occur during the month of October.
20
21   Though less pronounced than with once-through cooling, the operation of ESP Unit 3 would
22   increase the duration of periods of low lake level during drought conditions when the Lake Level
23   Contingency Plan would be applied. Implementation of the Lake Level Contingency Plan
24   reduces flow from Lake Anna as the level in the lake declines to a minimum flow of 0.57 m^3/s
25   (20 cfs) (Louisa County 2001). Hanover County, one of four downstream counties, has
26   identified a need for additional water for future development (Hanover County Department of
27   Public Utilities 2004). To meet its future projected demand, Hanover County proposes to
28   withdraw 1.3 m^3/s (46 cfs) from the North Anna River (Dominion 2006). Resolution of any future
29   conflicts over water use would fall within the regulatory authority of the Commonwealth of
30   Virginia. There are three basic approaches considered by the staff to mitigate water conflicts
31   including (1) alternative design of the Unit 3 cooling system, (2) alternative operation of the
32   proposed Unit 3, and (3) alternative operating procedures for the North Anna Dam. Alternative
33   cooling system designs are discussed in Section 8.2 of this Supplement to the Draft EIS
34   (SDEIS). Dry cooling would eliminate the consumptive water loss associated with Unit 3.
35   If water conditions were severe the Commonwealth of Virginia maintains the regulatory authority
36   to require Dominion to derate or terminate operation of one or more of the North Anna units.
37   Finally, the release of water from Lake Anna is regulated by the Commonwealth of Virginia. The
38   Lake Level Contingency Plan is an explicit statement of the Commonwealth's policy to balance
39   the demands of lake and downstream water needs. The Commonwealth can alter the normal
40   pool elevation and the trigger elevation at which releases are reduced, and can specify the
41   timing, duration, and magnitude of discharge flows from the North Anna Dam.

1   Based on the staff's independent water budget assessment, which includes the cumulative
2   impact of the existing NAPS Units 1 and 2 and the proposed Unit 3, the staff concludes that the
3   water use impacts would be SMALL except in drought years when the impacts would be
4   MODERATE. In drought years, the Commonwealth of Virginia may determine to require
5   Dominion to derate or cease operation. Water quality impacts are anticipated to be small.
6   However, because specific bounding water quality parameters were not provided in the PPE for
7   all discharge streams, the water quality impact is unresolved.
8
9   ## 7.4 Terrestrial Ecosystem
10
11 For purposes of this analysis, the geographic area in which adverse cumulative effects on
12 terrestrial resources, such as wildlife populations and habitat areas could occur, includes the
13 areas around Lake Anna, within the North Anna ESP site, and within the existing transmission
14 line rights-of-way.
15
16 Although the rate of housing and recreational development around Lake Anna has been
17 relatively high, the habitats at the North Anna ESP site and in the vicinity of Lake Anna are
18 common in central Virginia, and are not considered critical for the survival of any threatened or
19 endangered species. Therefore, the staff concludes that the contribution of development at the
20 North Anna ESP site to the cumulative habitat loss in the region would be SMALL.
21
22 There are no important terrestrial species (e.g., threatened or endangered species) or important
23 habitats (e.g., critical habitats, wildlife sanctuaries) in the vicinity of the site or transmission line
24 rights-of-way, and wildlife has adapted to the noise levels from the existing Units 1 and 2. The
25 nearest bald eagle (*Haliaeetus leucocephalus*) nest is 4.2 km (2.6 mi) west of the site and would
26 not be affected by noise from the NAPS site. Therefore, cumulative noise effects on wildlife are
27 expected to be minimal.
28
29 The combination wet and dry cooling system, as proposed for Unit 3, and dry cooling towers, as
30 proposed for Unit 4, include elevated structures that could pose a risk of avian collisions. In
31 NUREG-1437, *Generic Environmental Impact Statement for License Renewal of Nuclear Plants*,
32 the staff reviewed the issue of avian collisions with elevated structures at nuclear power plants
33 in the United States and concluded that cooling towers pose a very small hazard for birds (NRC
34 1996). Therefore, impacts to birds from collisions with heat dissipation structures of the
35 proposed Unit 3 and Unit 4 cooling systems and existing facilities at the NAPS site are expected
36 to be minimal. The North Anna ESP site is in an area with relatively few tall facilities or features
37 that would pose collision hazards to birds and additional industrial development is not likely in
38 the foreseeable future. Therefore, cumulative effects on birds resulting from collisions would be
39 expected to be minimal.
40

1   Because there would be no new transmission lines, transmission line operation and
2   maintenance, or alterations of rights-of-way, no changes to the level of impact on terrestrial
3   resources are expected to occur if additional power is transmitted through this system. Also,
4   the addition of the combination wet and dry cooling system for proposed Unit 3 and the dry
5   cooling system for proposed Unit 4 are not expected to adversely impact avian populations in
6   the vicinity of the North Anna ESP site. The staff concludes that the potential regional
7   cumulative impacts on terrestrial ecology contributed by the construction and operation of
8   Units 3 and 4 would be SMALL, and mitigation is not warranted.
9

7.5   Aquatic Ecosystem

11
12   The construction and operation of Units 3 and 4 were evaluated to determine if the potential
13   exists for interactions with past, present, and future actions that could contribute to adverse
14   cumulative impacts to aquatic resources. For the purpose of this analysis, the geographic area
15   of interest is the Lake Anna reservoir, the Waste Heat Treatment Facility (WHTF), and the
16   portion of the North Anna River downstream of Lake Anna dam. Environmental stressors
17   contributing to cumulative aquatic impacts include the operations of NAPS (with or without the
18   addition of Units 3 and 4), anthropogenic activities not directly related to NAPS (e.g., increased
19   urban development and recreational activity in or near the lake and river), and natural
20   environmental stressors (e.g., short- or long-term changes in precipitation or temperature, and
21   the resulting response of the aquatic community). The staff considered all of these sources of
22   impacts when evaluating the cumulative aquatic ecology impacts of Dominion's ESP
23   application.
24
25   The studies conducted by Dominion to establish compliance with sections 316(a) and 316(b) of
26   the Clean Water Act demonstrate that impingement and entrainment associated with the
27   operation of Units 1 and 2 have not resulted in a significant detectible adverse impact to fish
28   communities of Lake Anna. Thermal discharges from Units 1 and 2 have increased the overall
29   temperature of the WHTF and Lake Anna; however, heat-sensitive species are normally able to
30   find refuge in deeper parts of the lake, and heat shock or cold shock events have not resulted in
31   detectible changes to resident or stocked fish populations. Biocide releases from Units 1 and 2
32   currently comply with the Commonwealth of Virginia NPDES permit requirements, and will
33   continue to be monitored in the future. The addition of Units 3 and 4 is not expected to
34   noticeably contribute any aquatic impacts beyond those related to the operation of Units 1
35   and 2.
36
37   The closed-cycle cooling system proposed for Unit 4 uses very little water and would not result
38   in measurable impingement, entrainment, or discharge-related impacts. The combination wet
39   and dry cooling system proposed for Unit 3 would increase overall impingement and
40   entrainment at NAPS by approximately 1 percent. This is not expected to result in detectible
41   changes in the resident and introduced fish community in Lake Anna.

1   The thermal impacts of Unit 3 to the WHTF, Lake Anna, and the North Anna River are expected
2   to be minor because the temperature of the small volume of blowdown water released to the
3   WHTF is within the temperature range currently observed in the discharge canal. Operation of
4   Unit 3 would result in a decrease in the level in Lake Anna that would affect the downstream
5   release of water over the Lake Anna dam, especially during the summer months and drought
6   years. The decreases in lake level would not be expected to adversely impact aquatic organism
7   in the lake. However, the consumptive water use by Unit 3 may have some adverse impact on
8   aquatic communities downstream of the North Anna dam during summer months or drought
9   years. However, the aquatic communities of the North Anna River have adapted over time to
10   changes in stream flow, and it is expected that the overall impacts associated with Units 3 and 4
11   would be negligible.

12

13   Anthropogenic stressors not directly associated with NAPS activities may contribute to the
14   potential for adverse cumulative impacts to the lake and river. These impacts include habitat
15   loss and nonpoint pollution related to increased urbanization along the shores of the reservoir
16   and river, increased recreational use of the North Anna reservoir, impacts to the lake fishery
17   from changes in management practices or increased fishing pressure, and potential
18   downstream impacts from increased consumptive water use for human needs. Because the
19   Lake Anna reservoir is essentially a closed system that is regulated by Federal, State, and local
20   resource agencies, the staff assumes the cumulative impacts to the lake and river associated
21   with urbanization and increased resource use would be managed within the existing system of
22   rules and guidelines to ensure the aquatic communities are protected and the fishery resources
23   continue to be sustainable.

24

25   The presence of natural environmental stressors (e.g., short- or long-term changes in
26   precipitation or temperature) would contribute to the cumulative environmental impacts to Lake
27   Anna and the North Anna River. Because these impacts are not related to NAPS activities and
28   are difficult to predict, it is not possible to determine their contribution to cumulative impacts in
29   the study area. It is likely, however, that at certain times of the year, NAPS operations, other
30   anthropogenic stressors, and climactic events would combine to adversely impact the aquatic
31   populations of Lake Anna and the North Anna River. The staff expects that these events would
32   be of short duration and that the impacted resources would quickly recover after normal
33   patterns resume.

34

35   At present, Lake Anna represents a balanced community and supports a thriving population of
36   game fish and the forage species that support them. These communities either exist naturally
37   or are a product of management (stocking) strategies by State agencies. A diverse community
38   is also present in the North Anna River below the dam. Long-term monitoring has shown that
39   the lake and river communities are capable of adapting to changing environmental conditions
40   resulting from natural or anthropogenic sources of impact, and this adaptation is expected to

1   continue to occur with or without the addition of Units 3 and 4. Accordingly, the construction
2   and operation of proposed Units 3 and 4 is not expected to change the overall aquatic impacts
3   of NAPS.
4
5   Based on 25 years of aquatic monitoring data conducted by Virginia Power, there is no
6   evidence that Federally or State-listed threatened or endangered aquatic species are present in
7   Lake Anna or the North Anna River. Although aquatic species listed could occur in counties
8   adjacent to NAPS, there is no evidence that they have been observed or collected in those
9   locations. Based on this assessment, the staff concludes the cumulative impacts to threatened
10   or endangered aquatic species from the construction and operation of Units 3 and 4 are
11   SMALL, and mitigation is not warranted.
12
13   Consumptive water use for Unit 3 may exacerbate low-water conditions in Lake Anna and the
14   North Anna River during the summer months or droughts, but the overall impacts of Units 3
15   and 4 are not expected to be environmentally detectible or to contribute significantly to the
16   cumulative aquatic impacts that currently exist. The presence of anthropogenic or natural
17   stressors unrelated to NAPS operations currently influence the aquatic resources of Lake Anna,
18   the WHTF, and the North Anna River, and will continue to do so. These impacts would be
19   considered by Federal, State, and local regulatory agencies, and the various management
20   plans for the lake would be modified, as necessary. Based on the foregoing, the staff concludes
21   that the cumulative impacts of adding Units 3 and 4 would be SMALL, and mitigation is not
22   warranted.
23

24 ## 7.6 Socioeconomic, Historic and Cultural Resources,
25   **Environmental Justice**
26
27   Much of the analyses of the socioeconomic impacts presented in Sections 4.5 and 5.5 already
28   incorporate cumulative impact analysis because the metrics used for analysis only make sense
29   when placed in the total or cumulative context. The geographical area of the cumulative
30   analysis varies depending on the particular impact considered, and may depend on specific
31   boundaries, such as taxation jurisdictions, or distance, as in the case of environmental justice.
32   The construction and operation of Units 3 and 4 would not add any cumulative socioeconomic
33   impacts beyond those already evaluated in Sections 4.5 and 5.5. The staff concludes that
34   construction impacts would generally be SMALL, but there could be greater impacts if more
35   workers than expected settle in Louisa and Orange Counties, in which case MODERATE
36   impacts may be reached for physical impacts on roads, housing, and some public services.
37   In addition, during times of severe drought, the impacts to aesthetics and recreation during
38   operations may also reach MODERATE levels and there could be periodic MODERATE
39   aesthetic impacts from cooling tower plumes. In terms of beneficial effects, the impact on
40   regional economies and tax revenues would be beneficially SMALL to LARGE.

1 With regard to historic and cultural resources, construction and operation of Units 3 and 4 would
2 not add to any cumulative impacts to these resources beyond those identified in Sections 4.6
3 and 5.6. Dominion would implement the existing NAPS procedures to ensure that either known
4 or newly discovered potential historic and cultural sites would not be inadvertently impacted
5 during onsite activities that involve land disturbances (Dominion 2006). The staff concludes that
6 the cumulative impacts of construction and operation on historic and cultural resources would
7 be SMALL, and mitigation is not warranted.
8
9 The staff found no unusual resource dependencies or practices through which minority or
10 low-income populations would be disproportionately affected. As a result, cumulative impacts of
11 environmental justice would be SMALL.
12
13 Based on the above considerations, the staff concludes that under some circumstances,
14 construction and operation of Units 3 and 4 could make a detectable adverse contribution to the
15 cumulative effects associated with some socioeconomic issues under certain circumstances,
16 including aesthetics and recreation. The individual impacts range from MODERATE ADVERSE
17 to LARGE BENEFICIAL.
18

19 ## 7.7 Nonradiological Health
20
21 The cumulative impacts of construction and operation of the existing NAPS Units 1 and 2 and
22 the proposed North Anna Units 3 and 4 on the ambient temperature of Lake Anna with regard to
23 potential formation of thermophilic microorganisms was evaluated in Section 5.8.1. The
24 evaluation showed that the addition of two new units would not increase the temperature in
25 Lake Anna, and existing temperatures are not high enough to create an environment conducive
26 to the optimal growth of thermophilic organisms. Further, health risks to workers can be
27 expected to be dominated by occupational injuries at rates below the average U.S. industrial
28 rates and the impact of construction and operation of the Unit 3 cooling system would not
29 materially change the health risks. Noise, dust emissions, and acute and chronic
30 electromagnetic fields effects were also evaluated and found to have small impacts. The staff
31 concludes that the cumulative impacts resulting from construction and operation of Units 3 and
32 4 on nonradiological health would be SMALL, and mitigation is not warranted.
33

34 ## 7.8 Radiological Impacts of Normal Operation
35
36 The proposed increase in the power level from 4300 MW(t) to 4500 MW(t) per unit resulted from
37 the increase in power level for the ESBWR reactor design from 4000 MW(t) to 4500 MW(t). The
38 increase in the power level to 4500 MW(t) per unit results in a small increase to the liquid and
39 gaseous radiological effluent source terms for normal operation. The revised liquid and

1   gaseous effluent source terms result in small increases to the doses to the maximally exposed
2   individual as shown in revised Tables 5-7 and 5-8.
3
4   The dose from the existing Units 1 and 2 are well below regulatory limits and did not change as
5   a result of the power level increase for proposed Units 3 and 4. There is a small increase to the
6   maximally exposed individual as a result of the power level increase. This small increase to the
7   maximally exposed individual and the population dose estimates did not change the staff
8   conclusion that the cumulative radiological impacts are SMALL.
9
10   The radiological exposure limits and standards for the protection of the public and for
11   occupational exposures have been developed assuming long-term exposures, and therefore
12   incorporate cumulative impacts. As described in Section 5.9, the public and occupational doses
13   predicted from the operation of Units 3 and 4 would be well below regulatory limits and
14   standards. Specifically, the site boundary dose to the maximally exposed individual from the
15   existing Units 1 and 2 and the proposed Units 3 and 4 combined would be well within the
16   regulatory standard of 40 CFR Part 190. For purposes of this analysis, the geographical area is
17   the area included within an 80-km (50-mi) radius of the North Anna ESP site.
18
19   As stated in Section 2.5, Dominion has conducted a radiological environmental monitoring
20   program (REMP) around NAPS since 1976. The REMP measures radiation and radioactive
21   materials from all sources, including NAPS. The Commission would regulate any reasonably
22   foreseeable future actions that could contribute to cumulative radiological impacts. Therefore,
23   the staff concludes that the cumulative radiological impacts of operation of the proposed ESP
24   Units 3 and 4 and the existing operating NAPS Units 1 and 2 would remain SMALL, and
25   mitigation is not warranted.
26

7.9   Fuel Cycle, Transportation, and Decommissioning

28
29   *This section has been added since the Draft EIS was published and reflects the increase in fuel*
30   *use as presented in Revision 6 of the ER.*
31
32   The addition of the Units 3 and 4 on the North Anna ESP site would result in the need for
33   additional fuel. The impacts of producing this fuel include mining of the uranium ore, milling of
34   the ore, conversion of the uranium oxide to uranium hexafluoride, enrichment of the uranium
35   hexafluoride, fuel fabrication where the uranium hexafluoride in converted into uranium oxide
36   fuel pellets, and disposition of the spent fuel in a proposed Federal waste repository. As
37   discussed in Section 6.1 of the Draft EIS (NRC 2004), the environmental impacts of fuel cycle
38   activities for the proposed units would be a maximum of four times those presented in
39   Table S–3 of 10 CFR 51.51. Table S–3 provides the environmental impacts from uranium fuel
40   cycle operations for a model 1000-MW(e) LWR operating at 80 percent capacity with a
41   12-month fuel loading cycle and an average fuel burnup of 33,000 MWd/MTU. Per

1 10 CFR 51.51(a), the staff considers the impacts in Table S–3 to be acceptable for the 1000-
2 MW(e) reference reactor. As discussed in Section 6.1.1 of the Draft EIS, advances in reactors
3 since the development of Table S–3 impacts will have the effect of reducing environmental
4 impacts of the operating reference reactor. For example, a number of fuel management
5 improvements have been adopted by nuclear power plants to achieve higher performance and
6 to reduce fuel and separative work (enrichment) requirements. Fuel cycle impacts would occur
7 not only at the North Anna ESP site but would also be scattered through other locations in the
8 United States or, in the case of foreign-purchased uranium, in other countries. The staff
9 considers the cumulative fuel cycle impacts of operating NAPS Units 1 and 2 and the proposed
10 Units 3 and 4 for the 1000-MW(e) light-water scale model to be SMALL. Cumulative impacts for
11 other than light-water reactor designs are not resolved.
12
13 The addition of Units 3 and 4 would result in additional shipments of unirradiated fuel to the site
14 and additional shipments of spent fuel and waste from the site. Cumulative impacts would be
15 approximately twice that of the existing operating plants. Environmental impacts from
16 transportation of unirradiated fuel, spent fuel, and waste are found in Section 6.2 of this
17 environmental impact statement based on specific reactor types proposed for Units 3 and 4.
18 The following conclusions were derived from the staff's analysis of unirradiated fuel shipments:
19
20 • The number of unirradiated fuel shipments equates to less than one truck shipment per day
21   within criteria specified in Table S–4 of 10 CFR 51.52.
22
23 • Annual dose to workers and the public would be less than dose specified in Table S–4.
24
25 • Health impacts are projected to be small (i.e., less than 1×10^{-4} detriment/yr).
26
27 The following conclusions were derived from the staff's analysis of spent fuel: (1) after
28 accounting for conservative assumptions in the staff's evaluation, doses to the worker and the
29 public would be within criteria specified in Table S–4, and (2) health impacts from normal
30 conditions and accident conditions would be small (i.e., less than 0.1 detriment/yr). Regarding
31 transportation of waste shipments, the staff concluded that the normalized number of waste
32 shipments would be within the value specified in Table S–4 for the 1100-MW(e) reference
33 reactor. Cumulative impacts of transportation for operating both NAPS Units 1 and 2 and the
34 proposed Units 3 and 4 would be SMALL. Cumulative impacts for other than light-water reactor
35 designs are not resolved.
36
37 As discussed in Section 6.3 of this SDEIS and Section 6.3 of the Draft EIS (NRC 2004),
38 environmental impacts from decommissioning are expected to be small as the licensee would
39 have to comply with decommissioning regulatory requirements. In Supplement 1 to
40 NUREG-0586, *Generic Environmental Impact Statement on Decommissioning of Nuclear*
41 *Facilities*, the NRC found the impacts on radiation dose to workers and the public, waste

1 management, water quality, air quality, ecological resources, and socioeconomics to be small
2 (NRC 2002). However, because Dominion was not required to (and did not) submit information
3 regarding decommissioning in its ESP application, this issue is not resolved.
4

7.10 Staff Conclusions and Recommendations

7 The staff considered and evaluated the potential impacts resulting from construction and
8 operation of Units 3 and 4 together with past, present, and future actions in the North Anna ESP
9 site and surrounding area, including the changes presented in ER Revision 6. For several
10 impact areas, the staff concludes that the potential cumulative impacts resulting from
11 construction and operation are SMALL, and mitigation beyond the actions discussed in Sections
12 4.10 and 5.10 is not warranted. However, some areas have the potential for MODERATE
13 impacts, most of which would occur under temporary circumstances such as drought conditions
14 or as the result of a larger than expected concentration of construction workers settling near the
15 NAPS site. Further mitigation is not warranted because of the temporary nature of the impacts.
16

7.11 References

19 10 CFR Part 51. Code of Federal Regulations, Title 10, *Energy*, Part 51, "Environmental
20 Protection Regulations for Domestic Licensing and Related Regulatory Functions."
21

22 40 CFR Part 190. Code of Federal Regulations, Title 40, *Protection of Environment*, Part 190,
23 "Environmental Radiation Protection Standards for Nuclear Power Operations."
24

25 40 CFR Part 1508. Code of Federal Regulations. Title 40, *Protection of Environment*,
26 Part 1508, "Council on Environmental Quality, Terminology and Index."
27

28 Clean Water Act (also known as the Federal Water Pollution Control Act). 33 USC 1251 et seq.
29

30 Dominion Nuclear North Anna, LLC (Dominion). 2006. *North Anna Early Site Permit*
31 *Application – Part 3 – Environmental Report*. Revision 6, Richmond, Virginia.
32

33 Hanover County Department of Public Utilities. 2004. Letter to Chief, Rules and Directives
34 Branch, NRC., regarding Dominion's ESP application, January 7, 2004, Hanover, Virginia.
35

36 Louisa County. 2001. *The County of Louisa, Virginia Comprehensive Plan*. Comprehensive
37 Plan, Chapter V, September 4, 2001, Louisa, Virginia.
38

39 U.S. Nuclear Regulatory Commission (NRC). 1996. *Generic Environmental Impact Statement*
40 *for License Renewal of Nuclear Plants*. NUREG-1437, Volumes 1 and 2, Washington, D.C.

1   U.S. Nuclear Regulatory Commission (NRC). 2002. *Generic Environmental Impact Statement*
2   *on Decommissioning of Nuclear Facilities, Supplement 1 Regarding the Decommissioning of*
3   *Nuclear Power Reactors.* NUREG-0586, Washington, D.C.
4
5   U.S. Nuclear Regulatory Commission (NRC). 2004. *Draft Environmental Impact Statement for*
6   *an Early Site Permit (ESP) for the North Anna ESP Site.* NUREG-1811, Washington, D.C.

8.0 Impacts of the Alternatives

This chapter of the Supplement to the Draft Environmental Impact Statement was changed to reflect the higher power output level and an additional cooling system design alternative (i.e., once-through cooling) resulting from the cooling system changes proposed for Unit 3 in Revision 6 of the Environmental Report. In addition, this chapter is presented in its entirety to provide the analysis to support the comparisons of the proposed site to the alternative sites discussed in Chapter 9.

The purpose of this chapter of this Supplement to the Draft Environmental Impact Statement (SDEIS) is to examine the environmental impacts of alternatives to constructing and operating the proposed two nuclear units at the proposed North Anna early site permit (ESP) site. The U.S. Nuclear Regulatory Commission (NRC) staff considered the no-action alternative, system design alternatives, and the use of alternative sites. For the purposes of this SDEIS, the staff used the alternative sites selected by Dominion North Anna LLC (Dominion) in its ESP application (Dominion 2006). The results of the analysis described in this chapter were used and analyzed to determine whether any alternative site considered is obviously superior to the proposed site (described in Chapter 9).

Consideration of alternative sites involves a two-part examination as set forth in NUREG-1555, Section 9.3 (NRC 2000), in accordance with an NRC decision related to licensing the Seabrook Nuclear Power Plant (Public Service Company 1977). The first stage evaluates a full suite of environmental issues to determine whether any of the alternative sites is environmentally preferable to the proposed site. If not, then the evaluation of alternative sites ends at the first stage. If an alternative site appears environmentally preferable to the proposed site, the analysis proceeds to the second stage. The second stage of the test considers economic, technological, and institutional factors among the environmentally preferred sites to determine whether any alternative site that was considered is "obviously superior" to the proposed site. If there is no such obviously superior site, then the proposed site prevails; a staff conclusion that an alternative site is obviously superior to the proposed site would normally lead to a recommendation that the ESP application be denied.

Section 8.1 discusses the no-action alternative. Section 8.2 examines the station design alternatives. Section 8.3 reviews Dominion's region of interest (ROI) and examines the suitability of the ROI and Dominion's alternative site selection process, describing the method Dominion used to select the candidate and alternative sites. Section 8.4 examines issues that are common to all the sites, and addresses them collectively. Sections 8.5 through 8.7 individually evaluates the selected alternative sites. Section 8.8 provides a summary of alternative site impacts, and Section 8.9 cites the references relevant to this chapter.

8.1 No-Action Alternative

For this ESP application, the no-action alternative refers to a scenario in which the NRC would deny the ESP request. Upon such a denial, the construction and operation of new nuclear power reactors at the proposed North Anna ESP site in accordance with the Title 10 of the Code of Federal Regulations (CFR) Part 52 process would referencing an approved ESP not occur.

The no-action alternative consists of two parts. First, the no-action alternative would include a scenario in which the NRC would not issue the ESP. There are no environmental impacts associated with not issuing the ESP except that the impacts associated with site preparation and preliminary work, allowed pursuant to 10 CFR 52.17(c) and 10 CFR 52.25(a), would be avoided; changes to the Unit 3 cooling approach and the higher power level for proposed Units 3 and 4 (referred to hereafter as Units 3 and 4) would not change the impacts of the no-action alternative. Second, given that the EIS addresses the environmental impacts of construction and operation as directed by the Commission (10 CFR 52.18(a)(2)), the no-action alternative would result in no such construction and operation. Therefore, the impacts predicted in this EIS would not occur. Nonetheless, Part 52 does not require an ER or EIS for an ESP to include consideration of the benefits of construction and operation of a reactor or reactors at the ESP site (see 10 CFR 52.17(b)), nor does it require such an ER or EIS to include consideration of alternative energy sources (see Exelon Generation Co., LLC et al., CLI-05-17, 62 NRC 5 (2005)). Dominion did not include these matters in its ER, and the DEIS likewise did not consider them. Accordingly, should the NRC ultimately determine to issue an ESP for the North Anna ESP site, and a CP or COL application that references such an ESP is docketed, these matters will be considered in the EIS prepared in connection with the review of that CP or COL application.

However, the no-action alternative would not achieve the benefits intended by the ESP process, which would include (1) early resolution of siting issues prior to large investments of financial capital and human resources in new plant design and construction, (2) early resolution of issues on the environmental impact of construction and operation of reactors that fall within the site parameters, (3) the ability to bank sites on which nuclear plants may be located, and (4) facilitation of future decisions on whether to build new nuclear plants.

8.2 System Design Alternatives

Sections 8.2.1 through 8.2.3 contain information regarding alternative plant cooling systems for the proposed Unit 3 at the North Anna ESP site. Section 8.2.1 discusses once-through cooling system; Section 8.2.2 discusses wet cooling system heat-dissipation systems; and Section 8.2.3 discusses dry cooling system heat dissipation systems for Unit 3. A dry cooling tower has been proposed for Unit 4 at the North Anna ESP site. Water and energy balance studies of Lake Anna suggest that the lake would not support a once-through cooling system, a wet cooling tower heat dissipation system, or a combination wet and dry cooling system for Unit 4. Refer to Appendix K for more detail on the water budget analysis. Therefore, neither of these alternatives is considered for Unit 4 at the North Anna ESP site.

The purpose of the plant cooling system is to dissipate heat to the environment. The various cooling system options differ in how and where the heat transfer takes place and, therefore, have different environmental impacts. In the closed-cycle, combination wet and dry cooling system proposed for Unit 3, heat is transferred to the atmosphere through evaporation, long-wave radiation, and conduction. With the wet tower portion of the system, only a fraction of the water withdrawn from the lake is returned as blowdown, with the majority being evaporated. The dry tower portion of the system consumes a negligible amount of water.

8.2.1 Plant Cooling System: Unit 3 Once-Through Cooling System

A once-through cooling system for Unit 3 would transfer heat to the atmosphere and aquatic environment of the Waste Heat Treatment Facility (WHTF), Lake Anna, and the North Anna River downstream of the dam by convection, evaporation, long-wave radiation, and conduction. As described below, when compared to the proposed design, increased impingement, entrainment, circulation changes in the WHTF and Lake Anna, temperature in the aquatic environment, and consumptive use of water would result from the once-through design.

A once-through cooling system design would withdraw a larger volume of water from Lake Anna through the intakes, resulting in greater impingement and entrainment. The once-through design for Unit 3 was estimated to withdraw 71,900 L/s (1,140,000 gpm) compared to the maximum of 1405 L/s (22,269 gpm) for the proposed Unit 3 closed-cycle design operating in Energy Conservation (EC) mode. The additional recirculation from this flow combined with the existing recirculation in Lake Anna could further erode the limited water volume below the lake thermocline that may be an important refugium for certain fish populations in Lake Anna.

A once-through design would initially transfer all the reject heat to the aquatic environment. The increased heat load would push warm water out of the WHTF further into Lake Anna. The staff estimated that WHTF-type conditions would be extended into about 19 percent of the volume of Lake Anna. This could reduce the productivity of certain fish populations in Lake Anna that are

1 sensitive to temperature. Some of the heat entering the WHTF and Lake Anna would be lost to
2 the atmosphere through evaporation. This additional evaporation resulting from the increased
3 in lake surface temperatures would reduce the total water supply. Dominion had estimated that
4 induced evaporation from once-through cooling could result in water loss at a rate of 0.79 m^3/s
5 (28 cfs), whereas the design proposed in Revision 6 of the Environmental Report (ER) would
6 have induced evaporative losses at a lesser rate of about 0.57 m^3/s (20 cfs).
7
8 The staff evaluated mitigation for the once-through cooling system to reduce cooling water
9 discharge temperatures into Lake Anna. Wet mechanical draft cooling towers could be
10 employed as helper towers along with the once-through cooling design on an as-needed basis
11 during the late summer and early fall. Use of these towers in a helper mode would reduce the
12 station discharge temperature to Lake Anna but would result in an increase in consumptive
13 water use that would be greater than the combination wet and dry cooling system. Based on
14 the combination wet and dry cooling tower system's expected smaller impact on the aquatic
15 environment, the staff concludes that a combination wet and dry cooling system for Unit 3 would
16 be preferable to a once-through cooling system.
17
18 ## 8.2.2 Plant Cooling System: Unit 3 Wet Cooling System
19
20 Wet, mechanical, and natural draft cooling towers transfer heat to the atmosphere through
21 evaporation and conduction. Assuming all the heat transfer is through evaporation, a wet
22 cooling design would consume more water than either the once-through design or the
23 combination wet and dry cooling system proposed in ER Revision 6 (Dominion 2006). The
24 increased use of makeup water requirements for a wet cooling design would increase
25 impingement and entrainment slightly over the proposed design.
26
27 The use of a wet cooling tower design versus the proposed combination wet and dry cooling
28 system design for Unit 3 would increase water withdrawals from Lake Anna. The impact of the
29 increased evaporative losses of a wet cooling tower design would be particularly noticeable
30 during drought years. The results of water balance calculations suggest that the use of an wet
31 cooling tower system for the 2001 through 2003 critical water period would have resulted in an
32 additional 1.0 m (3.4 ft) drawdown of the lake in September 2002. In comparison, use of the
33 proposed combination wet and dry cooling would only have drawn the lake down by an
34 additional 0.5 m (1.6 ft). The use of a wet cooling tower design would also prolong the duration
35 of low-flow conditions downstream of the dam. The staff concludes that based on the expected
36 smaller impact on the lake level and downstream flows, a combination wet and dry cooling
37 system design for Unit 3 is preferable to a wet cooling tower design.
38

8.2.3 Plant Cooling System: Unit 3 Dry Cooling System

The use of a dry cooling design versus the proposed combination wet and dry cooling system design for Unit 3 would largely eliminate the impacts on aquatic biota in Lake Anna and the North Anna River downstream. The lake would not be heated by rejected heat from Unit 3, and there would be no additional consumptive water use.

A dry cooling tower designed to dissipate heat may reduce water-related impacts of operating Unit 3, but it also has some disadvantages. In particular, dry cooling systems are more expensive to build and are not as efficient as wet cooling systems. To achieve the necessary cooling, dry systems move of a large amount of air through a heat exchanger, and the fans that force the air through the heat exchanger use a significant amount of power. Dominion estimates that the power needed to operate dry cooling towers would be 8.5 to 11 percent of the plant power output (Dominion 2006). The power needed to operate a dry tower for Unit 3 would be about 150 MW(e). This power demand reduces the net power output of the plant. The power needed for operating the combination wet and dry cooling system would be 1.7 to 4 percent. The fans and the large volume of air required for cooling result in elevated noise levels. Nevertheless, the noise levels at the site boundary associated with the dry towers would likely be bounded by noise levels resulting from the operation of the North Anna Power Station (NAPS). The dry cooling tower would also occupy more land than a once-through or wet tower cooling system.

The staff concludes that based on its analysis that Lake Anna could support Unit 3 using a combination wet and dry cooling system and given the environmental impact of increased use of resources needed by using a less efficient dry cooling system, a combination wet and dry cooling system is preferable to a dry cooling system for Unit 3.

8.3 Alternative Sites, Region of Interest, and Selection and Evaluation Process

The power level increase affects the North Anna ESP site and the alternative sites equally. The cooling system design change affects only the North Anna ESP site. NRC regulations require that the ER submitted in conjunction with an application for an ESP include an evaluation of alternative sites to determine whether there is an obviously superior alternative to the site proposed (10 CFR 52.17(a)(2)). This section includes subsections discussing Dominion's Region of Interest (ROI) for selecting alternative sites and its alternative site-selection process. The three alternative sites examined in detail in this SDEIS are Dominion's Surry Power Station (Surry) site in Surry County, Virginia; the U.S. Department of Energy's (DOE) Portsmouth Gaseous Diffusion Plant (Portsmouth) site in Pike County, Ohio; and DOE's Savannah River Site, which is in Aiken and Barnwell Counties, South Carolina.

1 Dominion stated that the two DOE sites were selected as candidate sites because:
2
3 • The sites represent valuable national assets with prior or existing nuclear energy
4   potential.
5
6 • New nuclear power facilities would represent potentially promising new missions for
7   these sites.
8
9 • The sites have the potential to support reactor demonstrations and/or commercial
10   reactor development.
11
12 • There is extensive site information and an available infrastructure that could help to
13   reduce site development costs.
14
15 • The partially or fully developed site environment and the available infrastructure reduces
16   the incremental environmental impacts associated with the new plant construction and
17   operation on land use, ecological resources, aesthetics, and local transportation
18   networks.
19
20 • The sites are not in proximity to major population centers (Dominion 2006).
21
22 The Surry site was selected by Dominion as an ESP candidate site because:
23
24 • The existing environmental conditions and the environmental impacts are known from
25   data collected during years of monitoring air, water, ecological, and other parameters.
26
27 • Construction of new transmission line rights-of-way may potentially be avoided if the
28   existing transmission system (lines and rights-of-way) can accommodate the increased
29   power generation.
30
31 • No additional land acquisitions would be necessary if a new transmission line can be
32   avoided, and the resulting land-use impacts of the new plant would be small.
33
34 • The Surry site was recently subjected to an environmental review process during its
35   license renewal review.
36
37 • The Surry site had extensive environmental studies performed during the original
38   site-selection process, which could be updated and used for new units.
39
40 • Site physical criteria, including primarily geologic/seismic suitability, have been
41   characterized.
42

- Plant construction, operation, and maintenance costs would be reduced because of existing site infrastructure (e.g., roads, transmission line rights-of-way, water source, and intake/discharge system).

- The Surry site has nearby power markets.

- The Surry site has local community acceptance and support (Dominion 2006).

NRC's environmental review guidance for alternative nuclear plant sites recognizes that there will be special cases in which the proposed site was not selected on the basis of a systematic site-selection process, but was selected on the basis of environmentally acceptable operating experience at the site or because the site was previously found acceptable on the basis of a National Environmental Policy Act of 1969 (NEPA) review. In such cases the NRC will analyze the applicant's site-selection process only as it applies to the alternate sites. The site comparison may then be restricted to a site-by-site comparison of the alternate sites with the proposed site (NRC 2000).

8.3.1 Dominion's Region of Interest

The ROI is the geographic area considered in searching for candidate ESP sites. More specifically, the ROI is:

The geographical area initially considered in the site selection process. This area may represent the applicant's system, the power pool or area within which the applicant's planning studies are based, or the regional reliability council or the appropriate subregion or area of the reliability council (NRC 1999, 2000).

In its ESP application, Dominion selected its ROI for examining potential sites as the Mid-Atlantic, Northeast, and Midwest regions of the United States. These regions were selected because of Dominion's interest in continuing to grow and operate deregulated marketplaces in the region (Dominion 2006). The staff determined that Dominion's basis for deferring its ROI did not arbitrarily exclude desirable candidate areas. Within this ROI, Dominion used the candidate site criteria identified by NRC (NRC 1999) to identify candidate sites (Dominion 2006). The staff concludes that the ROI used by Dominion in its ESP application is appropriate for consideration and analysis of potential ESP sites because it consistent with the major load centers to be supplied by the proposed plant and that desirable candidate area have not been excluded on the basis of an arbitrarily defined ROI.

8.3.2 Dominion's Alternative Selection Process

Dominion evaluated its proposed North Anna ESP site and the three alternative sites using 45 site suitability/screening criteria (Dominion 2006). Dominion reviewed the alternative site

1 selection evaluation to determine the impact of the North Anna site merit score. Dominion
2 determined that the changes to the cooling system design had minimal impact on the North
3 Anna site ranking versus the alternative sites and did not affect its overall conclusion
4 (Dominion 2006). The increase in power level affects all sites equal and did not affect the
5 relative ranking of sites. The criteria were grouped into four major categories: (1) economic,
6 (2) engineering, (3) environmental, and (4) socioeconomic (see Table 8-1). The economic
7 category was given a relative weight of 40 percent by Dominion, and the other three categories
8 were weighted 20 percent each. A ranking or score for each of the 45 criteria was assigned by
9 Dominion (from 0 to 5, with 5 being the most favorable). The relative importance of each
10 criterion to the overall evaluation was established by assigning weights that reflected the
11 collective judgment of Dominion's experts involved in the process. The sum of the weighted
12 scores for all criteria represented a total site merit score. The preferred site was chosen based
13 on the highest site merit score. Based on its study, Dominion found the North Anna site to be
14 the preferred ESP site followed by the Savannah River, Portsmouth, and Surry sites,
15 respectively (Dominion and Bechtel 2002). (In this report, Dominion and Bechtel evaluated a
16 fourth site, the Idaho National Engineering and Environmental Laboratory (now known as the
17 Idaho National Laboratory) near Idaho Falls, Idaho, but Dominion did not consider it in this ESP
18 action because it was outside the ROI.) Accordingly, Dominion submitted its ESP application
19 for the North Anna site. However, Dominion concluded that all four sites are suitable locations
20 for deployment of new nuclear power plants.
21
22 Among the issues reviewed by Dominion in its site selection process were cooling water use,
23 ground water, aquatic and terrestrial resources transmission lines, socioeconomic, land use, air
24 quality, and population density. The range of issues evaluated by Dominion in its site selection
25 process was sufficient for the staff to use to determine that Dominion had employed a
26 reasonable site selection processthat resulted in identifying reasonable alternative sites within
27 Dominion's ROI.
28
29 ### 8.3.3 NRC's Evaluation of Alternative Sites
30
31 The staff independently performed an evaluation of the alternative sites identified by Dominion,
32 i.e., the Surry, Portsmouth, and Savannah River sites. Because the sites were evaluated at an
33 overview level using readily available information rather than using the more detailed approach
34 applied to the North Anna ESP site, which included independent analysis and modeling as
35
36 necessary, this evaluation is viewed as a "reconnaissance" level evaluation. All three
37 alternative sites previously had been characterized by their operators, but the basis for these
38 characterizations was not specific to an ESP action. In the *Study of Potential Sites for the*
39 *Deployment of New Nuclear Plants in the United States*, Dominion and Bechtel evaluated DOE
40 sites at Portsmouth, Savannah River, and Idaho Falls along with the Dominion Surry site
41
42

Table 8-1. Dominion Site Screening Criteria

Economic	Engineering	Environmental	Socioeconomic
Electricity Projections	Site Size	Terrestrial Habitat	Present/Planned Land Use
Transmission System	Site Topography	Terrestrial Vegetation	Demography
Stakeholder Support	Environmentally Sensitive Areas	Aquatic Habitat/ Organisms	Socioeconomic Benefits
Site Development Costs	Emergency Planning	Groundwater	Agricultural/Industrial
	Labor Supply	Surface Water	Aesthetics
	Transportation Access	Population	Historic/Archaeological
	Security		Transportation Network
	Hazardous Land Use		Environmental Justice
	Ease for Decommissioning		
	Water Rights and Air Quality Permits		
	Regulatory		
	Schedule		
	Geologic Hazards		
	Site-Specific Safe Shutdown Earthquake		
	Capable Faults		
	Liquefaction Potential		
	Bearing Material		
	Near-Surface Material		
	Groundwater		
	Flooding Potential		
	Ice Formation		
	Cooling Water Source		
	Temperature and Moisture		
	Winds		
	Rainfall		
	Snow		
	Atmospheric Dispersion		

with relation to economic, engineering, environmental, and sociological factors using preliminary advanced reactor design information (Dominion and Bechtel 2002). This report, funded by DOE, concluded that the three DOE sites and the two Dominion sites (North Anna and Surry) were suitable for potentially siting new nuclear power plants. This report also concluded that the North Anna site ranked highest overall of the sites evaluated and scored slightly higher with regard to environmental issues than the Surry, Portsmouth, and Savanna River sites.

1 Dominion reviewed the alternative site selection evaluation to determine the impact of the
2 change to the cooling system on North Anna site merit score. Dominion determined that the
3 changes to the cooling system design had minimal impact on the North Anna site ranking
4 versus the alternative sites and did not affect its overall conclusion (Dominion 2006). The
5 increase in power level affects all sites equally and did not affect the relative ranking of sites.
6
7 In its evaluation of the alternative sites, the staff toured each of sites and discussed topics
8 specifically relevant to the ESP evaluation process including potential sources of cooling
9 water, transmission line access, local ecology, and socioeconomics with site experts. The
10 staff relied primarily on direct observation, information provided by the site experts, DOE
11 environmental reviews on the Portsmouth and Savannah River Federal sites, and site-specific
12 information provided in the Dominion and Bechtel report for DOE (Dominion and Bechtel
13 2002). In the case of the Surry alternative site, the staff relied on the *Generic Environmental*
14 *Impact Statement for License Renewal of Nuclear Plants: Supplement 6 Regarding Surry*
15 *Power Station, Units 1 and 2* for much of the site background documentation (NRC 2002).
16 The staff also reviewed reports from relevant State and Federal agencies and other regional
17 information sources in evaluating the alternative sites.
18
19 ### 8.3.4 Greenfield and Brownfield Alternative Sites
20
21 Dominion also considered other existing nuclear power plant, greenfield, and brownfield sites
22 within the ROI. In as much as sites of current nuclear facilities have space for additional units,
23 the greenfield and brownfield sites were determined not to be environmentally preferable
24 because of the large land area that would need to be disturbed to build a new plant and to
25 support necessary transmission line rights-of-way. The associated land use, ecological
26 resource impacts, and the aesthetic impacts were determined to be large in comparison to
27 impacts at alternative sites with existing nuclear power plants.
28
29 The staff reviewed Dominion's alternative site-selection process as it applies to greenfield and
30 brownfield sites and concludes that the approach used and the findings of impacts are
31 reasonable.
32
33 ## 8.4 Generic Issues Consistent Among Alternative Sites
34
35 The power increase for Units 3 and 4 is consistent among the proposed and alternative sites.
36 In evaluating the alternative sites, the NRC staff found that certain impact areas would not
37 vary significantly among sites, and as a result, would not affect the evaluation of whether an
38 alternative site is environmentally preferable to the proposed site. These impact areas include
39 air quality, nonradiological health, and radiological health during construction and operations
40 for members of the public and workers, and radiological health during construction and
41 operations for biota. The staff evaluated the fuel cycle impacts for the proposed and alternate

sites in Chapter 6 and Appendix G of this EIS and found the impacts to be SMALL at all sites. Decommissioning impacts were analyzed in Chapter 6 for all sites and were determined to be unresolved, because the reactor design has not been selected at the ESP stage. The impacts from decommissioning are likely to affect all sites equally. In Chapter 5 the staff concluded that severe accident mitigation alternatives (SAMAs) are unresolved for North Anna ESP site, because the reactor design is not known at the ESP stage. SAMAs are also unresolved at all the alternative sites for the same reason. The analysis of SAMAs is likely to show the same result at all sites. In addition, the impacts to public service facilities (e.g., schools, water, and wastewater treatment) would not materially impact whether an alternative site is selected or not. As a result, air quality, health impacts, and radiation exposures are not evaluated as part of the site-specific alternatives analysis, but rather are discussed generically in the following sections.

8.4.1 Air Quality Impacts

During construction at any of the proposed alternative ESP sites, it is expected that some minor air quality impacts would occur in terms of fugitive dust emissions from general construction activities and the potential for elevated ambient levels of criteria pollutants caused by automotive emissions from the workforce traffic and emissions from construction equipment. The criteria pollutants of concern would be particulate matter less than 10 microns in diameter (PM_{10}), reactive organic gases, oxides of nitrogen, carbon monoxide, and sulfur dioxides from combustion engines of the construction equipment.

Air pollutants and fugitive dust would be emitted from operations of construction equipment and earth-moving and material-handling activities, respectively. In addition, operation of other equipment for hauling debris, equipment, and supplies on unpaved roads will produce fugitive dust emissions. Estimation of direct and indirect emissions is beyond the scope of the reconnaissance-level information. However, all activities would be conducted in accordance with State air quality agency requirements for visible and fugitive dust emissions as well as emission standards for mobile sources. If the Surry site were chosen as the alternative site, the same requirements set forth for the North Anna ESP site would apply because both are in the Commonwealth of Virginia. If the Savannah River Site were chosen, then requirements established by the South Carolina Department of Health and Environmental Control would apply. The Ohio Environmental Protection Agency would be consulted if the Portsmouth site were chosen. In addition, if construction activities include burning of construction materials, a permit would need to be secured from the State, and Dominion would need to contact local county officials to determine which local ordinances, if any, must be followed.

Dominion estimated that during construction activities, approximately 5000 workers would be divided between two 10-hr shifts (Dominion 2006). Using an assumption of 1.8 workers per vehicle, this would represent 2800 additional vehicles per day traveling on roads into and out of the proposed site (Dominion 2006). For any of the proposed alternative sites, the estimate

1   of work is similar to that proposed for the North Anna ESP site. Some roadways leading into
2   the site chosen may or may not experience congestion. This situation will impact the local
3   ambient air quality because of emissions from vehicles both during normal operation and
4   during congestion periods when vehicles are idling. However, because the current ambient
5   air quality pollutant levels at the proposed alternative sites are well below current national
6   standards, the resulting impact is estimated to be insignificant and would not create an air
7   quality impact. Therefore, the staff concludes that air quality impacts from construction at any
8   of the alternative sites would be SMALL, temporary and similar to those at the proposed site.
9

10   The meteorological and air quality impacts would be limited to additional nonradiological
11   pollutants during the operation of the wet cooling portion of the combination wet and dry
12   cooling system, auxiliary boilers, emergency generators, and emissions from onsite service
13   vehicles. The amount of pollutants emitted to the atmosphere is anticipated to be less than
14   [metric equivalent] (100 tons/yr) for any alternative site (Dominion and Bechtel 2002) and is
15   considered insignificant. However, Dominion would require approval under the existing
16   Federal, State, or local air quality laws and regulations on new sources for any activities
17   undertaken.
18

19   The current status of compliance regarding criteria pollutants in the regions surrounding the
20   alternative sites is the following (EPA 2005; 70 FR 30396, 70 FR 33771):
21

22   • Various areas around the Surry site have been designated as non-attainment areas
23   regarding the new U.S. Environmental Protection Agency (EPA) 8-hr ozone level
24   requirements. This includes James City and Isle of Wight Counties along with the City
25   of Williamsburg.
26

27   • No non-attainment areas were identified in the counties near the Portsmouth site.
28

29   • None of the counties around the Savannah River Site have been designated as
30   non-attainment. However, an Early Action Compact (EAC) was developed among the
31   counties in the region called the Lower Savannah Area; this includes Aiken, Allendale,
32   and Barnwell Counties that surround the Savannah River Site as well as Columbia and
33   Richmond counties in Georgia. The concept of EACs was developed by the EPA for
34   those areas that were classified as non-attainment with regard to the new 8-hr ozone
35   criteria pollutant level, to delay official designation as non-attainment and allow the
36   areas to develop their own means to achieve attainment. Some EACs, such as the
37   Lower Savannah Area EAC, were established and are participating in the EAC review
38   and evaluation process to demonstrate their support of cleaner air statewide including
39   ozone pollutant levels.
40

41   Although there are some existing air quality issues near the Surry and Savannah River
42   alternative sites, this is not expected to be a limiting factor in considering these sites, and the

staff concludes that the air quality impacts from operation at any of the alternative sites would be SMALL and similar to those at the proposed site.

8.4.2 Nonradiological Health Impacts

Nonradiological health impacts from construction of the proposed nuclear power plants on the construction workers at all the alternative sites would be similar to those evaluated in Section 4.8. They would include noise, odor, vehicle exhaust, and dust emissions. During the plant construction phase, activities would comply with applicable State regulations regarding fugitive dust emissions and air pollution control. The alternative sites considered by Dominion are in rural areas, and construction impacts on the surrounding population would be minimal. Accordingly, the staff concludes that health impacts to construction workers resulting from the construction of two new units at any of the alternative sites would be SMALL.

Occupational health impacts to operational employees would be expected to be the same for all the alternative sites. Thermophilic microorganisms would not be a concern at alternative sites for any facilities using either a wet or a combination wet and dry cooling process because the temperatures in the water bodies receiving the cooling system discharges are below those known to be conducive to the optimal growth and survival of thermophilic pathogens. Health impacts to workers from noise and electromagnetic fields would be similar among the sites. Noise and electromagnetic fields would be monitored and controlled in accordance with applicable Occupational Safety and Health Administration regulations. Based on the foregoing, the staff concludes that the occupational health impacts to construction or operations employees of proposed units at any of the alternative sites would be SMALL.

With respect to transmission systems, the potential exists for impacts to members of the public from operation of the transmission system in terms of electrical shock, electromagnetic field exposure, noise, and aesthetics. The impacts at the alternative sites are expected to be similar to those evaluated in Section 5.8.

8.4.2.1 Acute Effects of Electromagnetic Fields

All transmission lines, either constructed or used as part of an existing nuclear site, are designed to standards established by the most current version of the National Electrical Safety Code (NESC) (IEEE 2001), which is the standard that is applicable to the systems and equipment operated by utilities. The areas of particular concern are (1) the potential to create an electric shock that could disrupt the operation of pacemakers and health assistance devices and (2) the potential for chronic exposure to electromagnetic fields associated with the transport of electric current through large conductors, such as high-voltage transmission lines.

1   Currently, to limit the potential for electric shock, NESC requires transmission lines to be
2   designed so that electrostatic effects from operation do not create a steady-state current that
3   exceeds 5 mA root mean square. For the alternative sites considered, it is likely that NESC
4   requirements for preventing electric shock from induced current would be met, and the impact
5   to the public would be insignificant.
6
7   **8.4.2.2 Chronic Effects of Electromagnetic Fields**
8
9   There has been considerable debate in scientific circles regarding the potential impact from
10   exposure to 60-Hz electromagnetic fields resulting from energized transmission lines. The
11   potential for chronic effects from these fields continues to be studied and consensus results
12   are still outstanding. The National Institute of Environmental Health Sciences (NIEHS) directs
13   related research through the DOE. A recent NIEHS report contains the following conclusion
14   (NIEHS 1999):
15
16   The NIEHS concludes that ELF-EMF (extremely low frequency-electromagnetic field)
17   exposure cannot be recognized as entirely safe because of weak scientific evidence that
18   exposure may pose a leukemia hazard. In our opinion, this finding is insufficient to
19   warrant aggressive regulatory concern. However, because virtually everyone in the
20   United States uses electricity and is exposed to ELF-EMF, passive regulatory action is
21   warranted such as a continued emphasis on educating both the public and the regulated
22   community on means aimed at reducing exposure. The NIEHS does not believe that other
23   cancers or non-cancer health outcomes provide sufficient evidence of a risk to currently
24   warrant concern.
25
26   This statement is not sufficient to cause the staff to consider the potential impact as significant
27   to the public. In any event, the impacts would be similar at the proposed site and any of the
28   alternative sites.
29

30   **8.4.3 Radiological Health Impacts**
31
32   Exposure pathways for gaseous and liquid effluents from the proposed new Units 3 and 4 at
33   the North Anna ESP site would be similar for the alternative locations. Gaseous effluent
34   pathways would include external exposure to the airborne plume, external exposure to
35   contaminated soil, inhalation of airborne activity, and ingestion of contaminated agricultural
36   products. Liquid effluent pathways would include ingestion of aquatic foods, ingestion of
37   drinking water, external exposure to shoreline sediments, and external exposure to water
38   through boating and swimming.
39

1 **8.4.3.1 Radiation Doses and Health Impacts to Members of the Public**
2
3 Section 5.9 of the SDEIS provides an estimate of doses to the maximally exposed individual
4 and the general population at the North Anna ESP site for both the liquid effluent and gaseous
5 effluent pathways during operation. The same bounding liquid and gaseous effluent releases
6 would be used to evaluate doses to the maximally exposed individual and the population at
7 each alternative site. However, there would be differences in the estimated doses at each of
8 the sites. The differences would result from the use of site-specific atmospheric and water
9 dispersion data, different exposure pathways, and site-specific population data for the dose
10 calculations.
11
12 Section 5.9 shows that the estimated dose to the maximally exposed individual at the North
13 Anna ESP site would be well within the design objectives (10 CFR Part 50, Appendix I).
14 Considering the differences in pathways analyzed, atmospheric and water dispersion factors
15 and population, doses estimated to the maximally exposed individual for the alternative sites
16 would also be expected to be well within the 10 CFR Part 50, Appendix I design objectives.
17 Population dose within 80 km (50 mi) of these alternative sites would be expected to be small
18 compared to the population dose from natural background radiation.
19
20 Based on the forgoing, the staff concludes that the proposed system would likely result in
21 annual doses to the public well within regulatory limits, and there would be no observable
22 health impact to the public from construction or normal operation of the proposed North Anna
23 ESP facility or from any of the alternative sites. Therefore, the staff concludes that radiation
24 doses and resultant health impacts from construction or operation of the proposed new
25 nuclear units at the alternative sites would be SMALL.
26
27 **8.4.3.2 Occupational Doses to Workers**
28
29 Doses to construction workers during construction of the two proposed units were estimated
30 and compared against the requirements in 10 CFR Part 20. These doses were well below
31 limits for members of the public. In addition, annual collective doses were estimated and
32 appeared realistic and reasonable. Occupational doses to workers during construction would
33 be expected to be approximately the same for the alternative sites as for the proposed North
34 Anna ESP site. Therefore, the staff concludes that health impacts from radiological doses to
35 construction workers would be SMALL.
36
37 Occupational doses to workers during operations would be expected to be approximately the
38 same for the proposed ESP facilities at the alternative sites as for the North Anna ESP site.
39 The same (accumulated) annual occupational dose estimates of 1.5 person-Sv
40 (150 person-rem) would be expected for all the proposed units regardless of the site location.
41 The staff concludes that the occupational radiation doses from operation of the proposed units
42 at the alternative sites would be SMALL.

8.4.3.3 Impacts to Biota

Table 5-12 provides the annual whole body dose estimates to surrogate biota species for the two proposed units at the North Anna ESP site. The staff reviewed the available information relative to the radiological impact on biota, other than humans, and performed an independent estimate of dose to the biota. The staff concludes that no measurable radiological impact on populations of biota would be expected from the radiation and radioactive material released to the environment as a result of the construction or routine operation of the proposed units, or of operation at any of the alternative sites. The staff concludes that the impacts to biota of radiation doses from the construction or operation of the proposed units at the alternative sites would be SMALL.

8.4.4 Postulated Accidents

A suite of design basis accidents (DBAs) has been considered for the new nuclear units at the North Anna ESP site. The evaluation involved calculation of doses for specified periods at the exclusion area and low population zone boundaries, and comparison of those doses with doses based on regulatory limits and guidelines. Similar analyses have not been conducted for the alternative sites. Had such evaluations been conducted, differences in the results would be expected to be caused by differences in meteorological conditions and distances to the site boundaries. The release characteristics would be similar at all sites because the reactor designs are the same.

For the North Anna ESP site meteorology, the doses for each accident sequence considered were well below the corresponding regulatory limits and guidelines. Because the general climatological conditions at the North Anna ESP site are sufficiently similar to the conditions at the alternative sites, it is highly unlikely that differences in local meteorological conditions would be sufficient to cause doses from DBAs for new nuclear units at any of the alternative sites to exceed regulatory limits or guidelines. Similarly, because each of the alternative sites is located at a nuclear facility (although not a necessarily a nuclear reactor site), it is unlikely that differences in distances to the exclusion area and low population boundaries would be sufficient to cause doses from DBAs for new nuclear units at any of the alternative sites to exceed regulatory limits or guidelines. Similarly, doses at the alternative site are unlikely to be significantly lower than doses estimated at the North Anna ESP site. Therefore, the staff concludes that for the purposes of consideration of alternative sites, the impact of DBAs at each of the alternative sites would be SMALL.

A detailed analysis of the potential consequences of severe accidents for the postulated plants has been conducted for the North Anna ESP site. Similar analyses have not been conducted for the alternative sites. Had such evaluations been conducted, the differences in the results would likely have been limited to site-specific factors such as meteorological conditions, population distribution, and land-use distribution. The release characteristics would be similar at all sites because the reactor designs would be the same.

The probability-weighted consequences estimated for severe accidents for the proposed units at the North Anna ESP site would be well below the consequences estimated for severe accidents at current generation reactors (see Section 5.10). This result suggests that, as at the North Anna ESP site, the consequences of severe accidents at any of the alternative sites would be less than the consequences of a severe accident for a current-generation operating plant at each site. These risks are well below the NRC safety goals. In addition, the Commission has determined that the probability-weighted consequences of severe accidents is SMALL for all existing plants (10 CFR Part 51, Subpart B, Table B-1). On this basis, the staff concludes that for the purposes of consideration of alternative sites, the impact of severe accidents at each of the alternative sites would be SMALL.

8.5 Evaluation of Surry Power Station Site

The Surry site, operated by Dominion, was recently evaluated in a supplemental EIS prepared in connection with a license renewal application (NRC 2002). The analysis of environmental impacts for this section of the SDEIS draws from the data and conclusions gathered in the licence renewal process, the analysis provided by Dominion and Bechtel (2002), and the staff's independent review.

The following assumptions were made by the staff in the review of the Surry site as an alternative to the proposed North Anna ESP site.

- The units would use closed-cycle cooling.

- Mechanical draft towers would most likely be employed to avoid visual aspects of natural draft towers.

- The existing intake structures would be used with possible modifications to accommodate the proposed additional two units.

- The existing discharge canal would be used for cooling-water discharge.

- The land for additional reactors would be within the existing Surry site.

No additional transmission lines are assumed to be needed for power transmission for the proposed alternative units.

The station is on the Gravel Neck Peninsula on the south side of the James River, in an unincorporated portion of Surry County, Virginia. The station is approximately 40 km (25 mi) upstream of the point where the James River enters Chesapeake Bay. The James River is

1 about 4 km (2.5 mi) wide at the Surry site. The Surry site, shown in Figure 8.1, occupies
2 approximately 340 ha (840 ac).
3
4 The Surry site is 10 km (7 mi) south of Colonial Williamsburg and 13 km (8 mi) east-northeast
5 of the town of Surry. Jamestown Island, part of the Colonial National Historic Park, is to the
6 northwest on the northern shore of the James River. The area within 16 km (10 mi) of the site
7 includes Surry, Isle of Wight, York, and James City Counties, and parts of the cities of
8 Newport News and Williamsburg. The counties surrounding the Surry site are predominantly
9 rural, characterized by farmland, woods, and marshy wetlands. East and south of the site, at
10 distances between 16 and 48 km (10 and 30 mi), are the urban areas of Hampton, Newport
11 News, Norfolk, and Portsmouth, Virginia, and others, collectively known as Hampton Roads.
12
13 The site has two Westinghouse-designed light-water reactors, each with a design rating for a
14 gross electrical power output of 855 megawatts-electric (MW(e)). The Surry site was originally
15 planned for four units. Construction permits were issued for Units 3 and 4; however, the units
16 were never built. Cooling for the existing Surry site reactors is provided by a once-through
17 cooling system to remove waste heat from the reactor-steam electric system. Cooling water is
18 withdrawn from and returned to the James River.
19
20 Distinctive features of the Surry site include the 40-m (135-ft)-diameter cylindrical containment
21 buildings with hemispherical domes and the cooling canal. When the plant was designed,
22 there was a concern about the containment structures being visible from historic Jamestown
23 Island; consequently, the containment buildings were designed so the elevation would be
24 sufficiently low so as to blend with the surrounding forested lands. In addition to the two
25 nuclear reactors and their turbine building, intake and discharge canals, and auxiliary
26 buildings, the Surry site is the location of Dominion's Gravel Neck Combustion Turbine
27 Station, a switchyard, and an independent spent fuel storage installation (ISFSI).
28
29 Gravel Neck Peninsula is at the upstream limit of saltwater incursion to the James River;
30 upstream of Gravel Neck is tidal river and downstream is an estuary. Surry extends as a band
31 across the peninsula. Steep bluffs drop to the river on either side and to the tip of the
32 peninsula. Hog Island Wildlife Management Area, a Commonwealth of Virginia wildlife
33 management area, is located on the tip of the Gravel Neck Peninsula, and contains primarily
34 tidal marshes. Areas within 16 km (10 mi) of the site to the west, south, and east are
35 predominantly rural, characterized by farmland, forests, and marshy wetlands. The tidal flats
36 and marshes of Hog Island State Wildlife Management Area provide habitat for large numbers
37 and numerous species of migratory shorebirds, wading birds, and waterfowl. It also provides
38 habitat for numerous amphibians, reptiles, mammals, and upland game birds.

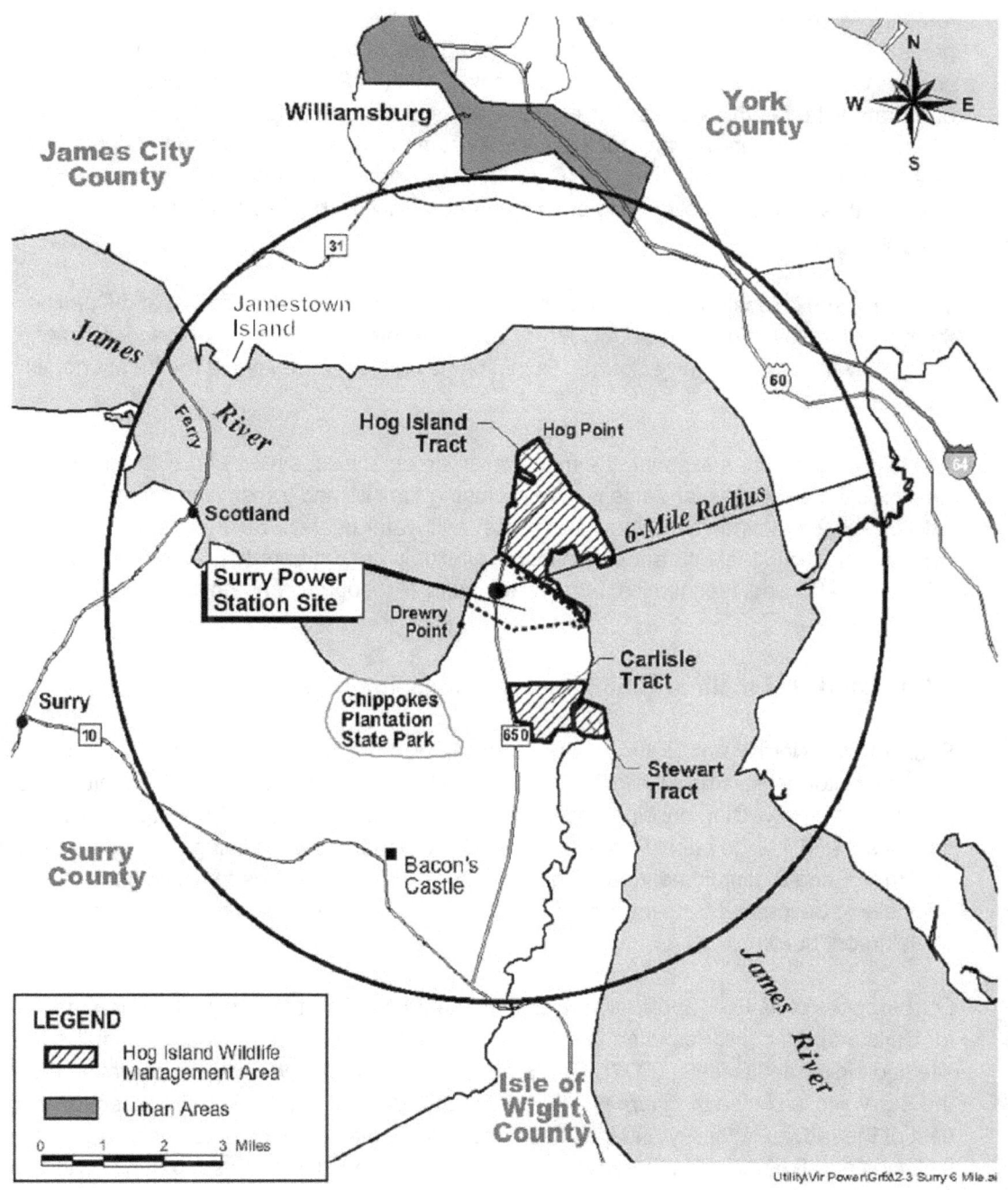

Figure 8-1. Surry Vicinity Map

1
2

1 The terrestrial community at the Surry site consists of remnants of mixed pine-hardwood
2 forests interspersed with early succession fields and developed areas. Wildlife species, found
3 primarily in the forested portions of the site, are those typically found in upland forests of
4 coastal Virginia. With the exception of the bald eagle (*Haliaeetus leucocephalus*) (Federally
5 and State-listed as threatened), terrestrial species that are Federally and/or State-listed as
6 endangered or threatened are not known to exist at the Surry site or along the rights-of-way of
7 its associated transmission lines (NRC 2002). The barking tree frog (*Hyla gratiosa*), State-
8 listed as threatened, is believed to be in the general vicinity but has not been observed at the
9 Surry site.
10
11 The Surry site is located in one of the strongest economic areas in Virginia, and Dominion is
12 the major employer in Surry County (NRC 2002). At present, because of the location of the
13 Surry Power Station in Surry County, Dominion has a significant impact on the economic well-
14 being of the county.
15
16 The following sections examine the major environmental issues reviewed by the staff.
17 Section 8.5.1 evaluates land-use issues, including the site and transmission lines.
18 Section 8.5.2 examines hydrology, water use, and water quality. Sections 8.5.3 and 8.5.4
19 evaluate the terrestrial and aquatic resources including endangered species, and Section
20 8.5.5 evaluates socioeconomics, historic and cultural resources and environmental justice
21 issues.
22

23 **8.5.1 Land Use Including Site and Transmission Lines**
24

25 Similar to the North Anna ESP site, the Surry site was originally designed for the construction
26 of four reactor units. Surry Units 3 and 4 were to be constructed to the east of Unit 2 where
27 the existing construction building and parking area are now situated. The original plans called
28 for Units 3 and 4 to be offset from Units 1 and 2, with the turbine building roughly in line with
29 the Units 1 and 2 containment buildings. The containment buildings for Units 3 and 4 were
30 originally to be located farther north of the intake canal than the existing Units 1 and 2
31 containment buildings.
32

33 For purposes of its ESP application, Dominion determined that the originally planned location
34 for Units 3 and 4 continues to be the best choice for two proposed nuclear units at the Surry
35 site (Dominion and Bechtel 2002). The proposed location of new nuclear generating units at
36 the Surry site is shown in Figure 8-2 (Dominion and Bechtel 2002). The proposed location is
37 east of the radwaste facility and includes construction, maintenance, and miscellaneous
38 buildings and the uncleared area west of the ISFSI. Relocation of these existing buildings to
39 another onsite location would therefore be necessary. The existing cleared area measures
40 approximately 300 m (900 ft) in the east-west direction. According to Dominion and Bechtel
41 (2002), an additional 300 m (900 ft) could be cleared to the east, while still maintaining

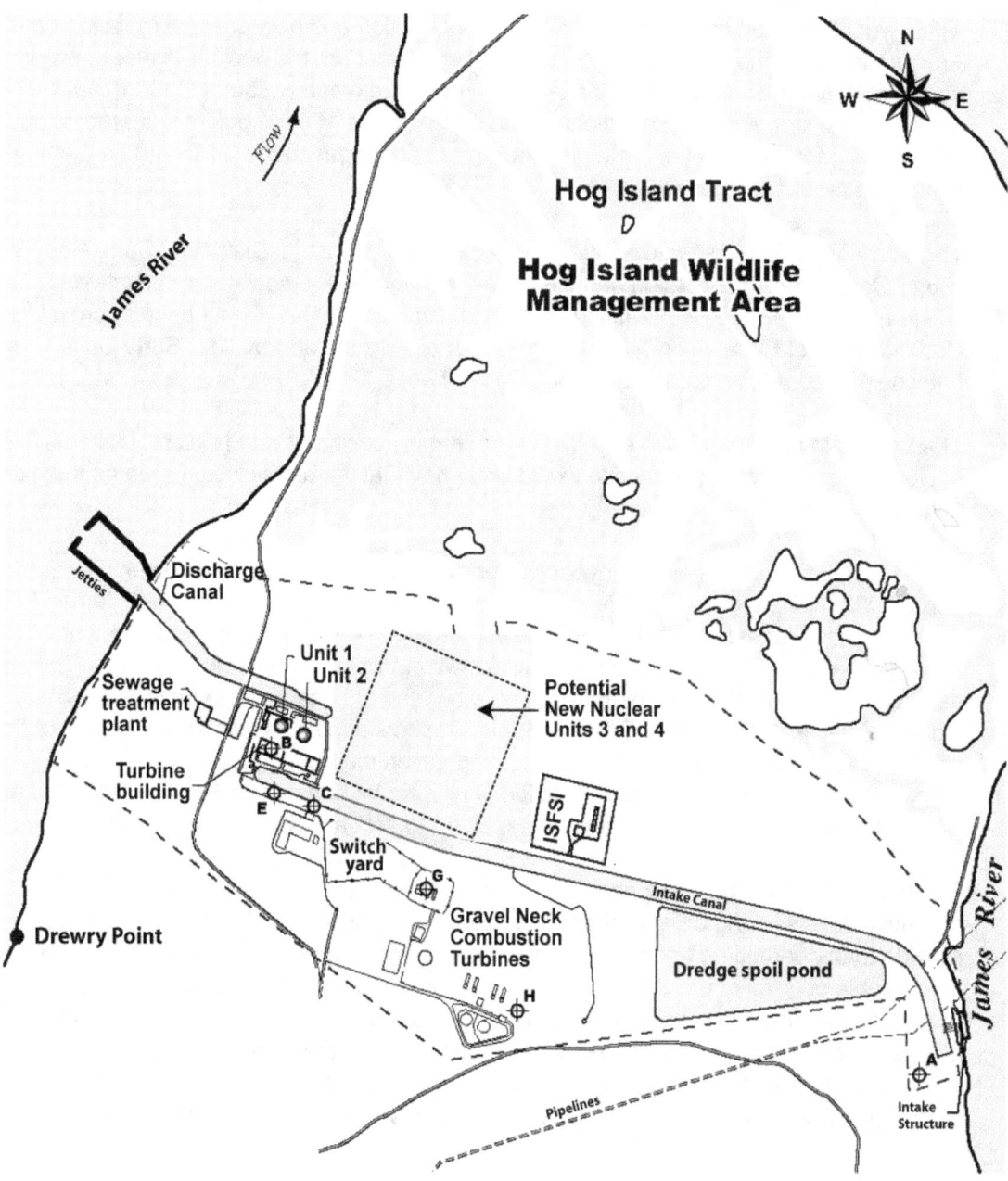

Figure 8-2. Surry Power Station Site

1 approximately 150 m (500 ft) to the ISFSI outer fence. An earthen berm around the
2 ISFSI would likely be constructed to reduce construction and occupational radiation doses.
3 Any expansion of the ISFSI would be to the east, away from the nuclear power units. In the
4 north-south direction, the cleared area measures approximately 350 m (1100 ft), including the
5 contractor parking area. An additional 30 to 45 m (100 to 150 ft) could be cleared without
6 encroaching too close to the north site boundary. The areas to the north and east of the ISFSI
7 could also be used, if needed.

9 The Surry site is in a district classified as M-2 General Industrial District by Surry County
10 (Surry County 1975). Location of nuclear power plants and associated radioactive
11 waste-handling facilities is permitted as a conditional use in this district upon approval by the
12 County Board of Supervisors. Dominion has received such approval for Surry Units 1 and 2,
13 but would need additional approval for proposed units.

15 The Surry site has an existing exclusion area that is consistent with NRC regulations. New
16 nuclear units sited at the Surry site would likely have the same exclusion area as the existing
17 units.

19 The residential locations of employees currently working at Surry are shown in Table 8-2
20 (NRC 2002). Approximately 60 percent of the employees live in Isle of Wight, James City, or
21 Surry Counties, or the City of Newport News, Virginia. The remaining 40 percent of
22 employees reside in other counties and cities within Virginia and adjacent states. The staff
23 assumes that the residences of the workforce needed to construct two units at the Surry site
24 would be similarly dispersed. Offsite land-use impacts associated with construction of the
25 proposed units would likely be relatively limited, given the temporary nature of the
26 construction (about 5 years). Construction of new rental housing and/or manufactured home
27 and recreational vehicle parks could be expected to accommodate construction workers.

29 Section 307(c)(3)(A) of the Coastal Zone Management Act (16 USC 1456(c)(3)(A)) requires
30 that applicants seeking a Federal permit to conduct an activity that affects a coastal zone area
31 provide to the permitting agency a certification that the proposed activity complies with the
32 enforceable policies of the State's coastal zone program. Surry is within the Virginia coastal
33 resources management area (VDEQ 2004). If construction of new nuclear units at Surry were
34 planned, Dominion would need to submit a certification to the Virginia Department of
35 Environmental Quality (VDEQ) stating that construction of the new units is consistent with the
36 Virginia Coastal Management Program. This submission would be reviewed by VDEQ.

Table 8-2. Surry Power Station, Units 1 and 2, Permanent Employee Residence by County/City

County/Independent City	Number of Personnel	Percentage of Total Personnel	Cumulative Percentage
Isle of Wight County	212	24	24
James City County	98	11	35
Newport News (city)	97	11	46
Surry County	90	10	57
Hampton (city)	71	8	65
Suffolk (city)	52	6	71
Chesapeake (city)	42	5	75
Chesterfield County	25	3	78
Portsmouth (city)	23	3	81
Virginia Beach (city)	21	2	83
York County	20	2	85
Prince George County	19	2	88
Sussex County	18	2	90
Southampton County	11	1	91
Others	79	9	100
Total	878	100	

Source: NRC 2002.

New land-use impacts associated with operation of new nuclear generating units at the Surry site would be expected to be limited. Some new housing in surrounding communities would likely be constructed to accommodate permanent workers at the new units. The incremental property tax revenue from the new units could affect future land use in Surry County as a result of infrastructure improvements made possible by the tax revenue. In the supplemental EIS related to renewal of the operating licenses for Surry Units 1 and 2, the staff determined that tax revenue impacts on land use during the 20-year license renewal term would be small (NRC 2002). The staff concludes that although the new units would be licensed for 40 years, the impacts would be similar. Based on the foregoing, the staff concludes that the land-use impacts on the site and vicinity of construction and operation would be SMALL.

The transmission line rights-of-way from the Surry site consist of three 500-kV transmission lines from the breaker, a 500-kV switchyard, and six 230-kV transmission lines from the 230-kV switchyard. The lines are not at or near capacity (Dominion and Bechtel 2002). For the addition of two advanced boiling water reactor-size units, it appears that the 500-kV transmission line rights-of-way would be able to transmit the new load; however, system-study (load-flow) modeling of the transmission lines and the new nuclear units would need to be performed to be certain whether any additional lines are required from the site. Based on the evaluation conducted by Dominion and Bechtel (2002), it is likely that no additional electrical

1 transmission line rights-of-way would be needed to transmit the power generated by additional
2 units at the Surry site to the regional power grid. This impact would be similar to land-use
3 impacts for construction and operation in the transmission line rights-of-way and offsite areas
4 associated with the North Anna ESP site. Therefore, based on the foregoing, the staff
5 concludes that the land-use impacts of transmission system construction and operation would
6 be SMALL.
7
8 ## 8.5.2 Water Use and Quality
9
10 The Surry site is located adjacent to the James River estuary. The consumptive use of water
11 to support mechanical-draft cooling towers for the proposed units would be undetectable
12 relative to the supply available in the estuary. Discharges to the James River could contain
13 water treatment chemicals that would be subject to regulation by the VDEQ to ensure
14 protection of the environment. The additional small amount of heat from blowdown water
15 would likely be undetectable if commingled with the once-through discharge of Surry Units 1
16 and 2. Therefore, based on the foregoing, the staff concludes that the impacts to water use
17 and water supply at the Surry site from construction and operation of two new units would be
18 SMALL.
19
20 ## 8.5.3 Terrestrial Resources Including Endangered Species
21
22 A maximum of approximately 200 ha (500 ac) of land would be disturbed to develop two
23 additional units with cooling towers at the Surry site (Dominion and Bechtel 2002). Much of
24 this area has been previously disturbed during development of the existing nuclear units and
25 associated facilities. Habitats in the area are a mixture of industrial areas, early successional
26 grasslands, and remnant mixed pine-hardwood forests. There are no threatened,
27 endangered, or other important terrestrial species known to exist within the Surry alternative
28 site, although Federally threatened bald eagles are known to nest near the Surry site.
29 Federally and State-listed threatened or endangered species reported to occur in Surry
30 County are listed in Table 8-3.
31
32 Potential construction impacts include erosion, dust generation, and noise which are typical of
33 large construction projects. These impacts could be mitigated using standard industrial
34 procedures or best management practices. Standard practices to limit potential construction
35 impacts including silt fences to control sedimentation and water sprays to limit dust generation
36 should protect wetlands and other ecological resources in the site vicinity.
37
38 Construction noise could affect bald eagles nesting in the vicinity of the site. Prior to initiation
39 of major construction activities, the presence and distribution of bald eagle nests in relation to
40 the location of planned facilities would need to be determined. If eagle nests are in the area,

1   **Table 8-3.** Federally and State-Listed Threatened or Endangered Terrestrial Species
2   Reported Within Surry County, Virginia
3

Scientific Name	Species	Federal Status	State Status
Birds			
Haliaeetus leucocephalus	bald eagle	T	T
Lanius ludovicianus[a]	loggerhead shrike	SC	
Falco peregrinus	peregrine falcon		T
Mammals			
Plecotus rafinesquii macrotis	eastern big-eared bat		E
Amphibians			
Hyla gratiosa	barking tree frog		T
Ambystoma mabeei	Mabee's salamander		T
Insects			
Speyeria diana	Diana fritillary	SC	
Vascular Plants			
Aeschynomene virginica	sensitive joint-vetch	T	T

18   T = threatened, E= endangered, SC = species of concern.
19   Sources: VDGIF 2004; VDCR 2004.
20   (a) The migrant subspecies *L.l. migrans* is a Federal species of concern; all loggerhead shrikes in Virginia are
21   State threatened.

23   mitigation measures would need to be developed. According to the Bald Eagle Protection
24   Guidelines for Virginia (FWS and VDGIF 2000), no major construction activities should occur
25   within 400 m (1300 ft) of an active eagle nest, and loud noises (such as blasting) should not
26   occur during the nesting/breeding season. No active nests currently exist within 400 m
27   (1300 ft) of the Surry construction site.

29   The staff evaluated the potential impacts of operation of new nuclear units, including operation
30   of the plants, cooling systems, and transmission systems on terrestrial threatened or
31   endangered species as follows. If new reactor units are constructed, very little usable habitat
32   would remain within the development area at the Surry site. Operation of additional units
33   would typically result in some noise generation, salt drift, icing, fogging, and bird collisions.
34   Noise would likely be typical of operating reactor units and cooling towers, which have been
35   found to be a SMALL impact (NRC 1996). However, it is possible that the such noise could
36   deter bald eagles from nesting near the site. There are no sensitive habitat areas adjacent to
37   the site that would be adversely affected by noise from plant operations. The terrestrial
38   vegetation in the immediate vicinity of he Surry site is not believed to be unusually sensitive to
39   salt drift, fogging, or icing (Dominion and Bechtel 2002). However, because the cooling tower
40   makeup water from the James River estuary is brackish (up to 17 parts per thousand of salt),
41   the cooling tower drift could have higher salt content than at freshwater sites. Because it is
42   likely that mechanical-draft towers with their much shorter towers would be used rather than

1 natural-draft towers, bird collisions would not be likely (NRC 1996). Based on the above
2 evaluation, the staff concludes that the impacts of operating two new units on terrestrial
3 threatened and endangered species would be SMALL.
4
5 According to the analysis performed by Dominion and Bechtel (2002), no additional
6 transmission lines would be needed to transmit electrical power generated by new units at the
7 Surry site to the regional distribution grid. The staff based its evaluation on the assumption
8 that new transmission lines would not be required to support two new units. Therefore,
9 maintenance and operation of the existing transmission line rights-of-way would likely not be
10 affected by two new nuclear units at the Surry site. NRC determined that the impacts of
11 continued operation of these transmission lines and maintenance of the rights-of-way would
12 have a small impact on terrestrial ecosystems (NRC 2002). However, some monitoring and
13 mitigation for nesting bald eagles might be warranted.
14
15 Based on the foregoing, the staff concludes that the overall impact to terrestrial ecological
16 resources of both construction and operation of two new units and associated cooling systems
17 and transmission line rights-of-way at the Surry site would be SMALL. However, some
18 monitoring and mitigation for nesting bald eagles might be warranted
19
20 **8.5.4 Aquatic Resources Including Endangered Species**
21
22 The aquatic environment near the Surry site is associated with the James River. The James
23 River rises in the Allegheny Mountains near the Virginia/West Virginia border and flows in a
24 southeasterly direction to Hampton Roads (that area of Virginia that includes Newport News,
25 Norfolk, Portsmouth, Hampton, and surrounding cities and towns), where it enters
26 Chesapeake Bay. The James River flows 692 km (430 mi) from its headwaters (the
27 confluence of the Cowpasture and Jackson Rivers) to Chesapeake Bay, crossing portions of
28 the Blue Ridge, Valley and Ridge, Piedmont, and Coastal Plain physiographic regions. The
29 river drains an area of 25,900 km^2 (10,000 mi^2), which is just over 25 percent of the total land
30 area of Virginia. Overall, about 71 percent of the basin is forested, 23 percent is agricultural,
31 and 6 percent is urban. The lower James River flows through the Coastal Plain of Virginia,
32 which is virtually flat in tidewater areas, generally ranging from 0 to 30 m (0 to 100 ft) above
33 mean sea level (MSL).
34
35 Two major tributaries enter the river between Richmond and Hampton Roads. The
36 Appomattox River enters the James River from the south, in the stretch of river between
37 Richmond and Petersburg. The Chickahominy River enters from the north, just west of
38 Williamsburg. Although the James River downstream of Richmond was severely polluted for
39 many years, the passage of the Clean Water Act in 1972 and implementation of associated
40 regulations, such as the National Pollutant Discharge Elimination System, have reduced the
41 flow of point-source pollutants into the James River watershed. Pollution prevention
42 measures and programs carried out by industrial entities in the area have further reduced

1 chemical discharges to the James River. At present, nutrients from sewage treatment
2 facilities, agricultural operations, and urban runoff and bacteria from combined sewer systems
3 (those that combine storm water and sewage) are considered the chief threats to the water
4 quality of the lower James River.
5
6 In the vicinity of the Surry site, the James River is approximately 4 km (2.5 mi) wide. Cobham
7 Bay lies west (just upstream) of the Gravel Neck Peninsula and represents the approximate
8 limit of saltwater incursion, effectively dividing the James River into a tidally influenced
9 freshwater river upstream (to the fall line at Richmond) and an estuary downstream. The U.S.
10 Army Corps of Engineers historically has dredged the main channel of the lower James River
11 so ocean-going vessels can proceed upriver as far as Hopewell, approximately 80 river km
12 (50 river mi) northwest of the Surry site.
13
14 The lower James River supports a diverse assemblage of finfish species, ranging from
15 exclusively marine species near Chesapeake Bay to exclusively freshwater species at the fall
16 line in Richmond. Approximately 80 fish species are known from the brackish portion of the
17 James River downstream of Surry, with another 40 or so species recorded from the tidally
18 influenced freshwater portion of the river upstream of the Surry site. Distributions and
19 abundances of particular species vary between seasons and years, depending on salinity
20 differences and natural fluctuations in fish populations.
21
22 Dominion's predecessors conducted extensive surveys of James River aquatic biota in the
23 1970s. While preparing its ER for the ESP application, Dominion contacted the Virginia
24 Institute of Marine Sciences for more recent information (Virginia Institute of Marine Sciences
25 2001). The following paragraphs describe the historic Virginia Electric and Power Company
26 (Virginia Power) data and the more recent data collected by the Virginia Institute of Marine
27 Sciences.
28
29 From 1970 to 1978, Virginia Power collected 63 fish species in monthly haul seine surveys
30 conducted to characterize fish populations of the shore zone in the vicinity of the Surry site.
31 Five species made up more than 75 percent of fish collected. These were the Atlantic
32 menhaden (*Brevoortia tyrannus*), blueback herring (*Alosa aestivalis*), inland silverside
33 (*Menidia beryllina*), bay anchovy (*Anchoa mitchilli*), and spottail shiner (*Notropis hudsonius*).
34 Over the same period, 42 fish species were collected in otter trawl samples that were intended
35 to characterize fish populations in deeper waters (the shelf zone) adjacent to the main river
36 channel. Five species comprised more than 80 percent of fish collected in trawl samples: the
37 hogchoker (*Trinectes maculatus*), spot (*Leiostomus xanthurus*), channel catfish
38 (*Ictalurus punctatus*), Atlantic croaker (*Micropogonias undulatus*), and bay anchovy.
39
40 Between 1996 and 2000, the Virginia Institute of Marine Sciences conducted approximately
41 350 deep-water ichthyoplankton trawl surveys in the James River in the vicinity of Hog Island.
42 In those collections, four species comprised more than 80 percent of the catch: hogchoker,

1 white perch (*Morone americana*), Atlantic croaker, and bay anchovy. Spot was the fifth most
2 abundant species. Salinity appears to be the most important factor influencing the relative
3 abundances of fishes between the two sampling periods.
4
5 In addition to finfish, several invertebrate aquatic species were found in the vicinity of the
6 Surry site. These include zooplankton (dominated by copepods), amphipods (notably the
7 scud [*Gammarus* spp.]), and a variety of benthic organisms (e.g., polychaetes and shellfish).
8 Shellfish formed the bulk of the benthic biomass from the transition zone in the vicinity of the
9 Surry site to Chesapeake Bay. The brackish water clam (*Rangia cuneata*), a species capable
10 of tolerating a wide range of salinities, dominated the benthic community in the vicinity of the
11 Surry site. Larval American oysters (*Crassostrea virginica*) occurred in the area as
12 meroplankton, but adults were uncommon. The more recent trawl survey collected American
13 oysters, blue crabs (*Callinectes sapidus*), spider crabs (*Libinia emarginata*), eight species of
14 shrimp (Penaeidae), and five species of clams (Bivalvia). The diversity of benthic
15 macroinvertebrate is usually low in a transition zone, increasing downstream to seawater and
16 upstream (moderately) to freshwater. A combination of physical, chemical, and biological
17 factors influence the distribution of benthic organisms, but as with the finfish, salinity appears
18 to exert the greatest influence.
19
20 No areas designated by the U.S. Fish and Wildlife Service (FWS) as critical habitat for
21 endangered species exist in the James River (Virginia Natural Heritage Program 2003).
22 Virginia Power and its contractors conducted extensive surveys of fish and aquatic
23 invertebrates in the lower James River in the vicinity of the Surry site in the 1970s. Based on
24 these historical surveys and a review of the scientific literature, no Federally listed aquatic
25 species is found in the lower James River. On Virginia's endangered species list, Jenkins and
26 Burkhead (1994) identify only one threatened or endangered fish species in the entire James
27 River drainage, the orangefin madtom (*Noturus gilberti*), which occurs in the headwaters of
28 the river, several hundred miles upstream of the Surry site.
29
30 The Atlantic sturgeon (*Acipenser oxyrhynchus*), a candidate for Federal listing, was reported
31 in the vicinity of the Surry site in the early 1970s and was subsequently collected in research
32 and monitoring studies conducted by Virginia Power and Virginia Power-funded entities in the
33 mid-to-late 1970s. A number of authorities on the fishes of Virginia and the mid-Atlantic coast
34 also list this species as occurring in the lower reaches of the James River. The blackbanded
35 sunfish (*Enneacanthus chaetodon*), listed as endangered by the Commonwealth of Virginia, is
36 reported to occur in Prince George, Surry, and Sussex Counties west of the Surry site.
37 However, this sunfish primarily inhabits thickly vegetated ponds, swamps, and pools and is
38 not reported to occur in the James River Drainage (Jenkins and Burkhead 1994.)
39
40 Although not recorded in Virginia for more than 100 years, the shortnose sturgeon
41 (*Acipenser brevirostrum*) is on the Commonwealth's list of rare animal species. This listing is
42 based on the fact that the species occurs in major river systems north and south of the

1 Chesapeake Bay, is presumed to have spawned in the four major estuarine drainages of the
2 Chesapeake Bay (including the James River) in Virginia as late as the 19th century, and may
3 reappear in the future if restoration efforts are successful. At present, the shortnose sturgeon
4 is listed as endangered by the National Marine Fisheries Service and by Virginia. It also
5 appears on the Virginia Department of Cultural Resources list of "Extinct and Extirpated
6 Animals of Virginia."
7
8 The staff evaluated the potential impacts of operating the proposed new nuclear units,
9 including operating the plants, cooling systems, and transmission systems on aquatic
10 threatened and endangered species. Based on this evaluation, the staff concludes that the
11 impacts of operating the proposed new units on aquatic threatened and endangered species
12 at the Surry site would be SMALL.
13
14 The potential for impingement and entrainment of aquatic resources is expected to be minimal
15 because of closed-cycle cooling. The potential impacts of heated water would be expected to
16 be mitigated by the placement of the discharge structures. The overall impact on aquatic
17 ecological resources of construction and operation of two new units and associated cooling
18 towers and transmission facilities at the Surry site would be SMALL.
19
20 **8.5.5 Socioeconomics, Historic and Cultural Resources, Environmental Justice**
21
22 In evaluating the socioeconomic impacts of construction at the Surry site, the staff used the
23 license renewal supplemental EIS information (NRC 2002) and Dominion and Bechtel's
24 (2002) analysis of potential sites. The staff also conducted a reconnaissance survey of the
25 site using readily obtainable data from the internet or published sources. No new data were
26 collected. The socioeconomic sections follow the organizational structure of the
27 socioeconomic discussions in Sections 2.8, 4.5, and 5.5. Both construction and station
28 operation impacts are addressed.
29
30 **8.5.5.1 Physical Impacts**
31
32 Construction activities can cause temporary and localized physical impacts such as noise,
33 odor, vehicle exhaust, vibration, shock from blasting, and dust emissions. The use of public
34 roadways, railways, and barges would be necessary to transport construction materials and
35 equipment. Dominion anticipates that the roadways could need some minor repairs or
36 upgrading, such as patching and filling potholes, to allow safe transport of these materials and
37 equipment. However, no extensive work is planned to the existing roads or railways, and no
38 new routes would be needed (Dominion and Bechtel 2002). All construction activities would
39 occur within the existing Surry site. Offsite areas that would support construction activities
40 (e.g., borrow pits, quarries, and disposal sites) are expected to be already permitted and
41 operational. Impacts on those facilities from construction of the new units would be small
42 incremental impacts associated with their normal operation.

1   Potential impacts from station operation include noise, odors, exhausts, thermal emissions,
2   and visual intrusions. Noise would be produced by the operation of pumps, cooling fans,
3   transformers, turbines, generators, switchyard equipment, and traffic. Dominion states in its
4   ER that any noise coming from the North Anna ESP site would be controlled in accordance
5   with applicable local county regulations. By inference, this is also expected to apply to the
6   Surry site. Virginia has no regulations or guidelines regarding noise limits. Commuter traffic
7   would be controlled by speed limits. Good road conditions and appropriate speed limits would
8   minimize the noise level generated by the workforce commuting to the Surry site
9   (Dominion 2006).
10
11 The new units would have standby diesel generators and auxiliary power systems. Permits
12 obtained for these generators would ensure that air emissions comply with regulations. In
13 addition, the generators would be operated on a limited short-term basis. During normal plant
14 operation, the new units would not use a significant quantity of chemicals that could generate
15 odors exceeding odor threshold values. Good access roads and appropriate speed limits
16 would minimize the dust generated by the commuting workforce (Dominion and Bechtel
17 2002).
18
19 Construction activities would be temporary and would occur mainly within the boundaries of
20 the Surry site. Offsite impacts would represent small incremental changes to offsite services
21 supporting the construction activities. During station operations, noise levels would be
22 managed in accordance with local ordinances. Air quality permits would be required for the
23 diesel generators, and chemical use would be limited, which should limit odors. Based on the
24 foregoing, the staff concludes that the physical impacts of construction and operation would
25 be SMALL.
26

27 **8.5.5.2 Demography**
28

29 The population base is considered to be the population of significant population centers within
30 a 80-km (50-mi) radius of the Surry site. The combined population of the Richmond-
31 Petersburg and Norfolk-Virginia Beach and Newport News, Virginia Metropolitan Statistical
32 Area is 2,566,050 (USCB 2000a). The estimated population within an 80-km (50-mi) radius of
33 the Surry site is 2,387,353 and is projected to grow by approximately 41 percent to 3,365,040
34 by 2030 (NRC 2002).
35

36 Most of the construction workforce is expected to come from within the region, and those who
37 might relocate to the region would represent a small percentage of the larger population base.
38 While the station operation workforce is expected to relocate into the region, their numbers
39 are small (720 new operating employees and their families) when compared to the total base
40 population and their locations of residence would probably be scattered throughout the region.
41 Based on the foregoing, the staff concludes that any environmental impacts caused by
42 population increases within an 80-km (50-mi) radius of the site would be SMALL.

8.5.5.3 Community Characteristics

Economy

The Surry site is located in one of the strongest economic areas in Virginia. The Richmond-Petersburg area is the primary economic driving force in the area within an 80-km (50-mi) radius of the Surry site. The Norfolk-Virginia Beach-Newport News area is characterized by the U.S. Navy's significant presence in the area (NRC 2002). Hampton Roads relies heavily on defense-related industry, particularly shipbuilding. In recent years, the regional economy has become more diversified with major businesses, financial and health care components, and a growing "high-tech" sector. Regionally, the service sector now offers the most employment opportunities. The construction and operation of two new nuclear units at the Surry site would be expected to benefit the economy of the region, especially Surry County.

Based on the foregoing, the staff reviewed the impacts of station construction and operation on the economy of the region and concludes that the impacts would be SMALL everywhere in the region except Surry County, where the impacts could be beneficially MODERATE. The magnitude of the economic impacts would be diffused in the larger economic bases of the Norfolk-Virginia Beach-Newport News area, including Isle of Wight, Surry, York, and James City Counties. With Surry County's smaller economic base, the economic impacts would be more noticeable.

Availability of Workers

Dominion estimates it would take approximately 5000 construction workers more than 5 years to build two new nuclear units at the Surry site (Dominion 2006). Dominion is expected to be able to attract the necessary workforce for construction activities at the Surry site because of its proximity to the major population centers of Richmond-Petersburg and Norfolk-Virginia Beach Newport News. While the availability of craft workers for outages at Surry is reported as very limited, this can be attributed to the short duration of the projects. However, the availability of craft workers for regular construction projects of longer duration is reported to be good (Dominion and Bechtel 2002). The construction workforce within 80 km (50 mi) of the site were estimated to number approximately 98,000 (Dominion and Bechtel 2002).

Approximately 990 employees work at Surry Units 1 and 2, (about 110 contract employees and 880 permanent employees). The addition of the proposed new units would result in an increase in the operations workforce of 720 employees. In its ER, Dominion stated that it expected most of the operations workforce for the new units to relocate from outside the region. The ER does not address from where these employees would come (Dominion 2006). Some nuclear defense sites are reducing their workforces as they change missions (such as the Portsmouth and the Savannah River sites), and workers from these sites could be potential pools of labor for the operating workforce at Surry.

1   Based on the foregoing, the staff concludes that the impacts are SMALL because construction
2   labor would be available from within the region, and there would be little problem recruiting the
3   required labor skills to enable the construction of the nuclear units at the Surry site and the
4   operations workforce would relocate to the region.
5
6   *Transportation*
7
8   The area around the Surry site is served by several major freeways and State and Federal
9   highways (NRC 2002). The most direct vehicular access to the Surry site is from the more
10   populous cities and counties on the north bank of the James River via State Route (SR) 31
11   and the James River Ferry service, operated by the Virginia Department of Transportation.
12   The principal road access to the Surry site is via SR 650, which is a two-lane paved road. SR
13   650 carries a level-of-service (LOS) designation of "A," which reflects a free flow of traffic and
14   users unaffected by the presence of others.
15
16   The construction of new nuclear units would involve additions to the workforce. In addition,
17   construction materials, wastes, and excavated materials would be transported both to and
18   from the site. These activities would result in increases in operation of personal-use vehicles
19   by commuting construction workers, in commercial truck traffic, and in traffic associated with
20   daily operations. However, five of the seven reactor types under consideration for this project
21   are generally smaller and modular in nature. Consequently, transportation of plant equipment
22   could be less challenging and workforce needs are expected to be less than those for
23   conventional nuclear plants (Dominion and Bechtel 2002).
24
25   The LOS designation on SR 650 would likely be degraded from "A" to "C" (which reflects a
26   stable flow that marks the beginning of the range of flow in which the operation of individual
27   users is significantly affected by interactions with the traffic stream) during the peak
28   construction period for a new nuclear plant at the Surry site (Dominion and Bechtel 2002).
29
30   SR 650 intersects SR 10 approximately 8 km (5 mi) from the plant. SR 10 in the vicinity of the
31   site, from Surry County Courthouse to the divergence of the business and bypass north of
32   Smithfield, carries a LOS designation of "C." Portions of Highway 10 would receive
33   significantly more traffic during plant construction (NRC 2002; Dominion and Bechtel 2002).
34
35   No direct rail access is available to the Surry site, so large equipment would have to be
36   offloaded and transported by road and/or barge from the nearest rail access points in
37   Richmond or Norfolk. Surry has an excellent barge slip adjacent to the cooling water intake.
38   This slip was used for the transport of the replacement steam generators in the late 1970s and
39   is regularly used to receive spent fuel storage casks and other large loads (Dominion and
40   Bechtel 2002).
41

1   The Williamsburg-Jamestown Airport, Newport News/Williamsburg International Airport,
2   Norfolk International Airport, and the Richmond International Airport all serve the area. The
3   airports in Richmond and Norfolk provide regular freight and passenger jet services and are of
4   sufficient size to accommodate the relatively small air shipments normally associated with a
5   construction project (Dominion and Bechtel 2002).

7   The impact of station operation employees on the transportation system would be less than
8   the impact incurred during construction. There would be increases in operation of personal-
9   use vehicles by commuting operators of both the existing and new units and in traffic
10   associated with daily operations. Portions of SR 10 may be impacted by commuters to the
11   plant site, particularly during shift changes. During new plant operation, the LOS designation
12   on SR 650 may retain its "A" status or perhaps degrade to "B" designation, which reflects a
13   condition of stable flow instead of the free flow indicated under an "A" designation. This
14   change in designation indicates that the freedom to select speed is unaffected, but the
15   freedom to maneuver is slightly diminished (Dominion and Bechtel 2002).

17   Based on the foregoing, the staff concludes that the impacts of a construction workforce and
18   related transportation of construction supplies and materials on transportation infrastructure at
19   Surry would be SMALL to MODERATE (but temporary). Some of the local roads could have
20   their LOS degraded during construction to the point where operations of individual drivers
21   could be significantly affected by interactions with other traffic. This would be at LOS levels of
22   C or lower. Also it is possible that, given the heavy loads carried by vehicles transporting
23   construction materials to the Surry site, some of the roads may need repair to carry the
24   additional load. The impacts during operation would be SMALL.

26   *Taxes*

28   Construction and operations workers would pay income, sales, and use taxes to Virginia and
29   the local governments in the region where sales take place and property taxes to the counties
30   in which they own a residence. Sales and use taxes would be paid from the sales of
31   construction materials and supplies purchased for the project and on expenditures of both the
32   construction and operations workforce for goods and services. Dominion estimates that
33   about half of the day-to-day expenditures during construction would occur in the region
34   (Dominion 2006). Corporate income taxes on profits would also be paid by those companies
35   engaged in construction at the site.

37   There are two types of property taxes in Virginia. The first is the tangible personal property
38   tax paid by contractors during construction of the additional units. This tax is based on the
39   value of property owned by the contractors that acquire taxable status in Surry County during
40   the construction period. The second is the real property tax levied for the incremental
41   increase in value to the entire site from the operation of the additional units. It is expected

1   that Surry County would be the only beneficiary of this tax. Dominion has a significant impact
2   on the economic well-being of Surry County, with Dominion paying well over 70 percent of the
3   property taxes between 1996 and 2000 (NRC 2002).
4
5   Based on the foregoing, the staff concludes that the overall impacts from construction and
6   operation of taxes collected through the income, sales and use, and property taxes would be
7   SMALL (with the exception of Surry County for property taxes). The taxes paid, while
8   substantial, are nevertheless a small sum when compared to the total amount of taxes
9   collected by Virginia and local governments in the region. The staff concludes that the overall
10 impacts of the property taxes collected in Surry County would be beneficially MODERATE
11 (construction) and LARGE (operation) relative to the total amount of taxes the county collects
12 through property taxes.
13
14 *Aesthetics and Recreation*
15
16 Although the Surry site is clearly an industrial site, its current structures are not visually
17 intrusive from any vantage point, even from across the James River. However, Surry Units 1
18 and 2 are visible from the highest amusement rides at Busch Gardens (Dominion and
19 Bechtel 2002) and at certain points of Jamestown Island and the Colonial Parkway. The
20 reasons for the lack of visual intrusiveness are the general wooded habitat surrounding the
21 site and the fact that the Units 1 and 2 reactor containment buildings are sunk into the ground
22 to minimize visual obtrusion from offsite. The licensee for Units 1 and 2 took this unusual step
23 because of the highly sensitive nature of the historic resources across the James River
24 (i.e., Jamestown and Williamsburg) (VEPCo 1970). Not all of the seven new reactor
25 technologies being considered for the Surry site could be designed to allow the reactor
26 containment building to be placed lower in the ground. For example, the AP1000, which has a
27 reactor containment building approximately 71 m (234 ft) above grade, could potentially be
28 more easily seen from Williamsburg across the James River from the Surry site. The design
29 of this building includes a hatch that determines the height that it must be above ground.
30 Dominion states that redesigning the AP1000 would be prohibitively expensive to allow the
31 building to be placed lower in the ground. In its ER, Dominion did not address the feasibility of
32 adapting the remaining reactor technology to minimize visual intrusiveness (Dominion and
33 Bechtel 2002).
34
35 The Surry site is a minimum of 5 km (3 mi) from any point across the James River. Except for
36 the west side of the site, which is open to the James River, the dense tree stands surrounding
37 the site effectively screen the existing unit from all but a few locations. No parks or
38 recreational areas are within 3 km (2 mi) of the Surry site. The closest recreational park is
39 Chippokes Plantation State Park located 4 km (2.5 mi) to the southwest (NRC 2002). The
40 only distinguishable view of the transmission lines by offsite observers is available from the
41 James River (Dominion and Bechtel 2002).
42

1   The addition of new units at the site would likely involve the use of cooling towers. Given the
2   historical nature of some of the surrounding communities (e.g., Jamestown), Dominion would
3   be more likely to use mechanical-draft cooling towers rather than the taller, natural-draft
4   towers, which may be considered unacceptable. Traditionally, visible plumes generated by
5   the operation of cooling towers could cause a negative aesthetic effect. These plumes would
6   be visible a majority of the time.
7
8   Most construction activities would be screened from offsite viewing. The exception may be
9   the reactor containment building as it nears completion. The AP1000 at 71 m (234 ft) would
10 be visibly intrusive. During operations, visible plumes could be generated by the cooling
11 towers.
12
13 Concerns regarding the design and operation of additional units on the viewshed of the
14 Colonial National Historic Park were raised by the National Park Service. The Colonial
15 National Historic Park includes Jamestown, the Yorktown Battlefield, and the Colonial
16 Parkway. The staff and its contractor met with the Park Service prior to conducting a
17 supplemental visit to the Surry site to assess the potential adverse effects at Surry as an
18 alternative site to the proposal for the North Anna ESP site. The survey of the site visit is
19 contained in a trip report dated September 19, 2005 (ML061720366) (NRC 2005). In its letter
20 dated October 25, 2005, the National Park Service suggested that the physical size of the
21 units and the operational impacts, would have our adverse effect on the view shed from both
22 Jamestown and the Colonial Parkway (NPS 2005).
23
24 In its summary, the staff found:
25
26   The Colonial National Historical Park (NHP), particularly the Jamestown Unit and the
27   associated initial stretch of the Colonial Parkway that extends eastward along the
28   shoreline of the James River, and the Jamestown National Historic Site would be the most
29   directly visually impacted. The Colonial NHP is managed by the National Park Service
30   while the adjoining Jamestown National Historical Site is owned and managed by the
31   Association for the Preservation of Virginia Antiquities. In the words of the co-managers:
32   "Jamestown is a world-class cultural historic site that needs to be promoted, explored, and
33   fully presented to communicate its significance in history."
34
35 and
36
37   Based on the high level of historical significance attributed to the Jamestown historical
38   features and the fact that current views of the Surry Power Plant range from full to partial,
39   from both the island and the Colonial Parkway, an even more visible plant infrastructure
40   and the added cooling towers and condensation plumes would constitute a major visual
41   intrusion from this significant historic property. ... In the context of a new ESP at the Surry

1   site, the visual impacts to Colonial NHP and the Jamestown National Historical Site would
2   be considered significant.

3

4 Based on the foregoing, the staff concludes that the impacts of construction and operation on
5 aesthetics in the vicinity of the Surry site generally would be SMALL to MODERATE, but that
6 a LARGE impact could occur at historically important sites (Section 8.5.5.4).

7

8 *Housing*

9

10 A 10 percent vacancy rate out of a total 110,250 housing units currently exists in Isle of Wight,
11 Surry, and James City Counties and the city of Newport News Independent City. Surry
12 County has the highest vacancy rate at 20 percent (NRC 2002). Given the proximity of the
13 Surry site to four major metropolitan areas, housing for construction workers, most of whom
14 will be coming from within the region, and the operations workforce is expected to be
15 available.

16

17 Based on the foregoing, the staff concludes that the impacts of a construction and operations
18 workforce on the demand and housing availability would be SMALL. The conclusion is based
19 on approximately 10,000 vacant housing units in the region and the Surry site's proximity to
20 the larger metropolitan areas in the region.

21

22 *Public Services*

23

24   Water Treatment Facilities

25

26 Isle of Wight County has municipal water supply systems in the towns of Windsor, Smithfield,
27 and Franklin. Permitted groundwater wells supply these systems. Surry County has
28 municipal water supply systems in the towns of Claremont, Dendron, and Surry. The
29 municipal water supply for James City County is provided by the Newport News Waterworks
30 and the James City Service Authority. Newport News Waterworks is one of the top
31 100 largest water utilities in the United States and one of the three largest in Virginia. James
32 City Service Authority's water system consists of the central system with 29 well facilities and
33 six independent water systems with five well facilities servicing them. Public water supply for
34 Newport News is provided by the Newport News Waterworks. Water is supplied to nearly
35 400,000 residents of Poquoson, Hampton, and Newport News, and to portions of York and
36 James City Counties. The primary source of raw water is the Chickahominy River. Water
37 supply needs in the intermediate term are expected to be met, with all towns and cities in the
38 region having excess capacity (NRC 2002).

39

40 Water supply needs near the Surry site are not a current concern with all towns and cities in
41 the region having excess capacity. Most of the construction workforce would come from
42 within the region and, therefore, are already accounted for in the demands being placed on

1　the systems and their excess capacities. The station operating workforce, while relocating to
2　the region, would probably take up residence across the region, thus not particularly impacting
3　any one community or jurisdiction. Based on the foregoing, the staff concludes that the
4　impacts of construction and operation on water supply treatment facilities would be SMALL.
5
6　　Police, Fire, and Medical Facilities
7
8　In the larger metropolitan areas of Richmond and Henrico County and the cities of Norfolk,
9　Hampton Roads, and Virginia Beach, police, fire, and medical facilities would not be materially
10　impacted by an increase in the construction workforce. Many of the construction workers are
11　anticipated to already live in the region and would commute to the Surry site. As a result,
12　these workers are already served by existing services and facilities.
13
14　The operations workforce of about 720 workers and their families is anticipated to relocate to
15　the site from outside the region. Most likely they would locate in residences across the region
16　and would not concentrate in one location. As such, inordinate demands are not likely to be
17　placed on these services and facilities.
18
19　Most construction workers already live within the region. The incoming operations workforce
20　would likely have residences scattered across the region. As a result, there should be
21　minimal new demands placed on these services and facilities by either construction or
22　operations employees. Based on the foregoing, the staff concludes that the impacts of
23　construction and operations workforce on police, fire, and medical facilities would be SMALL.
24
25　　Social Services
26
27　Social services in the Commonwealth are provided in each county by the Virginia Department
28　of Social Services. The Department, which provides a variety of services to children and
29　adults, has 131 local departments located throughout Virginia (VDSS 2004). During the
30　construction phase at the Surry site, there may be an increased demand for social services.
31
32　Generally, construction and operation of new units on the Surry site would be viewed as
33　economically beneficial to the disadvantaged population segments served by the Department.
34　The workforce associated with the Surry site would be relatively higher paid than other
35　employment categories in the region. Construction and operation of the new units, through
36　the multiplier effect, may enable members of the disadvantaged population to improve their
37　social and economic position by moving up to higher paying jobs. At a minimum, the
38　expenditures of the construction and operations workforce in the counties for items such as
39　food and services could, through the multiplier effect, increase the number of jobs that could
40　be filled by the disadvantaged population.
41

1   Based on the foregoing, the staff concludes that the demand for social and related services as
2   a result of construction and operation of two new units would be SMALL. Construction and
3   operation of two new units would be expected to have a beneficial economic impact to the
4   economically disadvantaged population of the region, which should lessen the demand for
5   social services. There could be an initial increase in demand for social services at the
6   beginning of the construction period, but this is considered manageable and limited.
7
8   *Education*
9
10   The Surry County School system has just over 1200 students (Great Schools 2004). There
11   currently is no overcrowding in the system (NRC 2002). In the other counties and cities of the
12   region, it is anticipated that the construction and operations workforce would minimally impact
13   school infrastructure. Many construction workers already live within the region. Those that do
14   not live within the area are not expected to move their families into the area. This conclusion
15   is based on Dominion's assertions in its ER for the North Anna ESP site (Dominion 2006), and
16   by inference is applicable to the Surry site because of the geographical proximity of the two
17   sites. The operations workforce, while coming from outside and relocating into the region,
18   would probably be scattered throughout the region, placing little demand on school
19   infrastructure as a result.
20
21   It is anticipated that most of the construction workforce would come from within the area and
22   would not relocate their families. Those construction and operations workers potentially
23   relocating to the region would most likely be scattered throughout the region and, as a result,
24   would not be in sufficient concentrated pockets to place an undue burden on the existing
25   infrastructure. Based on the foregoing, the staff concludes that the impacts of the
26   construction and operations workforce on education facilities in Surry County and the area
27   would be SMALL.
28
29   **8.5.5.4 Historic and Cultural Resources**
30
31   Historic and cultural resources at the Surry site have been addressed in the recent
32   supplemental EIS relating to renewal of the existing operating licenses for Surry Units 1 and 2
33   (NRC 2002) and supporting information (Louis Berger Group, Inc. 2001). Associated with that
34   effort was consultation with eight Commonwealth of Virginia-recognized Native American
35   Indian Tribes: the Chickahominy Indian Tribe, Chickahominy Indians-Eastern Division,
36   Mattaponi Indian Tribe, Monacan Indian Nation, Nansemond Indian Tribe, the Pamunkey
37   Indian Tribe, the Rappahannock Tribe, and the Upper Mattaponi Indian Tribe.
38
39   While there are no currently recorded historic and cultural resources at the Surry site,
40   evaluations of the potential for such occurrences included acreages with the following
41   designations: No Potential (areas previously disturbed during initial construction of the plant),
42   Low Potential (areas that may or may not have been disturbed during construction and areas

1 with little potential for human occupation), and Moderate-to-High Potential (areas with little
2 past surface disturbance and with a likelihood for prehistoric and historic sites based on
3 regional comparative data). Should this alternate site be selected for an ESP, the staff
4 expects that Dominion would consult with the Virginia Department of Historic Resources
5 concerning the need for additional field inventory of acreage for historic and cultural resources
6 prior to undertaking any ground-disturbing activities.
7
8 As noted in Section 8.5.5.3, the existing Units 1 and 2 containment structures were originally
9 constructed below grade so as to minimize visibility of the Surry plant from historic Jamestown
10 Island, located across the James River about 4.8 km (3 mi) from the plant. In that new ESP
11 plant would involve construction of a taller containment structure and the addition of
12 mechanical draft cooling towers at Surry, the potential for visual impacts to the area's
13 significant historical resources would be increased through increased visibility of the new
14 structures and view of a condensation plume from the cooling towers during plant operation,
15 or both. In addition to the Colonial National Historical Park, which includes both Jamestown
16 Island and the Colonial Parkway, other historic sites/districts that could be affected include the
17 Williamsburg Historic District and Carter's Grove Plantation on the north side of the James
18 River. South of the river, other historic properties that could experience visual intrusions
19 include Bacon's Castle and Smith's Fort Plantation, both owned and operated as visitor
20 attractions by the Association for the Preservation of Virginia Antiquities, and the Chippokes
21 Plantation Historic District, a component of the Chippokes Plantation State Park managed by
22 the Virginia Department of Conservation and Recreation. Under certain weather conditions,
23 such as cooler winter days, taller plumes of condensation could possibly be seen from other
24 historic properties, such as some of the James River Plantations located upriver from the
25 Surry plant.
26
27 Based on the foregoing, the staff concludes that the potential impacts within the plant site
28 boundaries on historic and cultural resources from construction and operation of two new
29 nuclear units at the Surry site would be SMALL. However, the potential for visual intrusion at
30 the area's significant historic properties, districts and landscapes, the potential impacts could
31 range from SMALL to LARGE depending on distance and geographic orientation from the
32 Surry site. At the Colonial National Historic Park where there are vantage points from historic
33 Jamestown and the Colonial Parkway, the physical size of additional units and the presence
34 of condensation plumes from operation of units within the bounds of the PPE could be
35 considered a MODERATE to LARGE visual impact to the historic qualities of this nationally
36 significant cultural landscape.
37
38 **8.5.5.5 Environmental Justice**
39
40 As part of the evaluation of the potential environmental justice impacts related to the Surry
41 site, the staff used information from NRC's supplemental EIS for the license renewal of Surry
42 Units 1 and 2 (NRC 2002). The pathways through which the environmental impacts

1　associated with the construction of two additional new nuclear units at the Surry site could
2　affect human populations were ascertained. The staff then evaluated whether minority and
3　low-income populations could be disproportionately affected by these impacts. The staff
4　found no unusual resource dependencies or practices, such as subsistence agriculture,
5　hunting, or fishing, through which the populations could be disproportionately affected. In
6　addition, the staff did not identify any location-dependent disproportionate impacts affecting
7　these minority and low-income populations.
8
9　Based on the foregoing, the staff concludes that the offsite impacts of construction and
10　operation of the new units at the Surry site to minority and low-income populations would be
11　SMALL. No disproportionately high and adverse impacts affecting minority and low-income
12　populations were identified.
13

14　## 8.6　Evaluation of the Portsmouth Site

15
16　Dominion identified a 138-ha (340-ac) parcel of land in the northeastern portion of the
17　Portsmouth site in Ohio as a possible location for two commercial nuclear units (Dominion and
18　Bechtel 2002). The parcel has been evaluated by DOE and slated for transfer from DOE to
19　the Southern Ohio Diversification Initiative for possible reindustrialization (66 FR 64963).
20
21　For this evaluation, the following assumptions were made by the staff about locating the
22　proposed units at the Portsmouth alternative site:
23
24　• The units would use closed-cycle cooling.

25　• Natural- or mechanical-draft cooling towers would be employed.

26　• Groundwater would be the source of cooling water (ostensibly from the Scioto River).

27　• The plant would discharge blowdown water to the Scioto River.

28　• The land-area needed for the site would be approximately 140 ha (340 ac).

29　• Additional transmission lines would be needed.
30
31　The Portsmouth site is located in Pike County, Ohio, approximately 35 km (22 mi) north of the
32　Ohio River and 5 km (3 mi) southeast of the town of Piketon. The Portsmouth site vicinity is
33　shown in Figure 8-3.
34
35　Pike County's largest community, Waverly, is about 16 km (10 mi) north of the Portsmouth site
36　and has a population of about 4400 residents. The nearest residential center to the site is
37　Piketon, which is about 5 km (3 mi) north on U.S. 23; its population in 2000 was approximately
38　1900 people. Additional population centers within 80 km (50 mi) of the plant are Portsmouth
39　(population 20,909), 35 km (22 mi) south; Chillicothe (population 21,796), 43 km (27 mi) north;

and Jackson (population 6184), 29 km (18 mi) east (Bechtel Jacobs 2003). Approximately 90 percent of Portsmouth site workers reside in Jackson, Pike, Ross, and Scioto Counties.

In 2003, the estimated population of the four counties was 215,700 (DOE 2003). The primary facility at the Portsmouth site is the Portsmouth Gaseous Diffusion Plant, a gaseous diffusion uranium enrichment plant previously operated first by DOE until 1993 and since then by the United States Enrichment Corporation (USEC). Uranium enrichment operations were discontinued in May 2001, and the plant was placed in cold standby, a nonoperational condition in which the plant retains the ability to resume operations within 18 to 24 months.

In December 2002, USEC announced that the Portsmouth site will be the location for a Lead Cascade Demonstration Facility for advanced centrifuge enrichment technology (SAIC 2004). NRC has recently authorized possession and use of source and special nuclear material at the proposed enrichment facility (69 FR 3956). In addition, USEC announced on January 12, 2004, that the Portsmouth site was selected as the location for a new $1.5 billion advanced centrifuge commercial plant (referred to as the American Centrifuge Uranium Enrichment Plant), expected to be operational by the end of the decade (DOE 2003). USEC submitted a license application for this proposed facility to the NRC on August 23, 2004.

The Portsmouth site encompasses approximately 1500 ha (3714 ac), including a 320-ha (800-ac) fenced core area that contains the former production facilities. The 1180-ha (2914-ac) area outside the core area includes restricted buffers, waste management areas, plant management and administrative facilities, gaseous diffusion plant support facilities, and vacant land. The site is 3 km (2 mi) east of the Scioto River in a small valley that runs parallel to and approximately 37 m (120 ft) above the Scioto River floodplain (Bechtel Jacobs 2003). Wayne National Forest borders the plant site on the east and southeast, and Brush Creek State Forest is located to the southwest, slightly more than 1.6 km (1 mi) from the site boundary. On the basis of an analysis of Landsat satellite imagery from 1992, dominant land cover categories in Pike County include deciduous forest (64.6 percent), pasture/hay (21.6 percent), and row crops (10.3 percent) (DOE 2003).

Water for the Portsmouth site comes from an onsite water treatment plant that in turn draws water from offsite supply wells adjacent to the Scioto River. The Ohio Valley Electric Corporation supplies the site with electrical power.

The topography of the Portsmouth site area consists of steep hills and narrow valleys, except where major rivers have formed broad floodplains. The site is underlain by bedrock composed of shale and sandstone. The most common type of vegetation on the Portsmouth site is managed grassland, which makes up approximately 30 percent of the site or about 445 ha (1100 ac). Approximately 28 percent of the site is forested, predominately by stands of oak-hickory and mixed hardwood (DOE 2003).

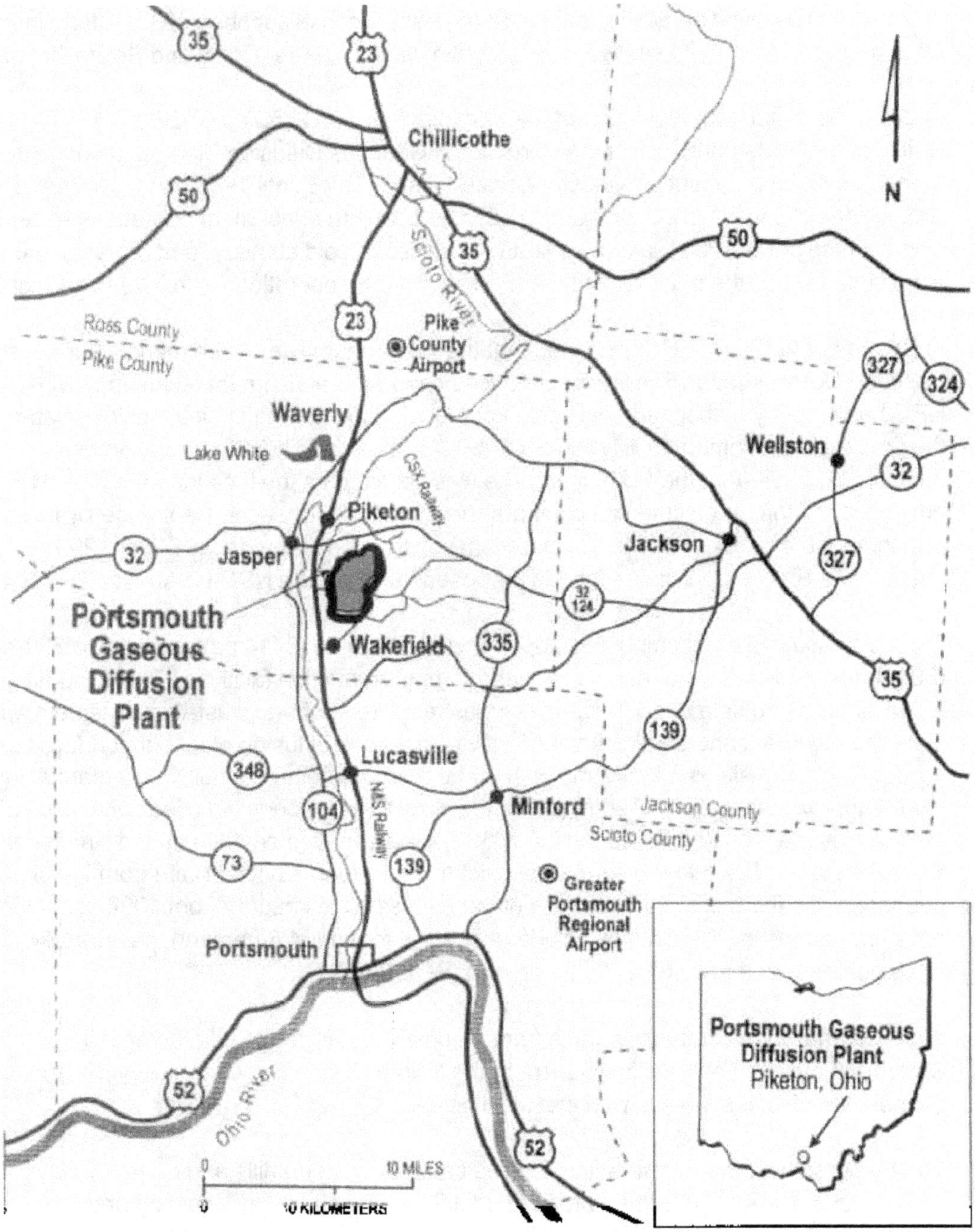

Figure 8-3. Portsmouth Gaseous Diffusion Plant Site Vicinity Map

1
2
3

1 The Portsmouth site is located in the humid continental climatic zone and has weather
2 conditions that vary greatly throughout the year. The site is in a rural setting, and no
3 residences or other sensitive locations (e.g., schools or hospitals) exist in the immediate
4 vicinity of the site. The Portsmouth site has direct access to major highway and rail systems,
5 a nearby regional airport, and barge terminals on the Ohio River. Use of the Ohio River barge
6 terminals requires transportation by public road to or from the Portsmouth site.
7
8 The site is located within the Western Allegheny Plateau ecological province (Omernik 1987).
9 The hilly, forested areas of the 138-ha (340-ac) site were harvested for timber before the
10 Portsmouth facilities were established. The eastern portion of the site has steep forested
11 slopes while the central and western areas are composed mainly of old fields and managed
12 grasslands that are not considered unique habitat or environmentally sensitive areas. Little
13 Beaver Creek runs through the southwestern part of the 138-ha (340-ac) alternative site area
14 and is identified as having riparian forest along its banks (DOE 2003). Oak-hickory forest
15 borders the riparian forest. Other than Little Beaver Creek, there is only about 1 ha (2 ac) of
16 wetlands within the alternative site. There is one Federally listed endangered species, the
17 Indiana bat (*Myotis sodalis*), and one non-listed Federal species of concern, (the timber
18 rattlesnake *Crotalus horridus*), that potentially could be found on the site.
19
20 The Portsmouth site itself has provided significant socioeconomic benefits for the surrounding
21 communities over the last 50 years, including jobs with above-average salaries. DOE does
22 not pay property taxes to the local communities around the Portsmouth site. However, it has
23 provided $12.9 million in grants to the Southern Ohio Diversification Initiative. Other economic
24 benefits include the collection of sales tax on uranium enrichment services. Sales, property,
25 and income taxes have been paid over the years to Ohio and local governments by
26 employees working at the site (Dominion and Bechtel 2002).
27
28 Major economic activities around Portsmouth consist mainly of farming, lumbering, and small
29 businesses. Other industries include a cabinet manufacturer and an automotive parts
30 manufacturer. The site itself has no prime agriculture lands. Sufficient public transportation
31 (rail and road) is present to support activities at the site (Dominion and Bechtel 2002).
32

8.6.1 Land Use Including Site and Transmission Lines

34
35 The alternative ESP parcel at the Portsmouth site is irregular in shape. At its widest points,
36 the parcel spans about 1737 m (5700 ft) in the north-south direction and about 1798 m
37 (5900 ft) in the east-west direction. The parcel is in a mostly undisturbed part of the site. No
38 hazardous substances have been stored, released, or disposed of on the parcel (Dominion
39 and Bechtel 2002). The closest disturbed land is used by security personnel for training and
40 as a firing range. The firing range is outside the 138-ha (340-ac) parcel, but is adjacent to the
41 parcel's boundary lines. The location of the 138-ha (340-ac) parcel is shown in Figure 8-4.
42 Two commercial nuclear units sited at the Portsmouth site would need to have an exclusion

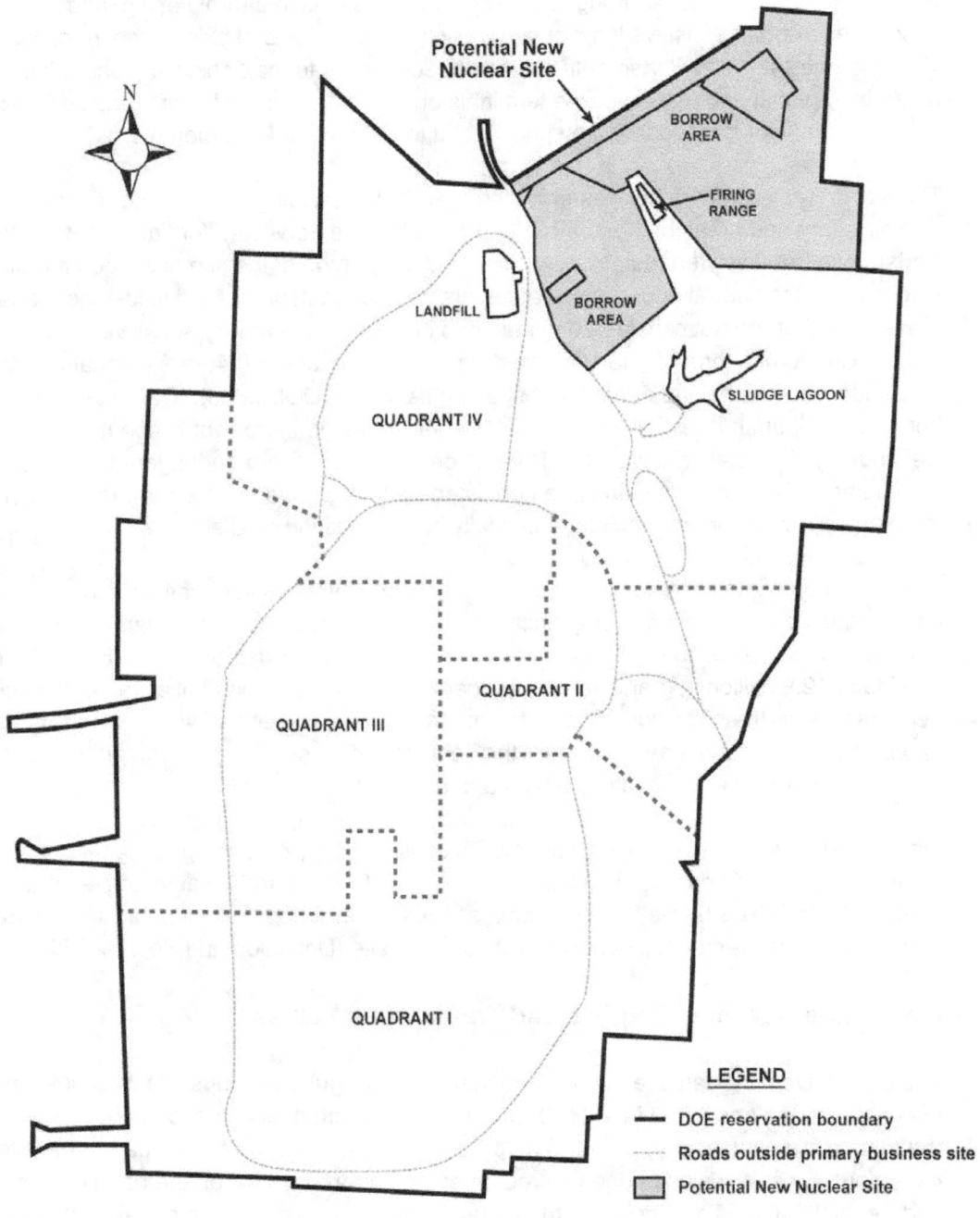

Figure 8-4. Potential New Nuclear Station Site at Portsmouth

1
2

1   area that meets NRC requirements (10 CFR 100.21(a)). The exclusion area is the area
2   surrounding the reactor within which the reactor licensee has the authority to determine all
3   activities, including exclusion or removal of personnel and property from the area
4   (10 CFR 100.3).
5

6   Six wetland areas comprise approximately 1 ha (2 ac) within the 138-ha (340-ac) parcel
7   (66 FR 64963). Five of these areas are ditches within a borrow area. The remaining wetland
8   is associated with a previously disturbed natural area. Dominion determined that construction
9   activities could take place without affecting the wetland areas (Dominion and Bechtel 2002).
10

11   Land within 8 km (5 mi) of the Portsmouth site is used primarily for farms, forests, and urban
12   or suburban residences (Dominion and Bechtel 2002). About 10,291 ha (25,430 ac) of
13   farmland, including cropland, wooded lots, and pasture, are within 8 km (5 mi) of the site. The
14   cropland is mostly found on or adjacent to the Scioto River floodplain and is farmed
15   extensively, particularly with grain crops. The hillsides and terraces are used as pasture for
16   both beef and dairy cattle. There are no state or national parks, conservation areas, wild and
17   scenic rivers, or other areas of recreational, ecological, scenic, or aesthetic importance within
18   the immediate vicinity of the Portsmouth site. There are approximately 9874 ha (24,400 ac) of
19   forestland within 8 km (5 mi) of the site. This land includes some commercial woodlands and
20   a small portion of Brush Creek State Forest. A relatively small area of urban land, about 206
21   ha (510 ac), is also within 8 km (5 mi) of the Portsmouth site. This land is situated primarily in
22   and around Piketon, approximately 5.6 km (3.5 mi) north of the center point of the site.
23

24   The Portsmouth site is about 113 km (70 mi) south of Columbus, Ohio, and 121 km (75 mi)
25   east of Cincinnati, Ohio, the two closest metropolitan areas. Huntington, West Virginia, is
26   approximately 140 km (87 mi) southeast of the site. The cities of Portsmouth, Jackson, and
27   Chillicothe, Ohio, are approximately 40 km (25 mi) from the facility (south, east, and north,
28   respectively). There are numerous small towns within 80 km (50 mi) of the site. Together,
29   these communities could supply an adequate workforce for construction of new generating
30   units and are within a 2-hour commuting distance via local transportation routes (Dominion
31   and Bechtel 2002). The regional transportation network consists of two major highways,
32   U.S. Route 23 and SR 32, and numerous state routes including SRs 35, 52, 124, and 139.
33   Offsite land-use impacts associated with construction of new nuclear generating units are
34   likely to be relatively limited, given the temporary nature of the construction. Construction of
35   new rental housing and/or new manufactured home and recreational vehicle parks could be
36   expected to accommodate construction workers.
37

38   The majority of current Portsmouth workers live in Scioto and Pike Counties and in the city of
39   Portsmouth, which is the county seat of Scioto County (DOE 2003). The staff assumed that
40   workers at two units located at the Portsmouth site would also primarily live in Scioto and Pike
41   Counties. Some new housing in Scioto and Pike Counties plus surrounding communities
42   would likely be constructed to accommodate permanent workers at the new units. The

1   property tax revenue from the new units could affect future land use in the area surrounding
2   the plant as a result of infrastructure improvements made possible by the tax revenue.
3
4   Based on the foregoing, the staff concludes that the land-use impacts on the site and in the
5   vicinity of construction and operation would be SMALL.
6
7   An extensive existing electric power transmission system serves the Portsmouth site. During
8   previous full-power operations of the enrichment facility, the site imported approximately
9   1900 MW(e) of power with a reported system capacity of approximately 2260 MW(e)
10   (Dominion and Bechtel 2002). One or more relatively short transmission lines (approximately
11   900 m [3000 ft]) could be used to connect to the existing transmission system serving the
12   Portsmouth site. Transmission lines that would be constructed to connect new power reactors
13   with the existing Portsmouth transmission system would primarily cross previously developed
14   industrial lands within the boundaries of the Portsmouth site (Dominion and Bechtel 2002).
15   Otherwise, relatively little land would be altered for the construction of the new lines.
16   Accordingly, the staff concluded that the overall impacts of constructing new lines and
17   operating and maintaining them would be SMALL.
18

19   ## 8.6.2 Water Use and Quality
20

21   Whether the consumptive water use for new nuclear units at the Portsmouth site would be
22   provided indirectly from the Scioto River aquifer via groundwater wells or directly from the river
23   itself, the consumptive water use for a power reactor at the Portsmouth site would impact the
24   Scioto River. Groundwater in the aquifer is directly connected to the Scioto River. The
25   aquifer is a major source of water to domestic, industrial, and agricultural users in the region.
26

27   The staff reviewed streamflow records reported by the USGS for stream gauge 03231500
28   (Scioto River at Chillicothe, Ohio). This gauge reflects runoff from a drainage of 9969 km^2
29   (3849 mi^2) and has data for the period from 1921 to present. Using these data, the staff
30   independently estimated the lowest 7-day discharge for low water condition that is estimated
31   every 10 years (7Q10) and the lowest 30 days of flow in an average year (30Q2) values. For
32   this gauge, the 7Q10 was estimated to be 5.72 m^3/s (202 cfs) and the 30Q2 was estimated to
33   be 11.4 m^3/s (403 cfs). The 7Q10 provides an estimate of the short-term, low-flow conditions
34   in a dry year. The 30Q2 provides an estimate of the intermediate-term low-flow conditions in
35   an average year.
36

37   The maximum make-up water flow rate for a single unit is estimated in the PPE as 2.78 m^3/s
38   (98.0 cfs); however, the portion of the flow not evaporated is ultimately returned to the river as
39   blowdown flow. Based on the PPE, the maximum evaporation for a single unit using
40   mechanical draft cooling towers is 1.23 m^3/s (43.5 cfs). For either one or two units, this
41   represents a significant fraction of both the 7Q10 and the 30Q2 values. Some mitigation,
42   such as aquifer recharge, offstream storage, etc., may be warranted to limit the impacts to

1 other water users in the region. Therefore, the staff estimates water use impacts at the
2 Portsmouth site during construction would be SMALL, and during operation would be SMALL
3 to MODERATE except during drought years when the impact is expected to be MODERATE.
4 Discharge of thermal and chemical effluents would be regulated by the State of Ohio's
5 National Pollutant Discharge Elimination System (NPDES) permitting process to limit water
6 quality impacts to the Scioto River, therefore water quality impacts during construction and
7 operation would be expected to be SMALL.
8
9 ### 8.6.3 Terrestrial Resources Including Endangered Species
10
11 Site preparation and construction of one or more nuclear reactor units at the Portsmouth
12 alternative site would result in the loss of wildlife habitat in the developed portion of the
13 approximately 140 ha (340-acre) site. In general, the types of habitat that would be lost are
14 relatively common in south-central Ohio, and are not considered to be unique or sensitive.
15 Less than 0.4 ha (1 ac) of wetlands would be disturbed for the development of this site
16 (Dominion and Bechtel 2002). Site development would have the potential to result in erosion,
17 dust generation, and noise impacts that are typical of large construction projects. These could
18 be mitigated using standard methods of erosion control, dust suppression, and noise
19 abatement.
20
21 Site development could result in the loss of habitat for several Federally or State-listed
22 species (Table 8-4). The Indiana bat (*Myotis sodalis*) could inhabit the riparian forest along
23 Little Beaver Creek, while the rough green snake (*Opheodrys aestivus*) and sharp-shinned
24 hawk (*Accipiter striatus*) may inhabit the forested portions of the Portsmouth site. The timber
25 rattlesnake (*Crotalus horridus*) has not been observed on the Portsmouth site, but is believed
26 to occur in the vicinity. DOE has indicated that additional field surveys and evaluations would
27 be needed prior to development of this location (Dominion and Bechtel 2002).
28
29 Development of the site could adversely affect at least two Ohio State-listed plant species, the
30 Virginia meadow-beauty (*Rhexia virginica*) and the Carolina yellow-eyed grass (*Xyris
31 difformis*). Evaluations for these species would be necessary prior to site development.
32 Pending site-specific surveys for Federally and State-listed species, site preparation and
33 construction within the Portsmouth alternative site would not result in noticeable or
34 destabilizing impacts to the terrestrial ecology of the site. Therefore, pending more detailed
35 evaluations of threatened and endangered species, the construction impacts would be
36 SMALL. If the Federally or State-listed animal species are observed at the site, the impacts
37 could be larger.

Table 8-4. Federally and State-Listed Threatened or Endangered Terrestrial Species Potentially Near the Portsmouth Alternative Site

Scientific Name	Species	Federal Status	State Status
Mammals			
Myotis sodalis	Indiana bat	E	E
Birds			
Accipiter stiatus	sharp-shinned hawk		E
Reptiles			
Crotalus horridus	timber rattlesnake		E
Opheodrys aistivus	rough green snake		S
Plants			
Rhexia virginica	Virginia meadow-beauty		P
Xyris difformis	Carolina yellow-eyed grass		E

E = endangered, S = Special concern, P = potentially threatened.
Source: DOE 2003.

If new reactor units were constructed at the Portsmouth alternative site, it would be expected that very little usable habitat would remain within the site. Operation of the facility would likely result in noise generation, and if wet cooling towers were employed, there could be impacts caused by salt drift, icing, fogging, and bird collisions. Noise is likely to be typical of operating reactor units and cooling towers, which has been found to have a minimal impact in most instances (NRC 1996). There are no sensitive habitat areas adjacent to the Portsmouth site that would be adversely affected by noise from plant operations. The terrestrial vegetation in the vicinity of the site is not believed to be unusually sensitive to salt drift, fogging, or icing (Dominion and Bechtel 2002). Bird collision impacts would not be expected to be different from most other power plants (NRC 1996). Based on the foregoing, the staff concludes that the impacts of operation of one or more reactor units at the Portsmouth alternative site on terrestrial systems would be SMALL.

Transmission lines that would be constructed to connect new power reactors with the existing Portsmouth transmission system would primarily cross either industrial lands or early-successional plant communities that are not considered to be unique in the region or otherwise sensitive. One exception is the potential for Indiana bats to inhabit the riparian zone along Little Beaver Creek. Additional evaluations and habitat preservation precautions may be necessary if the transmission lines cross this habitat. Otherwise, very little land would be altered for the construction of the new lines, and the overall impacts of constructing new lines would be minimal and would be similar to those of the construction of the reactor units.

Dominion has not indicated the specific maintenance procedures that would be followed for the new transmission lines, but they would likely include regular mowing and herbicide applications as needed. The primary area of potential concern is whether the lines must cross Little Beaver Creek, where precautions to protect Indiana bats and their habitat may be required. Additionally, there are at least two rare plant species (Table 8-4) that could occur within transmission lines rights-of-way and could therefore be affected by transmission line maintenance. Additional evaluations for the Indiana bat and rare plant species would be needed. Otherwise, because the transmission lines that would be needed are short, entirely within the bounds of the Portsmouth site and cross areas that have been previously disturbed, the impacts of operation and maintenance of the transmission line system on the terrestrial ecology would be expected to be minimal. No special mitigation measures would be warranted, except protection of the riparian zone along Little Beaver Creek and conservation of rare plant species, if present.

Based on the foregoing, the staff concludes that the overall impact on terrestrial ecological resources of the construction and operation of two commercial nuclear units and associated cooling systems and transmission facilities at the Portsmouth alternative site would be SMALL, pending additional surveys for threatened and endangered species within the construction area or transmission line rights-of-way. If such species were found, some mitigation measures may be warranted.

8.6.4 Aquatic Resources Including Endangered Species

The aquatic resources near the Portsmouth site would not be expected to be impacted by the construction and operation of two commercial nuclear units. The water used for cooling at Portsmouth would be withdrawn from groundwater wells. The cooling water would be expected to be discharged to the Scioto River. Discharge limits would be controlled by Federal and State regulations for protection of the river. Based on the foregoing, the staff concludes that the overall impact on aquatic ecological resources (including threatened and endangered species) of construction and operation of two commercial nuclear units and associated cooling towers and transmission facilities at the Portsmouth site would be SMALL.

8.6.5 Socioeconomics, Historic and Cultural Resources, Environmental Justice

The potential impacts on socioeconomics, historic and cultural resources, and environmental justice from construction and operation of two units at the Portsmouth site were evaluated using a reconnaissance survey of the site. That is, readily obtainable data from the Internet or published sources were used in the evaluation, and no new data were collected. The subsections that follow reflect the organizational structure of the socioeconomic discussions found in Sections 2.8.4, 4.5, and 5.5. The impacts resulting from both construction and operation of two units are addressed.

1 **8.6.5.1 Physical Impacts**

2

3 Construction activities can cause temporary and localized physical impacts due to matters
4 such as noise, odor, vehicle exhaust, vibration, shock from blasting, and dust emission. The
5 use of public roadways, railways, and barges would be necessary to transport construction
6 materials and equipment to the site. SR 32 and U.S. Route 23 appear to be well maintained
7 and have been used for transporting heavy loads in the past (Dominion and Bechtel 2002).
8 The staff expects that all construction activities would occur within the boundaries of the
9 Portsmouth site. Offsite areas that would support construction activities (e.g., borrow pits,
10 quarries, disposal sites) are expected to be already permitted and operational. Impacts on
11 those facilities from constructing two nuclear units would be small incremental impacts
12 associated with their normal operation.

13

14 Potential impacts from station operation could result from matters including noise, odors,
15 exhausts, thermal emissions, and visual intrusions. Noise would be produced from the
16 operation of cooling towers, pumps, transformers, turbines, generators, and switchyard
17 equipment, and from traffic. Dominion states in its ER that any noise coming from the North
18 Anna ESP site would be controlled in accordance with applicable local county regulations. By
19 inference, this is also expected to apply to the Portsmouth site. Good road conditions and
20 appropriate speed limits would minimize the noise level generated by the workforce
21 commuting to the Portsmouth site.

22

23 The nuclear units would be expected to have emissions from auxiliary power systems and
24 standby diesel generators. It is expected that the combined annual emissions of any pollutant
25 would be less than 91 MT/yr (100 tons/yr) (Dominion and Bechtel 2002). Air permits acquired
26 for these generators would ensure that air emissions comply with regulations. Paved access
27 roads and appropriate speed limits would minimize the amount of dust emissions generated
28 by the commuting workforce.

29

30 The nuclear facility with its two units and associated buildings and its cooling towers and
31 associated plume would change the landscape and would be visible from the sparsely
32 populated area north of the site.

33

34 Based on the foregoing, the staff concludes that the physical impacts of construction and
35 operation would be SMALL. Construction activities would be temporary and occur mainly
36 within the boundaries of the Portsmouth alternative site. Offsite impacts would represent
37 small incremental changes to offsite services supporting construction activities. During station
38 operations, noise levels would be managed to meet local ordinance requirements. Air quality
39 permits would be required for equipment such as diesel generators and auxiliary boilers,
40 which should limit air emissions and ensure that applicable standards are met.

41

8.6.5.2 Demography

The Portsmouth site is located in Pike County, Ohio, approximately 35 km (22 mi) north of the Ohio River and 5 km (3 mi) southeast of the town of Piketon. Pike County's largest community, Waverly, has a population of 4433 residents. The nearest residential center to the site is Piketon with a population of 1907. Additional population centers within 80 km (50 mi) of the plant are Portsmouth (population 20,909), 35 km (22 mi) south; Chillicothe (population 21,796), 43 km (27 mi) north; and Jackson (population 6184), 29 km (18 mi) east. Approximately 90 percent of Portsmouth site workers reside in Jackson, Pike, Ross, and Scioto Counties. The population for the four counties in 2000 was 212,876 (USCB 2000b).

Most of the construction and operations workforce are expected to come from within the region, and those who might relocate to the region would represent a small percentage of the larger population base. Those who do relocate to the region would most likely take up residency across the region. Based on the foregoing, the staff concludes that any environmental impacts caused by population increases within an 80-km (50-mi) radius of the Portsmouth site attributable to construction or operation of two units would be SMALL.

8.6.5.3 Community Characteristics

Economy

Economic activities near the Portsmouth site consist primarily of farming, timber harvesting and processing, and small businesses. The only significant industry in the vicinity is an industrial park south of Waverly. Industries include a cabinet manufacturer and an automotive parts manufacturer (Dominion and Bechtel 2002).

The unemployment rate in Ohio was 6.1 percent as of July 2004. At that time, the unemployment rates in Jackson, Pike, Ross, and Scioto Counties were 9.2, 8.3, 7.8 and 8.5 percent, respectively (ODJFS 2004). This data indicates that this area of Ohio has not fully recovered from the recession of 2001. The Portsmouth site itself has provided significant socioeconomic benefits for the surrounding communities over the last 50 years, providing jobs that paid above average wages and salaries. The overall economic impacts of constructing and operating two units at the Portsmouth alternative site would be beneficial to the local economy.

Based on the foregoing, the staff concludes that the beneficial impacts of station construction and operation on the economy of the region would be SMALL everywhere in the region except Pike County, where the beneficial impact level on the region would be MODERATE. The magnitude of the economic impacts would be diffused in the larger economic bases of the region; whereas, within the smaller economic base of Pike County and given the fact the new

1   units would be located in the county, the economic impacts could be more noticeable and
2   have a greater beneficial impact.
3
4   *Availability of Workers*
5
6   Dominion estimates it would take approximately 5000 construction workers more than 5 years
7   to build two nuclear units at the Portsmouth site. The Portsmouth site would draw its workers
8   from the tri-state area of southern Ohio, northern Kentucky, and western West Virginia. The
9   construction workforce in this region is estimated to be 491,265.[a]
10
11   Except for electricians, skilled craft workers in the Portsmouth area are reported as fully
12   employed. The concentration of industrial facilities within this region (e.g., oil refineries and
13   steel mills) provides yearly employment for the building trades. This could present significant
14   competition for manpower if this site were selected for construction of new nuclear units.
15   Moreover, this area has a reputation as a complicated labor environment, and the shutdown of
16   the Portsmouth enrichment facility operations has contributed to this climate (Dominion and
17   Bechtel 2002).
18
19   The Portsmouth site currently provides employment for more than 1800 people. The site
20   employs a highly skilled workforce with decades of nuclear-related experience. During the
21   last several years, Portsmouth has undergone a major downsizing. Dominion would need
22   approximately 720 new employees to operate the proposed new nuclear units. The addition
23   of commercial nuclear generation would be expected to add jobs of similar or higher quality to
24   the existing workforce, many of which could be filled by current or former Portsmouth site
25   employees (Dominion and Bechtel 2002).
26
27   Based on the foregoing, the staff concludes that there appears to be a large supply of
28   construction workers but a limited availability of skilled craft workers. Dominion may have to
29   recruit from outside the region to fill its requirements for skilled craft workers. Employees for
30   station operation would be expected to be available from within the region because of the
31   downsizing of the Portsmouth site workforce.
32
33   *Transportation*
34
35   Two major highways serve the Portsmouth site: U.S. 23 and SR 32. At their nearest points,
36   these highways run within 1.6 km (1 mi) of the site. Access to the site is by the main access
37   road, a four-lane interchange with U.S. 23, and the north access road, which initially is a

---

(a) The estimate is based on a methodology explained in Dominion and Bechtel (2002) and updated
     using 2000 census data.

1   two-lane road that transitions to four lanes with SR 32. SR 32 and U.S. 23 both appear to be
2   well maintained and have been used for the transport of heavy loads (Dominion and
3   Bechtel 2002).
4
5   As previously mentioned, constructing the two units would employ a construction workforce of
6   5000. Operating two nuclear units would employ an operations workforce of 720. During the
7   Portsmouth site operational period, between the 1970s and 2001, the total workforce
8   numbered about 5000 at its peak. Currently 1800 people work at the site. Based on this
9   previous peak, nearby access roads should be capable of supporting both construction and
10   operations commuter traffic at this level with some roadway upgrades and traffic signal
11   improvements. In addition, there are adequate transportation routes in the area to handle
12   transportation of bulk materials to and from the site (Dominion and Bechtel 2002).
13
14   Two major rail lines service the site: CSX and Norfolk and Southern. Both railways appear to
15   be in excellent condition. Approximately 35 km (22 mi) south of the Portsmouth site, two main
16   rail lines run east-west along the Ohio River. The river is used for barge transportation, so
17   materials could be off-loaded from barges onto rail cars, making transportation to the
18   Portsmouth site by either rail or road achievable (Dominion and Bechtel 2002).
19
20   Numerous airports are within 161 km (100 mi) of the site, including the airports at Columbus,
21   Cincinnati, and Dayton, Ohio, and Charleston, West Virginia. All these airports conduct
22   regular freight and passenger air services. In addition, there are numerous smaller airports in
23   the immediate vicinity. Thus, air passenger service and freight service for shipment of small
24   items via air are readily available (Dominion and Bechtel 2002).
25
26   The Portsmouth site is in a rural, low-population area. The regional transportation network is
27   adequate for commuter and transient traffic in the area. The transportation system around
28   the Portsmouth site was capable of handling 5000 workers during previous periods of peak
29   operations. Based on the foregoing, the staff concludes that the transportation impacts of a
30   construction or operating workforce on the transportation infrastructure would be SMALL.
31
32   *Taxes*
33
34   The State of Ohio has a 5-percent sales tax. In addition to the State sales tax, each county in
35   Ohio has a county sales tax. Jackson, Ross, and Scioto Counties each have a sales tax rate
36   of 1.5 percent, and Pike County has a sales tax rate of 1 percent (NRC 2004). Sales taxes
37   would be paid from the sales of construction materials and supplies purchased for the project.
38   The State of Ohio has a personal income tax rate with a top marginal rate of 5.2 percent for
39   incomes in excess of $40,000 (NRC 2004).
40
41   The average property tax rates for Ohio cities are divided into three separate classifications:
42   Class I Real (residential and agricultural); Class II Real (commercial, industrial, mineral, and

1   public utility); and Class III Tangible Personal (general and public utility) (NRC 2004). For
2   Waverly in Pike County, the rate was $0.07412 per $1000 for all three classifications in 2001;
3   for Portsmouth in Scioto County, the rate was $0.06013 per $1000 for all three classifications
4   in 2001; for Wellston in Jackson County, the rate was $0.05500 per $1000 for all three
5   classifications; and for Chillicothe in Ross County, the Class I rate was $0.05407, the Class II
6   rate was $0.05394, and the Class III rate was $0.05402 per $1000. Finally, because the units
7   would be built by a private company (Dominion) and not DOE, a property tax might be levied
8   on the value of the property that hosts the units as they are constructed and on the appraised
9   value of the units once construction is completed and the units are brought online. These
10   taxes would most likely go to Pike County.
11
12   DOE does not pay property taxes to the local communities around the Portsmouth site.
13   However, DOE has provided $12.9 million in grants to the Southern Ohio Diversification
14   Initiative. Other economic benefits include the collection of sales tax on uranium enrichment
15   services. Adding commercial nuclear capacity at the Portsmouth site would be expected to
16   increase the tax base for these localities for the life of the two units (Dominion and Bechtel
17   2002).[a]
18
19   Workers living outside Ohio and commuting to Portsmouth likely will have to pay income taxes
20   to their state of residence (West Virginia and Kentucky). They may also have to pay sales
21   taxes to the State and local governments in the region where sales take place and property
22   taxes to the counties in which they might own a residence.
23
24   Based on the foregoing, the staff concludes the overall beneficial impacts of construction and
25   operation of the new facility on taxes collected in the region through the income, sales and
26   use, and property taxes (except for Pike County) would be SMALL. The taxes paid, while
27   substantial, are nevertheless a small sum when compared to the total amount of taxes
28   collected by states and local governments in the region. For property taxes for Pike County,
29   the staff considers the overall beneficial impacts of the property taxes collected would be
30   LARGE[b] (operations) relative to the total amount of taxes the county collects through
31   property taxes.

(a) The proposed ESP site would not be on land owned by Dominion. Most likely, should the Portsmouth
     site be chosen for the new plant, the site for the new units would be leased to Dominion by DOE.
(b) The derivation of this impact is based on the fact that the fiscal year 2003 amount of property taxes
     collected in Pike County was $9,878,000 (Burton in Jaksch and Scott 2005). For comparison, NAPS
     Units 1 and 2 pay approximately $10 million in annual property tax to Louisa County (the actual
     amount that the proposed nuclear plant would pay to Pike County would depend on assessed value
     and millage rate per thousand of assessed value). On the assumption that there is a rough
     comparison between what Dominion pays to Louisa County and what they might pay to Pike County, it
     can be concluded that the potential percentage of the proposed facility's property taxes to the total of
     all property taxes paid in Pike County would be significant.

1　*Aesthetics and Recreation*

2

3　There are no significant recreational or residential areas within 3 km (2 mi) of the proposed
4　Portsmouth alternative site. Mechanical-draft cooling towers or natural-draft towers could
5　function as part of the new nuclear units' cooling system, which could produce visible plumes
6　offsite. Nearby trees may serve as a visual buffer for the transmission facilities. The preferred
7　location at the Portsmouth site is situated in an area with open terrain. Because the preferred
8　site is close to the northeast corner of the existing site boundary, it is possible that the
9　proposed units would have an identifiable nuclear power plant view offsite (Dominion and
10　Bechtel 2002), especially if natural draft towers were used for cooling.

11

12　Recreational facilities in the Portsmouth area include Brush Creek State Forest. Use of Lake
13　White State Park is occasionally heavy and is concentrated on the 43 ha (107 ac) of land
14　closest to the lake. The number of visitors in 1992 was 55,876 with a daily average of 153
15　(Dominion and Bechtel 2002).

16

17　The AP1000 reactor has a tall containment building, approximately 71 m (234 ft) above grade
18　level. Natural-draft cooling towers would be about 170 m (550 ft) above grade and
19　mechanical-draft towers would be about 18 m (60 ft) above grade. The design of the AP1000
20　containment building includes a hatch that determines the height that it must be above grade.
21　This building would be expensive to redesign to allow the building to be placed lower in the
22　ground. Thus, depending on the type of cooling tower chosen, the height of the AP1000
23　containment building could set the upper bound of what buildings would be visible from offsite.

24

25　There are no significant residential areas or recreational facilities within 3 km (2 mi) of the site.
26　Plumes from mechanical-draft cooling towers could be visible offsite. Trees ordinarily serve
27　as a visual buffer for the power transmission infrastructure. Based on the foregoing, the staff
28　concludes that the impacts of construction and station operation on aesthetics at a wooded
29　site would be SMALL. But the impacts could also be MODERATE, if the Portsmouth site is on
30　a mostly undisturbed part of the site and on open terrain, enabling the reactors and the
31　cooling towers and their plumes to be viewed from offsite.

32

33　*Housing*

34

35　In the four-county area of Jackson, Pike, Ross, and Scioto Counties, there were 89,026
36　housing units in 2000. Of these, 22,824 were rental units, 2150 of which were vacant, for an
37　8.6 percent vacancy rate (USCB 2000c). The Portsmouth site is about 113 km (70 mi) south
38　of Columbus, Ohio, (population in 2000 of 711,470), and 121 km (75 mi) east of Cincinnati,
39　Ohio, (population in 2000 of 331,285), which are the two closest metropolitan areas (USCB
40　2000b). Huntington, West Virginia (population 51,475), is approximately 140 km (87 mi) away
41　to the southeast. These three cities have a total housing stock of 519,075, of which 256,326
42　are renter-occupied and 24,868 units are available for rent for a vacancy rate of 8.8 percent

1   (USCB 2000c). Because it is not unusual for construction workers to commute up to 2 hours
2   (one way) per day, there appears to be enough vacant rental housing to house those who
3   might relocate to the region.
4

5   In the four-county area of Jackson, Pike, Ross, and Scioto Counties there were 58,246
6   owner-occupied housing units in 2000. In the four-county area, 1085 units were for sale, or
7   1.8 percent of the total of owner-occupied houses. In the three-city area of Columbus,
8   Cincinnati, and Huntington, there were 218,258 owner-occupied housing units and 4778 units
9   for sale, or 2.1 percent of total of owner-occupied houses (USCB 2000c). In both the local
10   and larger metropolitan areas (i.e., Columbus, Cincinnati, and Huntington), the percentage of
11   houses for sale in relation to owner-occupied housing is very low, indicating a fairly tight
12   housing market.
13

14   The operations workforce is expected to come from current or former employees at the
15   Portsmouth site. If, however, a substantial number of workers have to be recruited into the
16   area, upward pressure on housing values could emerge. This assumption is based on the low
17   number of homes for sale in the area and the fact that the workforce, which would be on the
18   higher end of the salary scale when compared to other job classifications in the area, may
19   tend to buy more upscale homes. In this case, the operational impacts could be moderate.
20

21   It is not unusual for construction workers to commute up to 2 hours (one way) per day to the
22   job site, and many of the construction workers are assumed to already live within the region.
23   There appears to be sufficient vacant rental housing to house those who might relocate to the
24   region. If, as expected, most of the operations workforce have residences already in the
25   region, then the impacts on housing because of station operation would be small. Based on
26   the foregoing, and assuming the operations workforce comes from within the region, the staff
27   concludes that the impacts to housing from construction and operation of the two units would
28   be SMALL, and mitigation is not warranted. If the operations workforce, the impacts would be
29   SMALL.
30

31   *Public Services*
32

33   <u>Water and Wastewater Treatment</u>
34

35   The capacity of communities to absorb an increase in population depends on the availability
36   of sufficient community resources, such as water and wastewater treatment. Two large
37   metropolitan areas (Columbus and Cincinnati) are within 145 km (90 mi) of the site.
38   Huntington, West Virginia, is approximately 140 km (87 mi) away. The cities of Portsmouth,
39   Jackson, and Chillicothe, Ohio, are within about 50 km (30 mi) from the facility (south, east
40   and north, respectively). There are numerous small towns within 80 km (50 mi) of
41   Portsmouth. All these towns and cities are within a 2-hour commuting distance via local

1 transportation routes of the site (Dominion and Bechtel 2002) and could provide public
2 services such as water and wastewater treatment to the construction and operations
3 workforce who might relocate to the area.
4
5 Many of the construction and operations workforce would come from within the region, and
6 those that choose to relocate to the region would most likely take up residence throughout the
7 region, thus placing minimal demands on the existing infrastructure. Based on the foregoing,
8 the staff concludes that the impacts of the construction and operations workforces on water
9 and wastewater treatment in the region would be SMALL.
10
11 Police, Fire, and Medical Facilities
12
13 The hospital nearest to the Portsmouth site is the Pike Community Hospital, located
14 approximately 12.1 km (7.5 mi) north of the facility on SR 104 south of Waverly. No other
15 acute-care facilities are located in Pike County. There is an urgent-care facility, Adena Health
16 Center, also on SR 104 near the hospital. In addition, two licensed nursing homes are located
17 near Piketon, and one nursing home is located in Wakefield; all are located within 8 km (5 mi)
18 of the Portsmouth site (NRC 2004). Other medical facilities exist in Jackson, Chillicothe, and
19 Waverly.
20
21 Several State, county, and local police departments provide law enforcement in the region
22 (NRC 2004). Any additional demands on law enforcement services could potentially be met
23 by the increased tax revenues available to support the services. There would most likely be a
24 time delay between the demand for the services and the collection of the tax revenues, which
25 could cause some short-term financial issues for the impacted jurisdictions.
26
27 Many of the potential construction and operations workforce probably already live within an
28 80-km (50-mi) radius of the region. There are a number of towns within a 2-hour commuting
29 distance of the site. Any new workers relocating to the area would most likely have places of
30 residency located throughout the region, which would not place an undue burden on any one
31 jurisdictional entity's infrastructure. Based on the foregoing, the staff concludes that the
32 impacts of the construction and operations workforce on police, fire and medical facilities in
33 the Portsmouth area would be SMALL.
34
35 *Social Services*
36
37 In Ohio, social services at the state level are overseen by the Ohio Department of Jobs and
38 Family Services. It develops and oversees programs and services designed to help Ohio
39 residents become independent through education, employment, job skills, and training. A
40 major responsibility of the department is to work with county departments of job and family
41 services, child support enforcement agencies, and public children's services agencies to
42 develop social service programs to strengthen families, protect children, and provide children

1   with an opportunity for a better life. The Department also administers the unemployment and
2   medicaid programs for Ohio. During construction, there could be increased demand for these
3   social services.
4

5   Generally, construction and operation of new nuclear units at the Portsmouth site would be
6   viewed as beneficial economically to the disadvantaged population segments served by the
7   Department. The workforce associated with the new units would most likely be better paid
8   than workers in other employment categories in the region. It is expected that, through the
9   multiplier effect, the number of jobs that could be filled by the disadvantaged population would
10   increase.
11

12   Construction and operation would have a beneficial economic impact to the economically
13   disadvantaged population in the region, which should lessen the demand for social services.
14   There could be an initial increase in demand for social services at the beginning of the
15   construction period, but this is considered manageable and limited. Based on the foregoing,
16   the staff concludes that the impacts of construction and station operation of two units at the
17   Portsmouth site on social and related services would be SMALL.
18

19   *Education*
20

21   Twenty-four public school districts provide public education for approximately 36,000 students
22   in the region. The two school systems nearest the Portsmouth site are in Pike and Scioto
23   Counties. In 2002, the combined enrollment of these schools was approximately 2387
24   (NRC 2004). Within the same area, three facilities provide daycare or schooling for
25   preschool-aged children and after-school care for school-aged children. Two of these
26   facilities accommodate 390 children (NRC 2004).
27

28   Many in the potential construction and operating workforce probably already live within the
29   region, and any new workers relocating to the area would most likely take up residency
30   throughout the region. Based on the foregoing, the staff concludes that the impacts of
31   construction and station operation on educational facilities and services as a result of
32   construction and operation of two units would be SMALL.
33

34   **8.6.5.4 Historic and Cultural Resources**
35

36   The area of southern Ohio where the Portsmouth site is located contains evidence from each
37   of the major prehistoric periods for eastern North America, including the Paleo-Indian, Archaic,
38   Woodland, and Fort Ancient periods. In early historic times, the area was occupied by the
39   Shawnee Tribe. The Euro-American historic period occupation in the vicinity began
40   about 1800. Doe completed archaeological and historic architectural surveys in 1996 at the
41   Portsmouth site. Consultation is ongoing with the Ohio State Historic Preservation Officer
42   concerning the results of these surveys (DOE 2003). Consultations with the Shawnee Indian

1 Tribe have not identified any traditional cultural properties or other resources of Native
2 American cultural value at the site.
3
4 The DOE survey found three archeological sites near the 138-ha (340-ac) parcel. The three
5 sites are northeast of the parcel (Dominion and Bechtel 2002). The closest site to the
6 Portsmouth parcel is the Holt Cemetery, which is located about 183 m (600 ft) from the
7 eastern boundary of the parcel. No national landmarks are near the Portsmouth site, and no
8 properties presently on the National Register of Historic Places are within the Portsmouth site.
9 The nearest National Register locations are Buzzardroost Rock and Lynx Prairie in Adams
10 County, about 48 km (30 mi) southeast of the Portsmouth site.
11
12 Based on the foregoing, the staff concludes that the potential impacts on historic and cultural
13 resources would be SMALL for construction at the Portsmouth site. Potential impacts from
14 operation of the proposed two units at the Portsmouth alternative site would also be SMALL,
15 because any such potential impacts would be identified and appropriate mitigation measures
16 could be effected during the construction phase.
17

8.6.5.5 Environmental Justice

19
20 DOE recently performed an environmental assessment (DOE 2001) for the Portsmouth site as
21 part of its winterization activities for placing the facility in cold standby. As part of that
22 assessment, an evaluation of potential environmental justice impacts was conducted. DOE
23 evaluated the distribution of minority populations in a four-county area around the Portsmouth
24 site. DOE defined a minority population as any area in which minority representation was
25 greater than the national average of 24.2 percent. In all four counties, minority populations
26 are smaller than the national average. Hence, using this definition, environmental justice was
27 not a concern (DOE 2001), nor is it a concern using the NRC criteria defined in 69 FR 52040
28 (which is different than that used by DOE – see Section 2.8.4). DOE then carried the analysis
29 a step further and examined the minority populations in the census tracts closest to the site.
30 None of the tracts closest to the site had minority representation greater than the national
31 average of 24.2 percent (DOE 2001).
32
33 Individuals with incomes below the poverty level were identified in the four-county region. A
34 low-income population included any census tract (1990 data) in which the percentage of
35 people with income below the poverty level was greater than the national average of
36 13.1 percent (Dominion and Bechtel 2002). Nearly all (41 of 48) of the census tracts in the
37 four-county area qualified as low-income populations, but none of the low-income populations
38 would suffer disproportionate impacts as a result of the construction and operation of new
39 nuclear units at the Portsmouth site.
40
41 Based on the foregoing, the staff concludes that the offsite impacts of construction and
42 operation of two units at the Portsmouth alternative site on minority and low-income

1 populations would be SMALL. No disproportionately high and adverse impacts were
2 identified.
3
4 ## 8.7   Evaluation of the Savannah River Site
5
6 Dominion selected a 100-hectare (250-acre) parcel of land in the northern portion of DOE's
7 Savannah River Site as a possible location for two commercial nuclear units (Dominion and
8 Bechtel 2002). For this evaluation, the following assumptions were made by the staff about
9 locating the proposed units at the alternative site at the Savannah River site.
10
11 • The units would use closed-cycle cooling.
12 • Natural- or mechanical-draft cooling towers would be employed.
13 • The Savannah River would be the source of cooling water.
14 • The existing intake structure is sufficient.
15 • Blowdown water would be discharged to the Savannah River or to Par Pond.
16 • The land area would be approximately 100 ha (250 ac).
17 • New transmission lines would be needed.
18
19 DOE's Savannah River Site occupies an area of approximately 800 km^2 (310 mi^2) adjacent to
20 the Savannah River, in Aiken and Barnwell Counties, South Carolina. The site is
21 approximately 40 km (25 mi) southeast of Augusta, Georgia, and 31 km (19.5 mi) south of
22 Aiken, South Carolina. The site is bounded along its southwest border by the Savannah River
23 for approximately 56 river km (35 river mi). The Savannah River Site vicinity is shown in
24 Figure 8-5.
25
26 The average population density in the counties surrounding the site is approximately 85
27 people per square mile, with the largest concentration in the Augusta metropolitan area.
28 Approximately 70 percent of the site employees live in South Carolina, primarily Aiken County,
29 and 30 percent live in Georgia (Westinghouse 2001).
30
31 The U.S. Atomic Energy Commission, predecessor agency to DOE, established the Savannah
32 River Site in the early 1950s. Historically, the mission of the site has been the production of
33 special radioactive isotopes to support national programs. DOE produced these isotopes in
34 five production reactors. After the material was produced at Savannah River Site, it was
35 shipped to other DOE sites for further processing.
36
37 Approximately 73 percent of the surface area of the Savannah River Site is composed of open
38 fields and upland forest. The forested areas consist primarily of upland pine and mixed
39 hardwoods. The remaining area consists of wetlands, streams, and reservoirs (22 percent)
40 and developed industrial and administrative areas (5 percent) (DOE 1999).

1

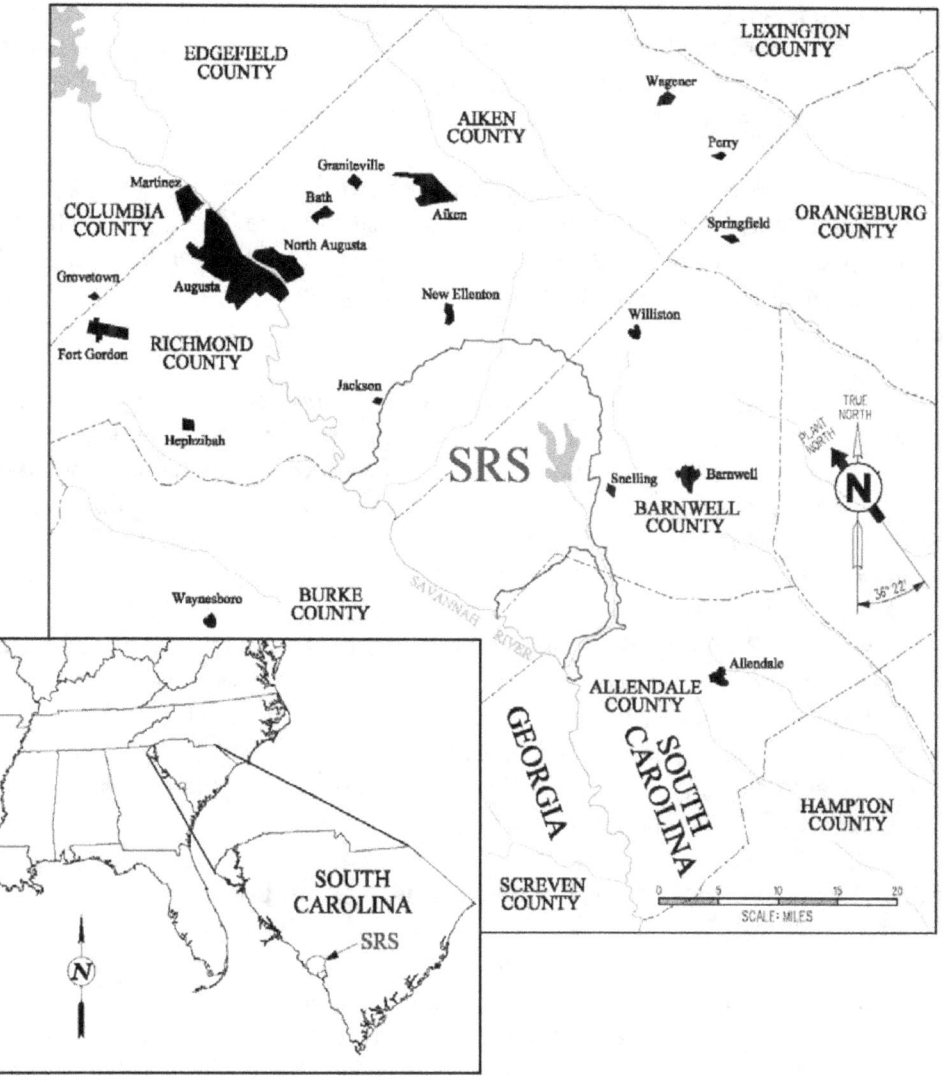

2
3   **Figure 8-5.** Savannah River Site Vicinity Map
4
5   The Savannah River is the principal surface water system associated with the Savannah River
6   Site. Five of its major tributaries (Upper Three Runs, Fourmile Branch, Pen Branch, Steel
7   Creek, and Lower Three Runs) flow through and drain the site. The Savannah River is a
8   domestic and industrial water source for the site and several downstream communities (the
9   cities of Port Wentworth and Savannah in Georgia and Beaufort and Jasper Counties in South
10   Carolina). In addition, the Vogtle Electric Generating Plant, located across the Savannah
11   River from the Savannah River Site, uses water from the river for its cooling system
12   (DOE 1999).

1   The southeastern United States has a humid subtropical climate characterized by relatively
2   short, mild winters and long, warm, humid summers. Summer-like weather typically lasts from
3   May through September. The humid conditions often result in scattered afternoon
4   thunderstorms. Average seasonal rainfall is usually lowest during the fall (DOE 1999).
5

6   The Savannah River Site is within the Southeastern Plains ecological province (Omernick
7   1987) near the transition between northern oak-hickory-pine forest and southern mixed forest.
8   Thus, species typical of both associations are found on the Savannah River Site (DOE 1995).
9   Farming, fire, soil, and topography have strongly influenced Savannah River Site vegetation
10   patterns.
11

12   The Savannah River Site currently provides employment for more than 13,000 people, many
13   of whom have nuclear facility training and who are highly skilled. Salaries are above average
14   for the area. During the last decade, the Savannah River Site has undergone a major
15   downsizing primarily because of the end of the Cold War. Because of downsizing, the
16   Savannah River Site has contracted many nonclassified operations to private companies for
17   support services (Dominion and Bechtel 2002).
18

19   The Savannah River Site itself has provided significant socioeconomic benefits for the
20   surrounding communities over the last five decades. The facility injects about $1.5 billion
21   annually into the economies of South Carolina and Georgia, the two states bordering the site.
22   The facility provides thousands of jobs with above-average salaries, conducts environmental
23   and nuclear technology research, and offers business development programs for local
24   communities (Dominion and Bechtel 2002).
25

26   8.7.1   Land Use Including Site and Transmission Lines
27

28   The Savannah River Site has extensive undeveloped land that is potentially suitable for
29   commercial nuclear power generation. DOE conducted a review of potential locations on the
30   site for possible location of an accelerator for the production of tritium (DOE 1999). DOE
31   identified six possible locations that satisfied its siting criteria. The preferred site from the
32   DOE review is approximately 10.4 km (6.5 mi) from the Savannah River Site boundary, 5 km
33   (3 mi) northeast of the Tritium Loading Facility, and north of Roads F and E (Dominion and
34   Bechtel 2002). The site, which is divided by the boundary line between Aiken and Barnwell
35   Counties, is bordered on the southwest by a 115-kV transmission line, a buried super-control
36   and relay cable, and Monroe Owens Road. Three other secondary roads cross the site. The
37   elevation of the site is 91 to 100 m (300 to 330 ft) MSL. Dominion has adopted the DOE site
38   for the accelerator as an alternative site for new nuclear generation. For this analysis, the site
39   boundaries are shown in the upper center portion of Figure 8-6, and the site is referred to as
40   the Savannah River alternative site.
41
42

1
2
3

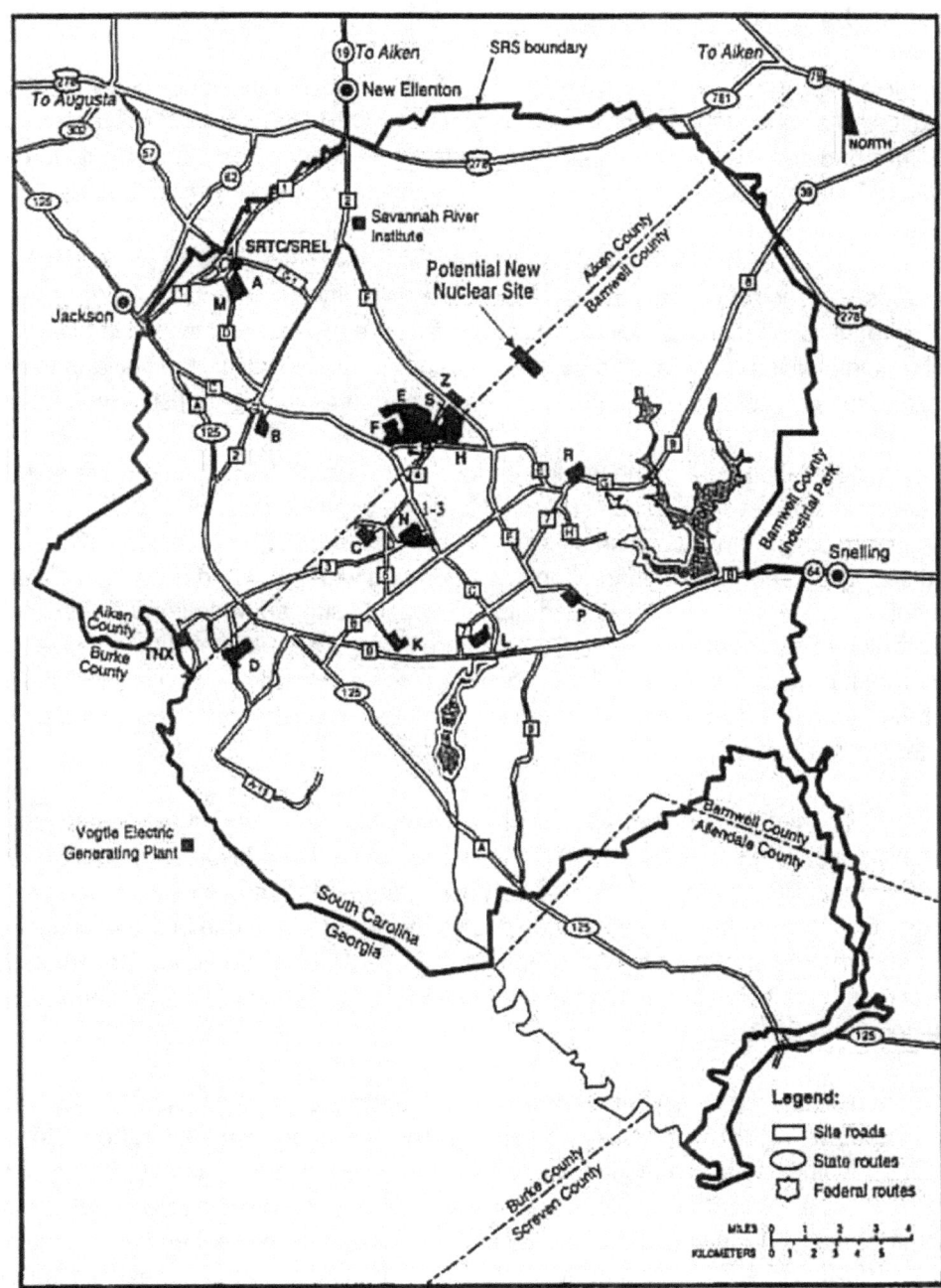

4
5 **Figure 8-6**. Potential New Nuclear Station Site within the DOE Savannah River Site

1   Dominion did not identify any current or possible future land-use restrictions that would prohibit
2   the construction of new units on the Savannah River alternative site (Dominion and
3   Bechtel 2002). DOE, however, would need to approve any such construction.
4

5   New nuclear generating units located at the Savannah River alternative site would need to have
6   an exclusion area that meets NRC requirements (10 CFR Part 100). The exclusion area is the
7   area surrounding the reactor within which the reactor licensee has the authority to determine all
8   activities, including exclusion or removal of personnel and property from the area.
9

10   Section 307(c)(3)(A) of the Coastal Zone Management Act (16 USC 1456(c)(3)(A)) requires that
11   applicants seeking a Federal permit to conduct an activity that affects a coastal zone area
12   provide to the permitting agency a certification that the proposed activity complies with the
13   enforceable policies of the state's coastal zone program. However, the Savannah River Site is
14   not within the coastal zone of South Carolina for purposes of the Act (SCDHEC 2004).
15

16   Approximately 90 percent of the workforce for the Savannah River Site lives in Aiken, Allendale,
17   Bamberg, and Barnwell Counties in South Carolina, and Columbia and Richmond Counties in
18   Georgia. There are numerous small towns within 80 km (50 mi) of the site. These communities
19   could supply an adequate construction and operating workforce and are within a 2-hour
20   commuting distance via local transportation routes. Offsite land-use impacts associated with
21   construction of two commercial nuclear units are likely to be relatively limited, given the
22   temporary nature of the construction. Some new rental housing or new manufactured home
23   and recreational vehicle parks would be expected to be constructed to accommodate
24   construction workers.
25

26   The staff assumed that workers at new units that would be located at the Savannah River
27   alternative site would live primarily in the aforementioned counties. Some new housing in these
28   counties would likely be constructed to accommodate permanent workers at the new units. The
29   property tax revenue from the new units could affect future land use in these counties as a
30   result of infrastructure improvements made possible by the tax revenue. Based on the
31   foregoing, the staff concludes that the land-use impacts of construction and operation are
32   expected to be SMALL.
33

34   The transmission system on the Savannah River Site consists of multiple 115-kV transmission
35   lines forming a ring network around the site. Three switching stations for the 115-kV
36   transmission lines exist around the site to feed the different area loads. Construction of one or
37   more new 500-kV transmission lines or several 230-kV transmission lines would be needed to
38   transmit power from new nuclear units located on the site to the regional grid (Dominion and
39   Bechtel 2002). Several options for transmitting the electrical output of new nuclear units were
40   evaluated by Dominion. The likely option would be to construct transmission lines either to the
41   west through the Savannah River Site, then cross the Savannah River to connect with the
42   existing system near the Vogtle Nuclear Power Plant in Burke County, Georgia, or to a

connection point approximately 97 km (60 mi) west of the Savannah River Site. Because the detailed routing of these transmission line rights-of-way are not known at this time, a detailed evaluation of the impacts to land use cannot be made. However, if a tie-in to the Vogtle Nuclear Power Plant is used, there would be minimal impacts to land use because most of the rights-of-way would be located on the Savannah River Site. The staff concludes that the impact of construction of new transmission capability at the Savannah River alternative site would likely be in the range of SMALL to MODERATE.

8.7.2 Water Use and Quality

The water consumed by the new units at the Savannah River alternative site would be pumped from Par Pond. In dry years, local inflows to Par Pond may be inadequate to offset consumption water demands for the new units. During such times, water to refill Par Pond would be pumped from the Savannah River. Therefore, new units at the Savannah River alternative site would impact the Savannah River. The staff reviewed streamflow records reported by the U.S. Geological Survey for stream gauge 02197000 (Savannah River at Augusta, Georgia). This gauge reflects runoff from a drainage of 19,450 km^2 (7508 mi^2) and has provided data for the period from 1884 to 2001. Using these data, the staff independently estimated the lowest 7-day discharge for low water condition that is estimated every 10 years (7Q10) and the lowest 30 days of flow in an average year (30Q2) values. For this gauge the 7Q10 discharge was estimated to be 60.8 m^3/s (2150 cfs) and 30Q2 was estimated to be 130 m^3/s (4600 cfs). The 7Q10 provides an estimate of the short-term, low-flow conditions in a dry year. The 30Q2 provides an estimate of the moderate-term, low-flow conditions in an average year.

The maximum makeup water flow rate for a single unit is estimated in the PPE as 2.78 m^3/s (98.0 cfs); however, the portion of the flow not evaporated is ultimately returned to the Par Pond as blowdown flow. Based on the PPE, the maximum evaporation for a single unit using mechanical-draft cooling towers would be 1.23 m^3/s (43.5 cfs). For either one or two units, this would represent a small fraction of both the 7Q10 and the 30Q2 values. Discharge of thermal and chemical effluents would be regulated by the State of South Carolina's NPDES permitting process to limit impacts to the Savannah River. Therefore, based on the foregoing, the staff concludes the impact of construction and operation of two units on water use and quality at the Savannah River alternative site would be SMALL.

8.7.3 Terrestrial Resources Including Endangered Species

The Savannah River alternative site is within the Southeastern Plains ecological province (Omernick 1987) near the transition between northern oak-hickory-pine forest and

1 southern mixed forest. Thus, species typical of both associations are found on the site
2 (DOE 1995). Farming, fire, soil, and topography have strongly influenced vegetation patterns
3 at the Savannah River Site.
4
5 A variety of plant communities occur in the upland areas. Typically, scrub oak communities are
6 found on the drier, sandier areas. Longleaf pine (*Pinus palustris*), turkey oak (*Quercus laevis*),
7 bluejack oak (*Q. incana*), and blackjack oak (*Q. marilandica*) dominate these communities,
8 which typically have understories of wiregrass (*Aristida stricta*) and huckleberry (*Vaccinium*
9 spp.). Oak-hickory communities are usually located on more fertile, dry uplands; characteristic
10 species are white oak (*Q. alba*), post oak (*Q. stellata*), red oak (*Q. falcata*), mockernut hickory
11 (*Carya tomentosa*), pignut hickory *(C. glabra)*, and loblolly pine (*P. taeda*), with an understory of
12 sparkleberry (*V. arboreum*), holly (*Ilex* spp.), greenbriar (*Smilax* spp.), and poison ivy
13 (*Toxicodendron radicans*) (DOE 1995).
14
15 Before the Federal government established, the Savannah River Site, the area was mainly
16 farmland that had been highly eroded. Approximately 90 percent of the site has been planted
17 with loblolly, slash pine (*P. elliottii*), and hardwood trees. The Savannah River alternative site
18 consists of mostly forested land, made up predominantly of loblolly and slash pine that have
19 been planted since the late 1950s. The site is part of a designated forest timber unit under the
20 Savannah River Site land-use system. The Savannah River Institute (formerly known as the
21 Savannah River Forest Station) coordinates the removal and sale of marketable timber from the
22 site (Dominion and Bechtel 2002).
23
24 The departure of residents in 1951 and the subsequent reforestation have provided the wildlife
25 of Savannah River Site with excellent habitat. The site has extensive, widely distributed
26 wetlands, most of which are associated with floodplains, creeks, or impoundments. In addition,
27 approximately 200 Carolina bays occur on the site (DOE 1995). Carolina bays are unique
28 wetland features of the southeastern United States.
29
30 Federally and State-listed rare, threatened, and endangered species, including the bald eagle,
31 wood stork (*Mycteria americana*), and red-cockaded woodpecker (*Picoides borealis*), reside
32 within the Savannah River Site. Federally and State threatened or endangered species
33 potentially occurring in Aiken or Barnwell Counties are listed in Table 8-5. In addition to the
34 species listed in Table 8-5, a large number of species, although not listed as threatened or
35 endangered, are still of concern or interest to the U.S. Fish and Wildlife Service (FWS)
36 (FWS 2004a, b) and/or the South Carolina Department of Natural Resources (SCDNR 2004).
37
38 Operation of the new units would likely result in noise generation, and if wet cooling towers were
39 employed, there could be impacts caused by drift, icing, fogging, and bird collisions. Noise
40 would likely be typical of operating reactor units and cooling towers, which has been determined
41 to be a SMALL impact in most instances (NRC 1996). There are no sensitive habitat areas
42 adjacent to the Savannah River alternative site that would be adversely affected by noise from

1 plant operations. The nearest bald eagles and wood storks are approximately 5 km (3 mi)
2 distant. Red-cockaded woodpeckers have not been observed at the Savannah River alternative
3 site, but could be deterred from using the area if there were increased noise and human activity.
4
5 The terrestrial vegetation in the vicinity of the Savannah River alternative site is not believed to
6 be unusually sensitive to drift, fogging, or icing (Dominion and Bechtel 2002). Bird collisions
7 would not be expected to be different from most other power plants (NRC 1996), and if
8 mechanical-draft towers are selected, bird strikes are likely to be very rare. Overall, it would be
9 expected that the impacts of operation of one or more nuclear units at the Savannah River
10 alternative site on terrestrial systems would be minimal.
11
12 The actual routes of transmission lines that would connect new units at the Savannah River
13 alternative site with the regional grid have not been determined (Dominion and Bechtel 2002).
14 Maintenance of the transmission line rights-of-way could impact wetlands, threatened or
15 endangered species habitat areas, or other sensitive ecological resources. Therefore, the
16

17 **Table 8-5.** Federally and State-Listed Threatened or Endangered Terrestrial Species
18 Potentially Occurring in Aiken and Barnwell Counties, South Carolina
19

Scientific Name	Species	Federal Status	State Status
Birds			
Haliaeetus leucocephalus	bald eagle	T	E
Mycteria americana	wood stork	E	E
Picoides borealis	red-cockaded woodpecker	E	E
Mammals			
Corynorhinus rafinesquii	Rafinesque's big-eared bat	SC	E
Amphibians			
Rana capito	gopher frog	SC	E
Reptiles			
Clemmys guttata	spotted turtle	--	T
Gopherus polyphemus	gopher tortoise	SC	E
Plants			
Trillium reliquum	relict trillium	E	E
Ptilimnium nodosum	harperella	E	E
Echinacea laevigata	smooth coneflower	E	E
Linderna melissifolia	pond berry	E	E
Oxypolis canbyi	Canby's dropwort	E	E
Schwalbea americana	American chaffseed	E	E

39 E = endangered, T = threatened, SC = species of concern.
40 Sources: FWS 2004a, b; SCDNR 2004.
41

1   potential impacts of construction, operation, and maintenance of the transmission line rights-of-
2   way on terrestrial ecosystems cannot be determined without more detailed information
3   concerning the location of the transmission line rights-of-way and the maintenance procedures
4   that would be employed. However, large impacts could be avoided by careful route selection;
5   therefore, the staff concludes the impact of construction on terrestrial resources (including
6   threatened and endangered species) would be SMALL to MODERATE depending on the
7   routing of the transmission line rights-of-way. Based on the foregoing, the overall impact on
8   terrestrial resources including threatened or endangered species of operating two units and
9   associated cooling systems at the Savannah River alternative site would be SMALL.
10 Depending on the location of transmission line rights-of-way, the construction and operational
11 impacts on threatened and endangered species could be SMALL to MODERATE.

12

13 ## 8.7.4 Aquatic Resources Including Endangered Species

14

15 The aquatic environment at the Savannah River Site is associated with the Savannah River.
16 The two main bodies of water onsite, Par Pond and L-Lake, were constructed to support site
17 operations. Par Pond, which was constructed to provide cooling water for, and to receive
18 heated cooling water from, P-Reactor and R-Reactor, has a surface area of about 1093 ha
19 (2700 ac). The 405-ha (1000-ac) L-Lake was constructed to receive heated cooling water from
20 L-Reactor. The Savannah River Site is bounded on its southwest border by the Savannah
21 River for about 56 river km (35 river mi). Five major streams from the Savannah River Site feed
22 into the river.

23

24 All the water for cooling is expected to be withdrawn from the Savannah River. The cooling
25 water blowdown would likely be discharged to Par Pond or the Savannah River. Because the
26 expected cooling system would be a closed-cycle system, the impacts to aquatic resources
27 would be expected to be minimal. The potential for impingement and entrainment of aquatic
28 resources would be expected to be mitigated by the current operation of the intake structure.
29 The potential impacts of heated water would be expected to be mitigated by the placement of
30 the discharge structures.

31

32 There are two endangered species in the Savannah River alternative site. They are the
33 shortnose sturgeon (*Acipenser brevirostrum*) and the fanshell (*Cyprogenia stegaria*). Both are
34 protected under current management practices used by DOE at the Savannah River Site. The
35 staff evaluated the potential impacts of operating the proposed new nuclear units, including
36 operating the plants, cooling systems, and transmission systems on aquatic threatened and
37 endangered species.

38

39 Based on the foregoing, the staff concludes that the overall impact on aquatic ecological
40 resources, including threatened and endangered species, of construction and operation of two
41 units and associated cooling towers and transmission facilities at the Savannah River
42 alternative site would be SMALL.

8.7.5 Socioeconomics, Historic and Cultural Resources, Environmental Justice

In evaluating the socioeconomic impacts of constructing and operating two units at the Savannah River Site, the staff undertook a reconnaissance survey of the site. That is, readily obtainable data from the Internet or published sources were used in the evaluation, and no new data were collected. The subsections that follow reflect the organizational structure of the socioeconomic discussions found in Sections 2.8.4, 4.5, and 5.5. The impacts resulting from both construction and operation of the two units are addressed.

8.7.5.1 Physical Impacts

Construction activities can result in temporary and localized physical conditions such as noise, odor, vehicle exhaust, vibration, shock from blasting, and dust emissions that affect the environment. Further, the use of public roadways, railways, and waterways would be necessary to transport construction materials and equipment to the site. There would, as a result, be increased use of these infrastructures, both in terms of increased volume and type of vehicular traffic.

Road access to the Savannah River Site is via SR 125. U.S. 278 cuts through a portion of the Savannah River Site. Easy access to the Savannah River alternative site could be accommodated by installing access roads from U.S. 278. Most roads leading to the site are two-lane roads, but appear to be kept in excellent condition (Dominion and Bechtel 2002).

The CSX Transportation, Inc., provides rail service to the Savannah River Site. Some upgrades would likely be needed to accommodate the large and heavy loads associated with construction of new nuclear units (Dominion and Bechtel 2002).

On the Savannah River, there is a barge slip situated on DOE property. This barge slip has been used in the past for heavy loads and large components such as steam generators. Shipment of heavy loads by barge to the Savannah River Site depends on the water level in the Savannah River. The Savannah River alternative site is on the opposite side of the property from the barge slip and some additional heavy-haul routes would need to be constructed to reach it (Dominion and Bechtel 2002).

All construction activities would likely occur within the boundaries of the Savannah River Site. Offsite areas that would support construction activities (e.g., borrow pits, quarries, disposal sites) are expected to be already permitted and operational. Impacts on those facilities from constructing new nuclear units are expected to be small incremental impacts associated with their normal operation. The alternative site is approximately 10.5 km (6.5 mi) from the Savannah River Site boundary (Dominion and Bechtel 2002).

1   Station operation could cause noise, odors, exhausts, thermal emissions, and visual intrusions.
2   Noise would be produced by operation of cooling towers, pumps, transformers, turbines,
3   generators, and switchyard equipment, and from traffic. The Savannah River Site is a Federal
4   Reservation. Dominion would need to comply with State and local ordinances which apply to
5   the Department of Energy, (NCA 1972). Good road conditions and appropriate speed limits
6   would minimize the noise level generated by the workforce commuting to the site. Nearby trees
7   would serve as a visual buffer, minimizing visual effects.
8
9   The new units would be expected to have emissions from auxiliary power systems, standby
10   diesel generators, and standby gas turbine generators (Dominion and Bechtel 2002). It is
11   expected that the combined annual emissions of any pollutant would be less than 91 MT/yr
12   (100 tons/yr) (Dominion and Bechtel 2002). Air quality permits acquired for these generators
13   would ensure that air emissions comply with regulations. Paved access roads and appropriate
14   speed limits would minimize the amount of dust generated by the commuting workforce.
15
16   Direct site-specific impacts from construction activities would be temporary and would occur
17   mainly within the boundaries of the Savannah River Site. Offsite impacts would represent small
18   incremental changes to offsite services supporting the construction activities. During station
19   operations, noise levels would be managed to comply with State and local ordinances as
20   required by Section 4(b) of the Noise Control Act (42 USC 4903). Air quality permits would be
21   required for the diesel generators, auxiliary boilers, and other equipment, which should limit air
22   emissions and meet applicable standards. Based on the foregoing, the staff concludes that the
23   physical impacts of construction and operation of new units would at the Savannah River
24   alternative site be SMALL.
25

26   **8.7.5.2 Demography**
27

28   The center of the Savannah River Site is approximately 40 km (25 mi) southeast of the city
29   limits of Augusta, Georgia. The population for Augusta-Richmond County, Georgia, was
30   195,182 in 2000 (USCB 2000d). The site is 161 km (100 mi) from the Atlantic Coast, and about
31   175 km (110 mi) south-southeast of the North Carolina border. The largest nearby population
32   centers are Aiken, South Carolina, with a population of 25,337 in 2000 (USCB 2000d), and
33   Augusta, Georgia. The only towns within 24 km (15 mi) of the center of the Savannah River
34   Site are New Ellenton, with a population of 2250; Jackson with a population of 1625; Barnwell
35   with a population of 5035; Snelling, with a population of 246; and Williston with a population of
36   3307 (USCB 2000d). All of these towns are in South Carolina.
37

38   Most of the construction and operations workforce are expected to come from within the region
39   (see more detailed discussion in Section 8.7.5.3), and those who might relocate to the region
40   would represent a small percentage of the larger population base. Those who do relocate to
41   the region would most likely take up residency across the region. Based on the foregoing, the
42   staff concludes that any environmental effects caused by population increases within an 80-km

1 (50-mi) radius of the Savannah River Site resulting from construction and operation of new units
2 would be SMALL.
3
4 **8.7.5.3 Community Characteristics**
5
6 *Economy*
7
8 The unemployment rate in the Augusta-Aiken Metropolitan Statistical Area was 5.7 percent in
9 June 2004 (Georgia Department of Labor 2004). This compares to a 4.6 percent
10 unemployment rate for Georgia (Georgia Department of Labor 2004) and a 6.6 percent
11 unemployment rate for South Carolina in June 2004 (South Carolina 2004a). Regional
12 unemployment statistics for selected South Carolina counties in the vicinity of the Savannah
13 River Site include Barnwell County at 12.3 percent unemployment in June 2004 and Aiken
14 County at 7.1 percent (South Carolina Department of Social Services 2004).
15
16 The Savannah River Site itself has provided significant socioeconomic benefits for the
17 surrounding communities over the last 50 years, and currently provides employment for more
18 than 13,000 people who are highly skilled workers, most of whom are college educated.
19 Salaries are above average salaries of the area. The site injects about $1.5 billion annually into
20 the economies of South Carolina and Georgia (Dominion and Bechtel 2002).
21
22 During the last decade, a major downsizing has occurred at the site because of the end of the
23 Cold War. Construction and operation of new units would increase employment at the site. These
24 jobs would provide economic benefits to the local communities (Dominion and Bechtel 2002).
25 The magnitude of the economic impacts would be diffused in the larger economic bases of
26 the region, whereas with the smaller economic base of Barnwell County and the higher
27 unemployment rate (compared to Aiken County and the State of South Carolina), the economic
28 impacts could be more noticeable and have a greater beneficial impact. Based on the
29 foregoing, the staff concludes that the beneficial impacts of construction and operation of two
30 units on the economy of the region would be SMALL everywhere in the region except Barnwell
31 County, where the beneficial impacts to the county could be MODERATE.
32
33 *Availability of Workers*
34
35 Dominion estimates it would take approximately 5000 construction workers over 5 years to build
36 two commercial nuclear units (Dominion 2006). As discussed in the previous section on the
37 economy, the Savannah River Site currently provides employment for more than 13,000 people.
38 However, during the last decade, some loss of jobs occurred because of the end of the Cold
39 War (Dominion and Bechtel 2002).
40
41 Construction of a nuclear generating facility would draw workers from South Carolina and
42 Georgia. The estimated number of construction workers in the two-state region is

1   approximately 459,725 (BEA 2000). With the extensive local transportation network in the area,
2   nearby cities could supply an adequate workforce and are well within a 2-hour commuting
3   distance of the Savannah River Site. Therefore, a minimal influx of project-related population
4   during plant construction and operation could be expected (Dominion and Bechtel 2002).
5
6   Dominion would need approximately 720 new employees to operate the proposed new facility.
7   The Savannah River Site currently provides employment for more than 13,000 people but also
8   has undergone downsizing. The addition of a new power generating facility would be expected
9   to add jobs for skilled craft workers (i.e., with skills comparable to or higher than skills of the
10   existing Savannah River Site workforce). Many of the jobs for skilled craft workers could be
11   filled by current or former Savannah River Site employees (Dominion and Bechtel 2002).
12
13   There appears to be a large supply of construction labor and skilled craft workers available.
14   Dominion may have to add incentives to draw craft workers from out-of-state with specific skills
15   to the area because of the lower prevailing wages when compared to other areas outside
16   Georgia and South Carolina, but it believes it can successfully manage this issue. The
17   unemployment rates in Aiken and Barnwell Counties are above the State of South Carolina
18   average unemployment rate. Likewise, the unemployment rate in Augusta is above the average
19   rate for the State of Georgia. Employees for station operation are expected to be available from
20   within the region because of the downsizing at Savannah River Site. Based on the foregoing,
21   the staff concludes that Dominion would be able to obtain a ready supply of construction and
22   operations labor for the new units.
23
24   *Transportation*
25
26   Two interstate highways serve the vicinity of the Savannah River Site. Several other highways
27   (U.S. Highways 221, 278, 301, 321, and 601) provide additional transport routes for the area.
28   Approximately 84 percent of the Savannah River Site workforce of 13,000 resides in Aiken and
29   Barnwell Counties in South Carolina and Columbia and Richmond Counties in Georgia
30   (Dominion and Bechtel 2002).
31
32   The regional transportation networks in the Savannah River Site vicinity serve Aiken, Allendale,
33   Bamberg, and Barnwell Counties in South Carolina and Columbia and Richmond Counties in
34   Georgia. Approximately 88 percent of the Savannah River Site commuter traffic originates from
35   these counties. On the site itself, there are more than 322 km (200 mi) of primary roads and
36   more than 1600 km (1000 mi) of unpaved secondary roads.
37
38   In general, heavy traffic occurs in the early morning and late afternoon when workers commute
39   to and from the Savannah River site. For the roads in the general region, the worst-case LOS is
40   associated with routes near the Savannah River bridges, including I-20 and U.S. 1 and urban
41   routes in North Augusta and Aiken, including South Carolina SRs 230, 25, 19, and 118.
42   Significant congestion occurs during peak traffic periods onsite on U.S. Highways 1-A and 278

1 and on SRs 19 and 125 at Savannah River Site access points. Long delays are also
2 experienced offsite along I-20 and U.S. Highways 25 and 1 where they cross the Savannah
3 River. The Savannah River Site has implemented changes to remedy the congestion at some
4 access points (Dominion and Bechtel 2002).
5
6 Other transportation in the area also includes a rail line for CSX Transportation, Inc. The
7 Savannah River Site has its own railroad system. Rail traffic is separated into two categories
8 depending on which track system it would use: CSX operations and the Savannah River Site
9 railway (Dominion and Bechtel 2002).
10
11 The nearest major airport to the Savannah River Site is in Atlanta, Georgia, and the closest
12 regional airport is in Augusta, Georgia. The Augusta airport conducts regular freight and
13 passenger airline services and is large enough to accommodate the relatively small air
14 shipments normally associated with a large construction project. The Atlanta airport can
15 accommodate large air shipments. Ground transportation from the Augusta airport takes
16 approximately 1 hour, and from the Atlanta airport, approximately 3 hours (Dominion and
17 Bechtel 2002).
18
19 During peak new plant construction, 5000 construction workers would be needed.[a] The units
20 operations workforce would be approximately 720. The extensive existing roadway network in
21 the area and the rail lines near the Savannah River alternative site are expected to be capable
22 of handling an additional 38 percent of the workforce commuting to the site during construction
23 and the transportation of bulk materials to and from the site. In addition, the workforce would
24 still be far below the peak levels of employment achieved in 1993. With implementation of
25 traffic mitigation measures, the construction of two units at the Savannah River alternative site
26 is expected to result in impacts that are manageable for traffic patterns, workforce commuter
27 traffic, and rail/truck delivery of materials (Dominion and Bechtel 2002).
28
29 The Savannah River alternative site is in a limited-access DOE site adjacent to a rural, low-
30 population area. The regional transportation network is adequate for commuter and transient
31 traffic in the area. Based on the foregoing, the staff concludes that the transportation impacts of
32 a construction workforce resulting in an approximately 38 percent increase to the existing
33 workforce at Savannah River Site would be SMALL. Because the increase in employment with
34 the operations workforce is less than 5.5 percent of the existing site workforce, the staff
35 concludes the transportation impacts of the operations workforce would be SMALL.
36

(a) Dominion and Bechtel (2002) states that 3000 to 3500 craft and an additional 800 to 1000 non-manual
    labor personnel (or 4500 workers) would be needed at the Savannah River Site. In its ER (Dominon
    2006), Dominion states that 5000 workers would be needed at the North Anna ESP site.

1 *Taxes*
2
3 In lieu of property taxes, DOE pays a fee to the localities bordering the site. For 2002, Barnwell
4 County received a fee of approximately $2 million, Aiken County approximately $800,000, and
5 Allendale County approximately $100,000 (Dominion and Bechtel 2002).
6
7 Construction and operation workers would pay personal income taxes to Georgia and South
8 Carolina, sales taxes to the State and local governments in the region where sales take place,
9 and property taxes to the counties in which they might own a residence. In addition, sales taxes
10 would be paid from the sales of construction materials and supplies purchased for the project.
11 Finally, because the units would be built by a private company (Dominion) and not DOE, a
12 property tax might be levied on the value of the property that becomes part of the additional
13 units as they are constructed. These taxes would most likely go to Aiken and Barnwell
14 Counties.[a] Georgia and South Carolina both have corporate income taxes, with the tax rates
15 being 6 and 5 percent, respectively (Federation of Tax Administrators 2004).
16
17 Based on the foregoing, the staff concludes that the overall beneficial impacts of taxes collected
18 in the region through the income, sales and use, and property taxes would be SMALL to
19 LARGE. The staff also concludes that the overall impacts of taxes collected through the
20 income, sales and use, and property taxes collected in the region would be SMALL for
21 jurisdictions other than Barnwell County. The taxes paid, while large in absolute value, are
22 nevertheless a small sum when compared to the total amount of taxes collected by State and
23 local governments in the region. For property taxes in Barnwell County, the staff concludes that
24 the overall beneficial impacts of the property taxes collected would be MODERATE
25 (construction) and LARGE (operation), relative to the total amount of property taxes the county
26 collects.[b]
27

(a) The proposed site of the new units would not be on land owned by Dominion. Most likely, should the
     Savannah River alternative site be chosen for the new power generating facility, the land for the new
     units would be leased from DOE by Dominion.
(b) The derivation of this impact is based on the fact the fiscal year 2003 amount of property taxes
     collected in Aiken and Barnwell Counties were $68,046,000 (Cornwell in Jaksch and Scott 2005) and
     $9,774,000 (Gibson in Jaksch and Scott 2005), respectively. For comparison, NAPS Units 1 and 2
     pay approximately $10 million in annual property tax to Louisa County (the actual amount that the
     proposed nuclear plant would pay to Barnwell County would depend on assessed value and millage
     rate per thousand of assessed value). On the assumption that there is a rough comparison between
     what Dominion pays to Louisa County and what they might pay to Barnwell County, it can be
     concluded that the potential percentage of the proposed facility's property taxes to the total of all
     property taxes paid in Barnwell County would be significant.

1 *Aesthetics and Recreation*
2
3 The preferred location for the two units on the Savannah River Site is more than 10 km (6 mi)
4 from the closest site boundary. There are no public amenity areas within 3 km (2 mi) of the site.
5 Most of the site is dense forest; therefore, nearby trees would provide a visual buffer for the
6 construction and operation of the units to the public. Because the location is at least 10 km
7 (6 mi) away from the existing site boundary, offsite observers would not have an identifiable
8 nuclear power plant view (Dominion and Bechtel 2002).
9
10 Cooling towers, could produce visible plumes offsite, are being proposed as part of the new
11 nuclear units cooling system. Dry cooling towers could be an alternative method for plant
12 cooling. Nearby trees would serve as a visual buffer for the transmission lines (Dominion and
13 Bechtel 2002).
14
15 The surrogate AP1000 reactor has a tall containment building that is approximately 71 m
16 (234 ft) above grade. The design of this building includes a hatch that determines the height
17 that it must be above ground. This building would be expensive to redesign to allow the building
18 to be placed lower in the ground. Thus, the height of the AP1000 containment building sets the
19 upper bound of what would be visible from offsite. In addition, if natural draft cooling towers are
20 used, the height of the towers is roughly 170 m (550 ft).
21
22 Prominent geographical features within 80 km (50 mi) of the Savannah River Site are Thurmond
23 Lake (formerly called Clarks Hill Reservoir) and the Savannah River. The principal surface-
24 water body associated with the Savannah River Site is the Savannah River, which flows along
25 the site's southwest border (Dominion and Bechtel 2002). The closest State park is Redcliffe
26 Plantation State Park, about 16 km (10 mi) northwest of the site location.
27
28 The proposed plant would be located about 10 km (6 mi) from the nearest site boundary and
29 would be screened by trees. There are no significant residential areas or recreational facilities
30 within 3 km (2 mi) of the site. Plumes from cooling towers could be visible offsite. Based on the
31 foregoing, the staff concludes that the construction and station operation impacts on aesthetics
32 and recreation would be SMALL.
33
34 *Housing*
35
36 In the four-county area of Richmond and Columbia Counties in Georgia and Barnwell and Aiken
37 Counties in South Carolina, there were 187,811 housing units in 2000. Of these units, 52,405
38 were rental units, 6424 of which were vacant (10.9 percent vacancy rate) (USCB 2000e). There
39 appears to be vacant rental housing units available for construction workers who might want to
40 relocate to the region.
41

1   In the same four-county area discussed above, there were 117,243 owner-occupied houses;
2   3089 of these houses were for sale (2.6 percent vacancy rate). The percentage of houses for
3   sale in relation to owner-occupied housing is low, indicating the market for resale housing is
4   tight.
5

6   The operations workforce is expected to come primarily from current or former employees at the
7   Savannah River Site. If, however, a substantial number of workers were recruited into the area,
8   there could be upward pressure on housing values. This assumption is based on the low
9   number of homes for sale in the area and the fact that the workforce, which would be on the
10   higher end of the salary scale when compared to other job classifications in the area, could tend
11   to buy more expensive homes.
12

13   It is not unusual for construction workers to commute up to 2 hours (one way) per day to the job
14   site. Many of the construction workers are assumed to already live within the region.
15   Therefore, there appears to be enough vacant rental housing to house those who might relocate
16   to the region. For the operations workforce, it is expected that most would already have
17   residences in the region, and few would relocate to the area given the potential supply of
18   workers in the region resulting from Savannah River Site downsizing. Based on the foregoing,
19   the staff concludes that the impacts to housing from construction and operation of two units at
20   the Savannah River alternative site would be SMALL.
21

22   *Public Services*
23

24   <u>Water and Wastewater Treatment</u>
25

26   Four major public sewage treatment facilities with a combined design capacity of 302.2 million
27   liters (79.8 million gallons) per day serve the six-county region composed of Aiken, Allendale,
28   Bamberg, and Barnwell Counties in South Carolina and Columbia and Richmond Counties in
29   Georgia. In 1989 (the latest year for which data were readily available), these systems were
30   operating at approximately 56 percent capacity, with an average daily flow of 170 million L/day
31   (44.9 MGD). Capacity utilization ranged from 45 percent in Aiken County to 80 percent in
32   Barnwell County (DOE 2000).
33

34   There are approximately 120 public water systems in the six-county area. About 40 of these
35   county and municipal systems are major facilities, while the remainder serve individual
36   subdivisions, water districts, manufactured home parks, or miscellaneous facilities. In 1989
37   (again the latest year for which data are readily available), the 40 major facilities had a
38   combined total flow of 576.3 million L/day (152.2 MGD). With an average daily flow rate of
39   approximately 268.8 million L/day (71 MGD), these systems were operating at 47 percent
40   capacity in 1989. Facility utilization rates ranged from 13 percent in Allendale County to
41   84 percent in the City of Aiken (DOE 2000).
42

1　The Savannah River alternative site is approximately 40 km (25 mi) southeast of Augusta,
2　Georgia, and 31.4 km (19.5 mi) south of Aiken, South Carolina. There are numerous towns and
3　cities within 80 km (50 mi) of the site, and all these towns and cities are within a 2-hour
4　commuting distance of the site via local transportation routes (Dominion and Bechtel 2002). In
5　addition, the utility infrastructures of the towns and cities could provide public services such as
6　water and wastewater treatment to members of the construction and operations workforce who
7　might relocate to the region. Based on the foregoing, the staff concludes that the impacts of the
8　construction and operations workforces on water and wastewater treatment in the region would
9　be SMALL.

10
11　<u>Police, Fire and Medical Facilities</u>
12
13　Eight general hospitals operate in the six-county region. Four of the eight general hospitals are
14　in Richmond County (Augusta), Georgia, while Columbia County, Georgia, has no hospital.
15　Aiken, Allendale, Bamberg, and Barnwell Counties in South Carolina each have one general
16　hospital (USHospital.info 2005).

17
18　Fifty-six fire departments provide fire protection in the region. Twenty-seven of these fire
19　departments are classified as municipal fire departments, but many provide protection to rural
20　areas outside municipal limits (DOE 2000).

21
22　County sheriff and municipal police departments provide most of the law enforcement in the
23　region. In addition, State law enforcement agents and State troopers assigned to each county
24　provide protection and assist county and municipal officers (DOE 2000).

25
26　Many of the potential construction and operations workforce probably already live within an
27　80-km (50-mi) radius of the region. There are a number of towns within a 2-hour commuting
28　distance of the site. Any new workers relocating to the area would most likely have places of
29　residency located throughout the region, which would not place an undue burden on the
30　infrastructure of any one jurisdictional entity. Based on the foregoing, the staff concludes that
31　the impacts of the construction and operations workforce on public facilities in the Savannah
32　River Site area would be SMALL.

33
34　*Social Services*
35
36　In Georgia, social services at the state level are overseen by the Department of Human
37　Resources. It oversees about 80 wide-ranging programs that include controlling the spread of
38　disease, enabling older people to live at home longer, preventing children from developing
39　lifelong disabilities, training single parents to find and hold jobs, and helping people with mental
40　or physical disabilities live and work in their communities (Georgia Department of Labor 2004).
41

1   In South Carolina, social services are overseen by the Department of Social Services, which
2   administers its programs through county offices. Services offered by the Department include
3   child care assistance to needy families, adult protective services, child protective services,
4   independent living, and emergency shelters food program, among other services (South
5   Carolina Employment Security Commission 2004). During construction of new nuclear units at
6   the Savannah River alternative site, there may be increased demand for social services from
7   the construction workforce and their dependents.
8

9   Generally, construction and operation of the new units at Savannah River alternative site would
10   be viewed as beneficial economically to the disadvantaged population segments served by the
11   Georgia Department of Human Resources and South Carolina Department of Social Services.
12   The workforce associated with construction and operation of two units at the Savannah River
13   alternative site would most likely receive higher wages than other employment categories in the
14   region. It is expected that through the multiplier effect, the number of jobs that could be filled by
15   members of the disadvantaged population would increase.
16

17   Construction and operation of two units would have a beneficial economic impact to the
18   disadvantaged population in the region, which should lessen the demand for social services.
19   There could be an initial increase in demand for social services at the beginning of the
20   construction period, but this is considered manageable and limited. Based on the foregoing, the
21   staff concludes that the impacts of construction and station operation of two units on social and
22   related services would be SMALL.
23

24   *Education*
25

26   Public education facilities in the six-county region (Aiken, Allendale, Bamberg, and Barnwell
27   Counties in South Carolina, and Columbia and Richmond Counties in Georgia) include
28   approximately 116 elementary or intermediate schools and 28 high schools (Great
29   Schools 2005). In addition to the public schools, there are approximately 50 private schools in
30   the region (NCES 2005). There are several local colleges, technical schools, and training
31   facilities available, such as the University of South Carolina Aiken, Augusta State University,
32   Paine College, Aiken Technical College, and Augusta Technical College.
33

34   Many of the potential construction and operating workforce probably already live within the
35   region, and any new workers relocating to the area would most likely take up residency
36   throughout the region. Based on the foregoing, the staff concludes that the impacts of
37   construction and operation of two units on educational facilities in the region would be SMALL.
38

39   **8.7.5.4 Historic and Cultural Resources**
40

41   Historic and cultural resources at the Savannah River Site are managed through a cooperative
42   agreement between DOE and the South Carolina Institute of Archaeology of the University of

1 South Carolina as the Savannah River Archaeological Research Program. Since 1974, more
2 than 60 percent of the 777-km^2 (300-mi^2) site has been inventoried for prehistoric and historic
3 sites and more than 1200 sites have been recorded, ranging in age from the Middle Archaic
4 prehistoric period to the 20th century (DOE 2002). Archaeological research has provided
5 considerable information about the distribution and content of historic and cultural sites on the
6 Savannah River Site.
7
8 Archaeologists have divided the Savannah River Site into three zones related to their potential
9 for containing sites with multiple archaeological components or dense or diverse artifacts, and
10 their potential for nomination to the National Register of Historic Places.
11
12 • Zone 1 is the zone of the highest archaeological site density with a high probability of
13   encountering large archaeological sites with dense and diverse artifacts and a high
14   potential for nomination to the National Register of Historic Places.
15
16 • Zone 2 includes areas of moderate archaeological site density. Activities in this zone
17   have a moderate probability of encountering large sites with more than three prehistoric
18   components or that would be eligible for nomination to the National Register of Historic
19   Places.
20
21 • Zone 3 includes areas of low archaeological site density. Activities in this zone have a
22   low probability of encountering archaeological sites and virtually no chance of
23   encountering large sites with more than three prehistoric components; the need for site
24   preservation is low. Some sites in the zone could be considered eligible for nomination
25   to the National Register of Historic Places.
26
27 The Savannah River alternative site parcel identified by Dominion lies in Zone 3. According to
28 Savannah River Site staff (Dominion and Bechtel 2002), no known historic and cultural
29 properties exist in the site.
30
31 In conjunction with previous studies, DOE solicited the concerns of Native American tribes
32 about traditional cultural values in the Central Savannah River Valley. Three Native American
33 groups, the Yuchi Tribal Organization, the National Council of Muskogee Creek, and the Indian
34 People's Muskogee Tribal Town Confederacy, expressed general concerns about the
35 Savannah River Site and the Central Savannah River Area but did not identify specific sites as
36 possessing religious significance. The Yuchi Tribal Organization and the National Council of
37 Muskogee Creek are interested in several plant species traditionally used in tribal ceremonies.
38
39 Based on the foregoing, the staff concludes that the potential impacts on historic and cultural
40 resources from construction and operation of two units at the Savannah River alternative site
41 would be SMALL.
42

1   **8.7.5.5 Environmental Justice**

2

3   DOE has performed an environmental assessment for the construction and operation of a linear
4   accelerator (since dropped from consideration in the general area of the Savannah River
5   alternative site) that would produce tritium (DOE 1999). As part of that assessment, an
6   evaluation of potential environmental justice impacts was conducted (Dominion and
7   Bechtel 2002).

8

9   DOE's environmental justice assessment evaluated whether minorities or low-income
10   populations could receive disproportionately high and adverse human health and environmental
11   impacts. Minority and low-income populations were identified by census tract. DOE's analysis
12   concluded that releases from the site would not disproportionally affect minority communities
13   (population equal to or greater than 35 percent of the total population) or low-income (equal to
14   or greater than 25 percent of the total population) within an 80-km (50-mi) radius of the region,
15   because the compared per capita doses did not vary significantly.[a] In addition, regarding
16   downstream communities, DOE evaluated doses to people using the Savannah River for
17   drinking water, sports, and food. Because the identified communities in the areas downstream
18   from the Savannah River Site are well distributed, DOE concluded there were no
19   disproportionate impacts among minority and low-income populations (Dominion and
20   Bechtel 2002).

21

22   Based on the foregoing, the staff concludes that the offsite impacts of construction and
23   operation of two units at the Savannah River alternative site on minority and low-income
24   populations would be SMALL. There are no disproportionately high and adverse impacts to
25   these populations.

26

27   # 8.8   Summaries of Alternative Site Impacts

28

29   Summaries of the impacts of construction and operation on each of the three proposed
30   alternative sites selected by Dominion are presented in Tables 8-6 and 8-7. Discussions of the
31   stated impacts are presented in the individual site sections (Sections 8.5 through 8.7). A
32   comparison of the alternative site impacts with impacts at the proposed North Anna ESP site is
33   presented in Chapter 9.

34

(a) NRC uses more complex threshold limits than DOE for defining whether minority or low-income
    populations exist within an 80-km (50-mi) radius of the Savannah River Site. See Section 2.8.4 for a
    more detailed discussion of the NRC criteria. However, the geographic distribution of minority
    low-income populations would be similar using either method for the region surrounding the Savannah
    River Site.

Table 8-6. Characterization of Construction Impacts at the Alternative ESP Sites

Category	Surry	Portsmouth	Savannah River
Land-use impacts	--	--	--
Site and vicinity	SMALL	SMALL	SMALL
Transmission line rights-of-way	SMALL	SMALL	SMALL to MODERATE
Air quality impacts	SMALL	SMALL	SMALL
Water-related impacts	--	--	--
Water use	SMALL	SMALL	SMALL
Water quality	SMALL	SMALL	SMALL
Ecological impacts	--	--	--
Terrestrial ecosystems	SMALL	SMALL	SMALL to MODERATE
Aquatic ecosystems	SMALL	SMALL	SMALL
Threatened and endangered species	SMALL	SMALL	SMALL to MODERATE
Socioeconomic impacts	--	--	--
Physical impacts	SMALL	SMALL	SMALL
Demography	SMALL	SMALL	SMALL
Social and economic[a]	SMALL BENEFICIAL to MODERATE BENEFICIAL	SMALL BENEFICIAL to MODERATE BENEFICIAL	SMALL BENEFICIAL to MODERATE BENEFICIAL
Infrastructure and community services	SMALL to MODERATE	SMALL to MODERATE	SMALL
Historic and cultural resources	MODERATE to LARGE	SMALL	SMALL
Environmental justice	SMALL	SMALL	SMALL
Nonradiological health impacts	SMALL	SMALL	SMALL
Radiological health impacts	SMALL	SMALL	SMALL

(a) Impacts of construction on the economy and increases in taxes collected are considered beneficial impacts. These beneficial impacts are discussed in the applicable sections.

Table 8-7. Characterization of Operational Impacts at the Alternative ESP Sites

Category	Surry	Portsmouth	Savannah River
Land-use impacts		--	--
The site and vicinity	SMALL	SMALL	SMALL
Transmission line rights-of-way	SMALL	SMALL	SMALL
Air quality impacts	SMALL	SMALL	SMALL
Water-related impacts	--	--	--
Water use	SMALL	SMALL to MODERATE	SMALL
Water quality	SMALL	SMALL	SMALL
Water use in drought year	SMALL	MODERATE	SMALL
Ecological impacts	--	--	--
Terrestrial ecosystems	SMALL	SMALL	SMALL
Aquatic ecosystems	SMALL	SMALL	SMALL
Threatened and endangered species	SMALL	SMALL	SMALL to MODERATE
Socioeconomic impacts	--	--	--
Physical impacts	SMALL to MODERATE	SMALL	SMALL
Demographics	SMALL	SMALL	SMALL
Social and economic[a]	SMALL BENEFICIAL to LARGE BENEFICIAL	SMALL BENEFICIAL to LARGE BENEFICIAL	SMALL BENEFICIAL to LARGE BENEFICIAL
Infrastructure and community services	SMALL to MODERATE[b]	SMALL to MODERATE	SMALL
Historic and cultural resources	MODERATE to LARGE[b]	SMALL	SMALL
Environmental justice	SMALL	SMALL	SMALL
Nonradiological health impacts	SMALL	SMALL	SMALL
Radiological health impacts	SMALL	SMALL	SMALL
Impacts of postulated accidents	SMALL	SMALL	SMALL
Fuel cycle impacts[c]	SMALL	SMALL	SMALL

(a) Impacts of operation on the economy and increases in taxes collected are considered beneficial impacts. The beneficial economic impacts are discussed in the applicable sections.

(b) Aesthetic impacts could be LARGE at historically important sites in the vicinity. This is captured in the historic and cultural resources evaluation. (Sections 8.5.5.3 and 8.5.5.4)

(c) Fuel cycle impacts are evaluated in Chapter 6

8.9 References

Note: Because the web pages cited in this document may become unavailable, the staff has entered the appropriate pages into ADAMS. The accession number of the package containing the Web-sites used as references in Chapter 8 of the North Anna ESP EIS is ML051580551.

10 CFR Part 20. Code of Federal Regulations, Title 10, *Energy*, Part 20, "Standards for Protection Against Radiation."

10 CFR Part 50. Code of Federal Regulations, Title 10, *Energy*, Part 50, "Domestic Licensing of Production and Utilization Facilities."

10 CFR Part 51. Code of Federal Regulations, Title 10, *Energy*, Part 51, "Environmental Protection Regulations for Domestic Licensing and Related Regulatory Functions."

10 CFR Part 52. Code of Federal Regulations, Title 10, *Energy*, Part 52, "Early Site Permits; Standard Design Certifications; and Combined Licenses for Nuclear Power Plants."

10 CFR Part 100. Code of Federal Regulations, Title 10, *Energy*, Part 100, "Reactor Site Criteria."

66 FR 64963. "Notice of Wetlands Involvement for the Portsmouth Gaseous Diffusion Plant Reindustrialization Program." *Federal Register.* Vol. 66, No. 242, December 17, 2001.

69 FR 3956. "Notice of Availability of Environmental Assessment and Finding of No Significant Impact for License Application for USEC Inc., Bethesda, MD." *Federal Register.* Vol. 69, No. 17, January 27, 2004.

69 FR 52040. "Policy Statement on the Treatment of Environmental Justice Matters in NRC Regulatory and Licensing Activities." *Federal Register.* Vol. 69, No. 163, August 24, 2004.

70 FR 30396. "Early Action Compact Areas for South Carolina/Georgia." *Federal Register.* U.S. Environmental Protection Agency, May 26, 2005.

70 FR 33771. "Correction to information on page 30404 of 70 FR 30396." *Federal Register.* U.S. Environmental Protection Agency, June 9, 2005.

Bechtel Jacobs Co. 2003. *U.S. Department of Energy Portsmouth Annual Environmental Report for 2002 Piketon, Ohio.* DOE/OR/11-3132&D1. Accessed at http://www.bechteljacobs.com/ports_reports.shtml on March 5, 2004.

1   Bureau of Economic Analysis (BEA). 2000. CA25 Total Full-time and Part-time Employment by
2   Industry. Accessed at http://www.bea.doc.gov/bea/regional/reis/ on March 30, 2004.
3
4   Clean Water Act (also referred to as the Federal Water Pollution Control Act). 33 USC 1251, et
5   seq.
6
7   Coastal Zone Management Act (CZMA). 16 USC 1451, et seq.
8
9   Dominion Energy, Inc. and Bechtel Power Corp (Dominion and Bechtel). 2002. *Study of*
10 *Potential Sites for the Deployment of New Nuclear Plants in the United States.*
11 U.S. Department of Energy Cooperative Agreement No. DE-FC07-02ID14313. Available at
12 www.ne.doe.gov/NucPwr2010/ESP_Study/ESP_Study_Dominion1.pdf
13
14 Dominion Nuclear North Anna, LLC (Dominion). 2006. *North Anna Early Site Permit*
15 *Application – Part 3 – Environmental Report.* Revision 6, Glen Allen, Virginia.
16
17 Federation of Tax Administrators. 2004. Range of State Corporation Income Taxes 2003
18 (For the tax year 2003 – as of January 1, 2003). Accessed at
19 http://taxpolicycenter.org/TaxFacts/TFDB/Content/PDF/state_corp_2003.pdf on
20 August 24, 2004.
21
22 Georgia Department of Human Resources. 2004a. Georgia – Online access to Georgia
23 Government. Accessed at
24 http://dhr.georgia.gov/02/dhr/home/0,2220,5696,00.html;jsessionid=73BA632F785E95B6CBCB
25 2081541FE127 on August 23, 2004.
26
27 Georgia Department of Labor. 2004. Unemployment – Augusta - Aiken Compared to Georgia
28 (June 2004). Accessed at http://www.dol.state.ga.us/pdf/pr/u0604augusta.pdf on August 22,
29 2004.
30
31 Great Schools. 2004. Surry County Public Schools, Virginia. Accessed at
32 http://www.greatschools.net/cgi-bin/va/district_profile/155 on August 19, 2004.
33
34 Great Schools. 2005. Welcome to GreatSchools.net. Accessed at http://www.greatschools.net/
35 on August 24, 2005.
36
37 Institute of Electrical and Electronics Engineers, Inc. (IEEE). 2001. *National Electrical Safety*
38 *Code – 2002 Edition.* Accredited Standard Committee C2-2002, New York, New York.
39

Jaksch J. and M. Scott. 2005. North Anna ESP Site Audit Trip Report – Socioeconomics. 12-8-2003 through 12-12-2003 with Additional Telephone Interviews 2-26-2003 through 12-12-2003 with Additional Telephone Interviews 2-26-2004 through 9-29-2004 and 7-11-2005 through 7-15-2005. Available at http://www.nrc.gov/reading-rm/adams.html, Accession No. ML052170374.

Jenkins R.E. and N.M. Burkhead. 1994. *Freshwater Fishes of Virginia.* American Fisheries Society, Bethesda, Maryland.

Louis Berger Group, Inc. 2001. *Cultural Resource Assessment, Surry Power Station, Surry County, Virginia.* Surry, Virginia.

National Center for Education Statistics (NCES). 2005. Search for schools, colleges, and libraries. Accessed at http://nces.ed.gov/globallocator on August 24, 2005.

National Environmental Policy Act of 1969 (NEPA). 42 USC 4321, et seq.

National Institute of Environmental Health Sciences (NIEHS). 1999. *NIEHS Report on Health Effects from Exposure to Power Line Frequency and Electric and Magnetic Fields.* Publication No. 99-4493, Research Triangle Park, North Carolina.

Noise Control Act of 1972 (NCR), 42 USC 4901, et seq.

Ohio Department of Job and Family Services (ODJFS). 2004. Table of County Unemployment Rates. Civilian Labor Force Estimates – For Counties and Cites with Population Over 50,000: July 2004. Accessed at http://jfs.ohio.gov/releases/unemp/OhioCivilianLaborForceEstimates.pdf on August 20, 2004.

Omernik J.M. 1987. "Ecoregions of the Conterminous United States. Map (Scale 1:7500000)." *Annals of the Association of American Geographers*, 77(1):118-125.

Public Service Company of New Hampshire. 1977. *Public Service Co. of New Hampshire (Seabrook Station, Units 1 & 2)*, CLI-77-8, 5 NRC 503, 526 (1977), affirmed, *New England Coalition on Nuclear Pollution v. NRC*, 582 F.2d 87 (1st Circuit 1978).

Science Applications International Corporation (SAIC). 2004. *Draft Risk-Based End State Vision and Variance Report for the Portsmouth Gaseous Diffusion Plant, Piketon, Ohio.* DOE/OR/11-3137&D0v1. Accessed at http://www.bechteljacobs.com/pdf/port/rbesd/RBES2-9-04-2ReportText.pdf on March 5, 2004.

1　South Carolina Department of Health and Environmental Control (SCDHEC).　2004.　"South
2　Carolina Coastal Zone."　Columbia, South Carolina.　Accessed at
3　http://www.scdhec.net/ocrm/images/map.jpg on January 20, 2004.
4
5　South Carolina Department of Natural Resources (SCDNR).　2004.　"South Carolina Rare,
6　Threatened, and Endangered Species Inventory."　Columbia, South Carolina.　Accessed at
7　http:www.dnr.state.sc.us/pls/heritage/county_species.select_county_map on July 16, 2004.
8
9　South Carolina Employment Security Commission (South Carolina).　2004a.　Workforce Trends
10　Newsline.　Accessed at http://www.sces.org/lmi/news/July_2004.pdf on August 22, 2004.
11
12　South Carolina Department of Social Services (South Carolina).　2004.　Welcome to the South
13　Carolina Department of Social Services Web Site.　Accessed at
14　http://www.state.sc.us/dss/index.html on August 23, 2004.
15
16　Surry County.　1975.　"Surry County Land Development Ordinance."　Surry County Planning
17　Commission and Board of Supervisors, Surry, Virginia.
18
19　U.S. Census Bureau (USCB).　2000a.　Table p1" Total Population (1) – Universe Total
20　Population.　Accessed at
21　http://factfinder.census.gov/servlet/DTTable?_bm=y&-context=dt&-ds_name=DEC_2000_SF1_
22　U&-mt_name=DEC_2000_SF1_U_P001&-CONTEXT=dt&-tree_id=4001&-all_geo_types=N&-g
23　eo_id=38000US5720&-geo_id=38000US6760&-format=&-_lang=en on March 28, 2004.
24
25　U.S. Census Bureau (USCB).　2000b.　American Fact Finder.　Highlights from the Census 2000
26　Demographic Profiles.　For Waverly, Piekton, Portsmouth, Chillicothe; Jackson, Ohio and
27　Jackson, Pike, Ross and Scioto Counties in Ohio and Columbus and Cincinnati, Ohio and
28　Huntington, West Virginia.　Accessed at
29　http://factfinder.census.gov/home/saff/main.html?_lang=en on August 19, 2004.
30
31　U.S. Census Bureau (USCB).　2000c.　American Fact Finder.　Quick Tables.　Table QT-H1.
32　General Housing Characteristics: 2000.　For Jackson, Pike, Ross and Scioto Counties in Ohio
33　and Columbus and Cincinnati, Ohio and Huntington, West Virginia.　Accessed at
34　http://factfinder.census.gov/servlet/SAFFHousing?_sse=on on August 20, 2004.
35
36　U.S. Census Bureau (USCB).　2000d.　American Fact Finder.　Highlights from the Census 2000
37　Demographic Profiles.　For Augusta-Richmond County (balance) Georgia; Aiken, New Ellenton,
38　Jackson, Barnwell, Snelling, and Willston, South Carolina; and Aiken and Barnwell Counties,
39　South Carolina.　Accessed at http://factfinder.census.gov/home/saff/main.html?_lang=en on
40　August 22, 2004.
41

1   U.S. Census Bureau (USCB). 2000e. American Fact Finder. U.S. Census Bureau (USCB).
2   2000c. American Fact Finder. Quick Tables. Table QT-H1. General Housing Characteristics:
3   2000. For Richmond County and Columbia Counties, Georgia and Aiken and Barnwell
4   Counties, South Carolina. Accessed at
5   http://factfinder.census.gov/servlet/SAFFHousing?_sse=on on August 23, 2004.
6
7   U.S. Department of Energy (DOE). 1995. *Savannah River Site Waste Management Final*
8   *Environmental Impact Statement.* DOE-EIS 0217, Washington, D.C.
9
10   U.S. Department of Energy (DOE). 1999. *Final Environmental Impact Statement Construction*
11   *and Operation of a Tritium Extraction Facility at the Savannah River Site.* DOE/EIS-0271,
12   Washington, D.C.
13
14   U.S. Department of Energy (DOE). 2000. *Savannah River Site Waste Management Final*
15   *Environmental Impact Statement.* DOE-EIS 0217, Washington, D.C. Accessed at
16   http://www.globalsecurity.org/wmd/library/report/enviro/eis-0217/index.html on August 23, 2004.
17
18   U.S. Department of Energy (DOE). 2001. *Environmental Assessment: Winterization Activities*
19   *in Preparation for Cold Standby at the Portsmouth Gaseous Diffusion Plant, Piketon, Ohio.*
20   DOE/EA-1392, Oak Ridge Operations Office, Oak Ridge, Tennessee. As cited in Dominion and
21   Bechtel. 2002.
22
23   U.S. Department of Energy (DOE). 2002. *Savannah River Site High-Level Waste Tank Closure*
24   *Final Environmental Impact Statement.* DOE/EIS-0303, Washington, D.C.
25
26   U.S. Department of Energy (DOE). 2003. *Draft Environmental Impact Statement for*
27   *Construction and Operation of a Depleted Uranium Hexafluoride Conversion Facility at*
28   *Portsmouth, Ohio, Site.* DOE/EIS-0360. Washington, D.C. Accessed at
29   http://tis.eh.doe.gov/nepa/docs/deis/eis0360/ on March 5, 2004.
30
31   U.S. Department of the Interior, National Park Services (NPS). 2005. Letter from S. Brooks,
32   NPS to J. Cushing, NRC, regarding the impacts of building a nuclear power plant at the Surry
33   site, October 25, 2005.
34
35   U.S. Environmental Protection Agency (EPA). 2005. 8-Hour Ozone Nonattainment Areas.
36   Accessed at http://www.epa.gov/oar/oaqps/greenbk/o8index.html on September 13, 2005.
37
38   U.S. Fish and Wildlife Service (FWS). 2004a. Letter from M. Knapp, FWS to P. T. Kuo, NRC.
39
40   U.S. Fish and Wildlife Service (FWS). 2004b. North Anna Power Station Alternate, Barnwell
41   County, South Carolina, FWS log No. 4-6-04-T-110. Letter from T. Hall (FWS) to P. T. Kuo
42   (NRC). January 15, 2004.

1   U.S. Fish and Wildlife Service and Virginia Department of Game and Inland Fisheries (FWS and
2   VDGIF). 2000. Bald Eagle Protection Guidelines for Virginia. Accessed at
3   http://www.dgif.state.va.us/wildlife/publications/EagleGuidelines.pdf on March 23, 2004.

4

5   USHospital.info. 2005. United States Hospitals and Medical Centers. Accessed at
6   http://www.ushospital.info/index.htm on August 25, 2005.

7

8   U.S. Nuclear Regulatory Commission (NRC). 1996. *Generic Environmental Impact Statement*
9   *for License Renewal of Nuclear Plant*, NUREG-1437, Volumes 1 and 2, Washington, D.C.

10

11   U.S. Nuclear Regulatory Commission (NRC). 1999. *Environmental Standard Review Plans for*
12   *Nuclear Power Plants.* NUREG-1555, Supplement, Office of Nuclear Reactor Regulation,
13   Washington, D.C.

14

15   U.S. Nuclear Regulatory Commission (NRC). 2000. *Standard Review Plans for Environmental*
16   *Reviews for Nuclear Power Plants.* NUREG-1555, Washington, D.C. Available at
17   http://www.nrc.gov/reading-rm/doc-collections/nuregs/staff/sr1437/supplement7.

18

19   U.S. Nuclear Regulatory Commission (NRC). 2002. *Generic Environmental Impact Statement*
20   *for License Renewal of Nuclear Plants, Supplement 6 Regarding Surry Power Station, Units 1*
21   *and 2.* NUREG-1437, Supplement 6, Washington, D.C.

22

23   U.S. Nuclear Regulatory Commission (NRC). 2004. *Environmental Assessment of the USEC*
24   *Inc. American Centrifuge Lead Cascade Facility at Piketon, Ohio.* Office of Nuclear Material
25   Safety and Safeguards Washington, D.C. Accessed at
26   http://www.eh.doe.gov/nepa/ea/EA1495/ea-1495.pdf on August 20, 2004.

27

28   U.S. Nuclear Regulatory Commission (NRC). 2005. September 19 through 22, 2005 Trip
29   Report Tour of the North Anna River, Lake Anna, and the Surry Alternative Site (ML061720366)
30   (NRC 2005).

31

32   Virginia Department of Conservation and Recreation (VDCR). 2004. Natural Heritage
33   Resources Factsheet. *Migratory Songbird Habitat in Virginia's Coastal Plain*. Accessed at
34   http://www.dcr.state.va.us/dnh/songfact.htm on March 4, 2004.

35

36   Virginia Department of Environmental Quality (VDEQ). 2004. "Virginia's Coastal Resources
37   Management Area." Accessed at http://www.deq.state.va.us/coastal/thezone.html on
38   January 8, 2004.

39

40   Virginia Department of Game and Inland Fisheries (VDGIF). 2004. Virginia Fish and Wildlife
41   Information Service. Accessed at http://vafwis.org/WIS/ASP/default.asp on March 1, 2004.

42

1   Virginia Department of Social Services (VDSS). 2004. Accessed at
2   http://www.dss.state.va.us/index.html on August 18, 2004.
3
4   Virginia Electric and Power Company (VEPCo). 1970. Surry Power Station, Units 1 and 2,
5   Environmental Report. December 31, 1970.
6
7   Virginia Institute of Marine Sciences. 2001. Letter from J. E. Olney to Tony Banks, VEPCo
8   related to "Evaluation of potential impacts of operation of Surry Power Station on Federally
9   Managed Species." April 4, 2001.
10
11   Virginia Natural Heritage Program. 2003. Online Information on Virginia Natural Communities,
12   Rare, Threatened and Endangered Animals and Plants. Accessed at
13   http://192.206.31.52/cfprog/dnh/naturalheritage/display_counties.cfm on August 29, 2005.
14
15   Westinghouse Savannah River Company (Westinghouse). 2001. *Savannah River Site*
16   *Environmental Report for 2000*. WSRC-TR-2000-00328, Aiken, South Carolina.

9.0 Comparison of the Impacts of the Proposed Action and Alternative Sites

This chapter of the Supplement to the Draft Environmental Impact Statement was changed to reflect the higher power level and the proposed cooling system approach for Unit 3 in Revision 6 of the Environmental Report. In addition, to compare the impact of the action at the proposed site to impacts at the alternative sites, the chapter is presented in its entirety.

The need to compare the proposed early site permit (ESP) site at the North Anna Power Station (NAPS) with alternative sites arises from the requirement in Section 102(2)(c)(iii) of the National Environmental Policy Act of 1969 (NEPA) (42 USC 4332(2)(c)(iii)) that environmental impact statements (EISs) include an analysis of alternatives to the proposed action. The U.S. Nuclear Regulatory Commission (NRC) criterion to be employed in assessing whether a proposed ESP site should be rejected in favor of an alternative site is whether the alternative site is "obviously superior" to the site proposed by the applicant (NRC 1977). An alternative site is "obviously superior" to the proposed site if it is "clearly and substantially" superior to the proposed site.

The standard of obvious superiority "...is designed to guarantee that a proposed site will not be rejected in favor of a substitute unless, on the basis of appropriate study, the Commission can be confident that such action is called for" (NRC 1978a). The "obviously superior" test is appropriate for two reasons. First, the analysis performed by NRC in evaluating alternative ESP sites is necessarily imprecise. Key factors considered in the alternative site analysis, such as population distribution and density, hydrology, air quality, aquatic and terrestrial ecological resources, aesthetics, land use, and socioeconomics, are difficult to quantify in common metrics. Given this difficulty, any evaluation of a particular site would necessarily have a wide range of uncertainty. Second, the applicant's proposed ESP site has been analyzed in detail, with the expectation that most adverse environmental impacts associated with the site have been identified. By design, the alternative sites have not undergone a comparable level of detailed study. For these reasons, a proposed ESP site may not be rejected in favor of an alternative site when the alternative is "marginally better" than the proposed site, but only when it is "obviously superior" (NRC 1978b). NEPA does not require that a nuclear plant be constructed on the single best site for environmental purposes. Rather, "...[a]ll that NEPA requires is that alternative sites be considered and that the effects on the environment of building the plant at the alternative sites be carefully studied and factored into the ultimate decision" (NRC 1978a).

The NRC staff's review of alternative sites consists of a two-part sequential test for obvious superiority (NRC 2000). The first part of the test determines whether there are "environmentally preferred"[a] sites among the candidate ESP sites. The staff considers whether the applicant

---

(a) An "environmentally preferred" alternative site is a site for which the environmental impacts are sufficiently less than the proposed site so that environmental preference for the alternative site can be established (NRC 2000).

1   has (1) reasonably identified alternative sites, (2) evaluated the likely environmental impacts of
2   construction and operation at these sites, and (3) used a logical means of comparing sites that
3   has led to the applicant's selection of the proposed site. Based on its independent review, the
4   staff then determines whether any of the alternative sites are environmentally preferable to the
5   applicant's proposed ESP site.
6
7   If the staff determines that one or more alternative sites is environmentally preferable, it would
8   then compare the estimated costs (e.g., environmental, economic, and time) of constructing
9   the proposed plant at the proposed site and at the environmentally preferable site or sites
10   (NRC 2000). To find an obviously superior alternative site, the staff must determine that (1) one
11   or more important aspects, either singly or in combination, of a reasonably available alternative
12   site are obviously superior to the corresponding aspects of the applicant's proposed site and
13   (2) the alternative site does not have offsetting deficiencies in other important areas. A staff
14   conclusion that an alternative site is obviously superior to the applicant's proposed site would
15   normally lead to a recommendation that the application for the ESP be denied.
16

9.1   Comparison of the Proposed Site with the Alternatives

18
19   The staff reviewed the Environmental Report (ER) submitted by Dominion Nuclear North Anna,
20   LLC (Dominion) (Dominion 2006a), the Dominion and Bechtel study for the U.S. Department of
21   Energy on potential sites for nuclear power plant development (Dominion and Bechtel 2002),
22   and supporting documentation. The staff also conducted site visits at the proposed North Anna
23   ESP site and the alternative sites. As discussed in Section 8.3, the staff concluded that
24   Dominion had reasonably identified alternative sites, evaluated the environmental impacts of
25   construction and operation of new nuclear power facilities at those sites, and used a logical
26   means of comparing the sites. As discussed in Section 8.4, some environmental impacts
27   considered for the North Anna ESP site and the alternative site are generic to all sites and,
28   therefore, do not influence the comparison of impacts between the North Anna ESP site and the
29   alternative sites. These generic environmental impacts common to all sites include air quality,
30   nonradiological and radiological health impacts, fuel cycle, impacts for light water reactors, and
31   environmental impacts from postulated accidents. Fuel cycle impacts for gas-cooled reactors
32   are unresolved for all sites, but are likely to be SMALL. Decommissioning impacts were
33   determined to be unresolved because the reactor design has not been selected at the ESP
34   stage. The impacts from decommissioning are likely to be SMALL and affect all sites in a
35   similar manner. While the probability-weighted consequences of severe accidents were
36   resolved for all sites, the severe accident mitigation alternatives were determined to be
37   unresolved because the reactor design has not been selected at the ESP stage. The
38   combination of population characteristics and dispersion potential are not significantly different
39   among the sites to differentiate one from another given the extremely low risk already.
40

1   The staff conducted its own evaluation of the sites at a reconnaissance level before writing the
2   *Draft Environmental Impact Statement (EIS) for an Early Site Permit (ESP) at the North Anna*
3   *ESP Site* (Draft EIS)(NRC 2004), touring the sites, and reviewing existing environmental and
4   socioeconomic assessments and relevant data maintained by State and Federal agencies for
5   information relevant to potential impacts at the alternative sites. Selected staff also revisited the
6   site during the analysis of the new cooling system for Unit 3 and the power increase. For this
7   Supplement to the Draft Environmental Impact Statement (SDEIS), the staff has evaluated the
8   new information presented in ER Revision 6 and Dominion's response to the NRC request for
9   additional information (Dominion 2006a, b) and determined expected environmental impacts at
10   the proposed North Anna ESP site and at the three alternative sites.

12   The staff's characterization of the expected environmental impacts of constructing and
13   operating two new nuclear units at the proposed ESP site and alternative sites within the
14   revised plant parameter envelope presented by Dominion in ER Revision 6 (Dominion 2006a)
15   are summarized in Tables 9-1 and 9-2. These tables include all of the impacts evaluated for
16   reference and context, not just the ones that have changed as a result of the changes to
17   ER Revision 6. For those impacts to environmental resources for which the staff was unable to
18   reach a significance level for the North Anna ESP site or the alternative sites as a result of
19   insufficient information, the most likely level of impact for the purposes of comparison to
20   alternative sites was identified and the staff assumed that impacts would affect all sites in a
21   similar manner. In the following analysis, the staff indicated a likely impact level for these
22   unresolved issues based on professional judgment, experience, and consideration of controls
23   likely to be imposed under required Federal, State, or local permits that would not be acquired
24   until an application for a construction permit or combined license were underway. These
25   considerations and assumptions were similarly applied at each of the alternative sites to provide
26   a common basis for comparison. These impact levels are, therefore, best estimates of impacts
27   that the staff used for its "obviously superior" determination. No new data were collected.

29   The environmental impact categories shown in Tables 9-1 and 9-2 have been evaluated using
30   NRC's three-level standard of significance – SMALL, MODERATE, or LARGE – developed
31   using the Council on Environmental Quality guidelines. The rationale for these significance
32   levels is outlined in the footnotes to Table B-1 of Title 10 of the Code of Federal Regulations
33   (CFR) Part 51, Subpart A, Appendix B:

35   SMALL – Environmental effects are not detectable or are so minor that they will
36   neither destabilize nor noticeably alter any important attribute of the resource.

38   MODERATE – Environmental effects are sufficient to alter noticeably, but not to
39   destabilize important attributes of the resource.

41   LARGE – Environmental effects are clearly noticeable and are sufficient to
42   destabilize important attributes of the resource.

Table 9-1. Comparison of the Construction Impacts at the Proposed ESP and Alternative Sites

Impact Area Category	North Anna ESP Site	Surry Site	Portsmouth Site	Savannah River Site
Land-use impacts	--	--	--	--
The site and vicinity	SMALL	SMALL	SMALL	SMALL
Transmission line rights-of-way	SMALL	SMALL	SMALL	SMALL TO MODERATE
Air quality impacts	SMALL	SMALL	SMALL	SMALL
Water-related impacts	--	--	--	--
Water use	SMALL	SMALL	SMALL	SMALL
Water quality	SMALL	SMALL	SMALL	SMALL
Ecological impacts	--	--	--	--
Terrestrial ecosystems	SMALL	SMALL	SMALL	SMALL TO MODERATE
Aquatic ecosystems	SMALL	SMALL	SMALL	SMALL
Threatened and endangered species	SMALL	SMALL	SMALL	SMALL TO MODERATE
Socioeconomic impacts	--	--	--	--
Physical impacts	SMALL	SMALL	SMALL	SMALL
Demography	SMALL	SMALL	SMALL	SMALL
Social and Economic[a]	SMALL BENEFICIAL to LARGE BENEFICIAL	SMALL BENEFICIAL to MODERATE BENEFICIAL	SMALL BENEFICIAL to MODERATE BENEFICIAL	SMALL BENEFICIAL to MODERATE BENEFICIAL
Infrastructure and community services	SMALL to MODERATE	SMALL to MODERATE	SMALL to MODERATE	SMALL
Historic and cultural resources	SMALL	MODERATE to LARGE	SMALL	SMALL
Environmental justice	SMALL	SMALL	SMALL	SMALL
Nonradiological health impacts	SMALL	SMALL	SMALL	SMALL
Radiological health impacts	SMALL	SMALL	SMALL	SMALL

(a) Impacts of construction on the economy and increases in taxes collected are considered beneficial impacts. These beneficial impacts are discussed in the applicable sections.

Table 9-2. Comparison of the Operational Impacts at the Proposed ESP and Alternative Sites

Impact Area Category	North Anna ESP Site	Surry Site	Portsmouth Site	Savannah River Site
Land-use impacts	--	--	--	--
The site and vicinity	SMALL	SMALL	SMALL	SMALL
Transmission line rights-of-way	SMALL	SMALL	SMALL	SMALL
Air quality impacts	SMALL	SMALL	SMALL	SMALL
Water-related impacts	--	--	--	--
Water use	SMALL	SMALL	SMALL TO MODERATE	SMALL
Water quality	*Unresolved, but likely SMALL	*Unresolved, but likely SMALL	*Unresolved, but likely SMALL	*Unresolved, but likely SMALL
Water use in drought year	MODERATE	SMALL	MODERATE	SMALL
Ecological impacts	--	--	--	--
Terrestrial ecosystems	SMALL	SMALL	SMALL	SMALL
Aquatic ecosystems	SMALL	SMALL	SMALL	SMALL
Threatened and endangered species	SMALL	SMALL	SMALL	SMALL TO MODERATE
Socioeconomic impacts	--	--	--	--
Physical impacts	SMALL to MODERATE	SMALL to MODERATE	SMALL	SMALL
Demographics	SMALL	SMALL	SMALL	SMALL
Social and economic	SMALL BENEFICIAL to LARGE BENEFICIAL	SMALL BENEFICIAL to LARGE BENEFICIAL	SMALL BENEFICIAL to LARGE BENEFICIAL	SMALL BENEFICIAL to LARGE BENEFICIAL
Infrastructure and community services	SMALL to MODERATE	SMALL to MODERATE	SMALL to MODERATE	SMALL
Historic and cultural resources	SMALL	MODERATE TO LARGE	SMALL	SMALL
Environmental justice	SMALL	SMALL	SMALL	SMALL
Nonradiological health impacts	SMALL	SMALL	SMALL	SMALL
Radiological health impacts	SMALL	SMALL	SMALL	SMALL
Postulated accidents	SMALL	SMALL	SMALL	SMALL
Fuel Cycle Impacts	SMALL	SMALL	SMALL	SMALL

The socioeconomic impact level category reflects both adverse and beneficial impacts. Positive impacts (e.g., tax receipts to local government) would occur but are not the determining factors in the analysis of an environmentally preferable or obviously superior site. For impact

1 categories in which no impact is predicted, the adverse impact level is shown as SMALL.
2 Within some impact categories, the impact levels varied. Professional judgments were made to
3 conclude, where possible, a single overall level of impact. In several cases, a range of probable
4 impacts is given.
5
6 The staff determined that the impact level from construction on most of the environmental
7 resources at most of the sites is SMALL, and was not affected by the proposed changes to the
8 Unit 3 cooling system or to the higher power level. In some cases, there are factors related to a
9 site that could cause the impact level to increase from SMALL to MODERATE. In one case, the
10 impact level category for an alternative site could be as high as LARGE. Impacts on the local
11 economy and tax base range from SMALL BENEFICIAL to LARGE BENEFICIAL at the various
12 sites. More detailed information on these cases is presented in Chapters 4, 5, and 6 for the
13 North Anna ESP site, and Chapters 6 and 8 for the alternative sites. The staff based its
14 analysis of the environmental impacts on the implementation of mitigation measures in
15 accordance with Federal, State, and local permit requirements and on the mitigation measures
16 identified in the ER. In its analysis of the alternative sites, the staff assumed that similar permit
17 requirements and mitigative measures would apply.
18
19 The staff determined that the impact from operation on most of the environmental resources at
20 most of the sites is SMALL. In some cases, there are factors related to a site that could
21 cause the impact level to range from SMALL to LARGE. Impacts on the local tax base range
22 from SMALL BENEFICIAL to LARGE BENEFICIAL at all sites. More detailed information on
23 these cases is presented in Chapter 5 for the North Anna ESP site and Chapter 8 for the
24 alternative sites.
25

26 ## 9.2 Environmentally Preferable Sites
27

28 ### 9.2.1 Construction
29

30 The impacts of construction at the North Anna ESP site are SMALL for most major impact
31 categories. However, as noted in Section 4.5, there are some impact subcategories under
32 infrastructure and community services (housing, public services, and education) for which the
33 impacts could be MODERATE if a larger number of construction workers than the staff assumed
34 relocate to Louisa or Orange Counties. The tax benefits to Louisa County could be LARGE
35 BENEFICIAL, and the impacts on jobs and the economy could be MODERATE BENEFICIAL.
36

37 The impacts of construction at the Surry alternative site are SMALL for all impact categories
38 except infrastructure and community services (economy and taxes) and historic and cultural
39 resources. As noted in Section 8.5, the impacts in this area are SMALL to MODERATE for

1　transportation and MODERATE to LARGE because of its potential effect on the Colonial
2　National Historic Park. Impacts on the economy and taxes may be SMALL BENEFICIAL to
3　MODERATE BENEFICIAL.
4
5　The impacts of construction at the Portsmouth alternative site are SMALL for all impact
6　categories except infrastructure and community services. As noted in Section 8.6, the impacts
7　in this area are SMALL to MODERATE for aesthetics. In addition, the impacts on the economy
8　and taxes are SMALL BENEFICIAL to MODERATE BENEFICIAL.
9
10　The impacts of construction at the Savannah River alternative site are SMALL for all impact
11　categories except terrestrial resources (including endangered species). As noted in
12　Section 8.7, the impacts on terrestrial resources are SMALL to MODERATE. The staff arrived
13　at this range of potential impacts because the routing for the new transmission line rights-of-way
14　that would be needed is not known with certainty and, consequently, neither are the impacts of
15　construction. In addition, the impacts on the economy and tax base are SMALL BENEFICIAL to
16　MODERATE BENEFICIAL.
17
18　While there are minor differences in most of the construction impacts at the four sites, none of
19　these differences is sufficient to determine that any of the alternative sites is environmentally
20　preferable to the proposed North Anna ESP site.
21
22　**9.2.2　Operations**
23
24　The impacts of operations at the North Anna ESP site are SMALL for all major impact
25　categories except water use and socioeconomic categories. As discussed in Section 5.3, the
26　impacts of Unit 3 operations on water use are SMALL most years. However, during a significant
27　drought, the impacts could be MODERATE. In addition, as discussed in Section 5.5, the
28　impacts for aesthetics and housing are SMALL to MODERATE, and impacts to recreation may
29　be MODERATE during drought years. The impacts on the economy and taxes are SMALL
30　BENEFICIAL to LARGE BENEFICIAL and impacts on community services would be SMALL to
31　MODERATE.
32
33　The impacts of operations at the Surry alternative site are SMALL for all impact categories
34　except community characteristics and historical and cultural resources. As noted in Section 8.5,
35　the impacts in this area are SMALL to MODERATE for aesthetics and infrastructure in the
36　vicinity and MODERATE to LARGE because the particular impacts that could be realized on the
37　Colonial National Historic Park's resources. The impacts on the economy and taxes are SMALL
38　BENEFICIAL to LARGE BENEFICIAL.
39
40　The impacts of operations at the Portsmouth alternative site are SMALL for all impact categories
41　except water use and socioeconomic categories. As noted in Section 8.6, the impacts of plant
42　operations on water use are SMALL to MODERATE most years. However, during a significant

1   drought, the impacts would be MODERATE. In addition, impacts under infrastructure and
2   community services are SMALL to MODERATE for aesthetics. The impacts on the economy
3   and taxes are SMALL BENEFICIAL to LARGE BENEFICIAL.
4
5   The impacts of operations at the Savannah River alternative site are SMALL for all impact
6   categories except threatened and endangered species and socioeconomics. As noted in
7   Section 8.7, the impacts to threatened and endangered species are SMALL to MODERATE
8   because the routing for the new transmission line rights-of-way that would be needed for plant
9   operation is not known and so the associated impacts of operation and maintenance must be
10   assigned a range of potential impacts. The impacts on the economy and taxes are SMALL
11   BENEFICIAL to LARGE BENEFICIAL.
12
13   In summary, although the water-use impacts at the North Anna ESP site are projected to be
14   MODERATE during years when there is a severe drought, this event is expected to be an
15   infrequent and temporary as are any associated impacts. Aesthetic impacts are expected to be
16   periodic and MODERATE. The operational impact of the units at North Anna is also expected
17   to have a MODERATE impact in the recreational subcategory of community characteristics
18   during a severe drought, at which time the lake marinas could be affected. The Portsmouth
19   alternative site has a similar water-use issue, and the Savannah River alternative site has
20   unknown impacts associated with the transmission line rights-of-way, which could range from
21   SMALL to MODERATE. The Surry alternative site has a cultural and historical impact that could
22   be LARGE. The impacts on economy and taxes are generally beneficial and similar across
23   sites, and the impacts on infrastructure and community services are similar and up to
24   MODERATE.
25
26   For those impacts to environmental resources for which the staff was unable to reach a
27   significance level for the North Anna ESP site or the alternative sites as a result of insufficient
28   information, the most likely level of impact for the purposes of comparison to alternative sites
29   was identified and the staff assumed that impacts would affect all sites in a similar manner. In
30   the following analysis, the staff indicated a likely impact level for these unresolved issues based
31   on professional judgment, experience, and consideration of controls likely to be imposed under
32   required Federal, State, or local permits that would not be acquired until an application for a
33   construction permit or combined license were underway. These considerations and
34   assumptions were similarly applied at each of the alternative sites to provide a common basis
35   for comparison. These impact levels are, therefore, best estimates of impacts that the staff
36   used for its "obviously superior" determination. No new data were collected. For example,
37   insufficient information was provided for operational water quality, gas-cooled reactor fuel cycle,
38   decommissioning and severe accident mitigation alternatives. While there is insufficient
39   information to reach a conclusion on the significance levels for the unresolved issues, the staff
40   does not expect, based on the information available, that there would be significant differences
41   in impact categories among the proposed and alternative sites. While there are some

differences in the environmental impacts of operation at the four sites, none of these differences is sufficient for the staff to determine that any of the alternative sites is environmentally preferable to the proposed North Anna ESP site.

9.3 Obviously Superior Sites

None of the alternative sites was determined to be environmentally preferable to the proposed North Anna ESP site. Therefore, the staff concluded that none of the alternative sites is obviously superior to the North Anna ESP site.

9.4 Comparison with the No-Action Alternative

The no-action alternative refers to a scenario in which the NRC would deny the ESP application. If the ESP application for the North Anna ESP site were denied, the impacts of the site preparation activities would not occur. Further, denial of the ESP application would prevent early resolution of safety and environmental issues for the site. These issues would have to be addressed during a future licensing action (ESP, construction permit, or combined license), should an applicant decide to pursue construction and operation activities for a nuclear facility at the site at a later time.

In the event of NRC's denial of the ESP application, Dominion could follow several paths to satisfy its electric power needs. The potential paths include (1) seeking an ESP for a different proposed site, (2) purchase of power from other electricity providers, (3) conservation and demand-side management programs, (4) construction of new generation facilities other than nuclear at the North Anna site, (5) construction of new generation facilities at other locations, (6) delayed retirement of existing generating facilities, and (7) reactivation of previously retired generating facilities. These paths could be pursued individually or in combination. Each of the paths would have associated environmental impacts. Nonetheless, since 10 CFR Part 52 does not require an ER or EIS for an ESP to include consideration of energy alternatives or the benefits of construction and operation of a reactor or reactors at the ESP site, Dominion did not addressed those matters in its ER, and this EIS does not consider such matters. Accordingly, should the NRC ultimately determine to issue an ESP for the North Anna ESP site, and a construction permit or combined license application that references such an ESP is docketed, these matters would be considered in the EIS on the CP or COL application.

The activities that may be permissible under an ESP are limited to the site preparation and limited construction activities enumerated in 10 CFR 50.10(e). Pursuant to 10 CFR 52.25, such activities are permissible only if the final environmental impact statement concludes that the activities would not result in any significant impacts that could not be redressed, and an ESP that incorporates the site redress plan is granted. The results of the staff's assessment of the site redress plan are discussed in Section 4.11 of the DEIS. As discussed in that section, the staff concludes that the potential site-preparation activities described in Dominion's redress plan would not result in any significant adverse impacts that could not be redressed. Because the site preparation and preliminary work could be redressed by the site redress plan described in Section 4.11 of the DEIS, the impacts of the proposed action and the no-action alternative would be similar.

9.5 References

10 CFR Part 50. Code of Federal Regulations, Title 10, *Energy*, Part 20, "Domestic Licensing of Production and Utilization Facilities."

10 CFR Part 51. Code of Federal Regulations, Title 10, *Energy*, Part 51, "Environmental Protection Regulations for Domestic Licensing and Related Regulatory Functions."

10 CFR Part 52. Code of Federal Regulations, Title 10, *Energy*, Part 52, "Early Site Permits Standard Design Certification and Combined License for Nuclear Power Plants."

Dominion Energy, Inc. and Bechtel Power Corporation (Dominion and Bechtel). 2002. *Study of Potential Sites for the Deployment of New Nuclear Plants in the United States.* U.S. Department of Energy Cooperative Agreement No. DE-FC07-02ID14313, Washington, D.C.

Dominion Nuclear North Anna, LLC (Dominion). 2006a. *North Anna Early Site Permit Application   Part 3   Environmental Report.* Revision 6, Glen Allen, Virginia.

Dominion Nuclear North Anna, LLC (Dominion). 2006b. Letter from E. Grecheck (DNNA) to NRC dated May 24, 2006, submitting additional information in response to an NRC request dated May 10, 2006, (ML061510131).

Dominion Nuclear North Anna, LLC (Dominion). 2006c. *North Anna Early Site Permit Application – Part 4 – Programs and Plans.* Revision 6, Glen Allen, Virginia.

National Environmental Policy Act of 1969 (NEPA). 42 USC 4321 et seq.

U.S. Nuclear Regulatory Commission (NRC). 1977. Public Service Co. of New Hampshire (Seabrook Station, Units 1 & 2), CLI-77-8, 5 NRC 503, 526 (1977), *aff'd, New England Coalition on Nuclear Pollution v. NRC*, 582 F.2d 87, 95-96 (1st Cir 1978).

U.S. Nuclear Regulatory Commission (NRC). 1978a. New England Coalition on Nuclear Pollution. *New England Coalition on Nuclear Pollution v. NRC,* 582 F.2d 87 (1st Circuit 1978).

U.S. Nuclear Regulatory Commission (NRC). 1978b. Rochester Gas & Electric Corp. (Sterling Power Project Nuclear Unit No. 1), ALAB-502, 8 NRC 383, 397 (1978), *aff'd,* CLI-80-23, 11 NRC 731 (1980).

1   U.S. Nuclear Regulatory Commission (NRC). 2000. *Standard Review Plans for Environmental*
2   *Reviews for Nuclear Power Plants.* NUREG-1555, Vol. 1, Washington, D.C.
3
4   U.S. Nuclear Regulatory Commission (NRC). 2002. *Generic Environmental Impact Statement*
5   *for License Renewal of Nuclear Plants, Supplement 6 Regarding Surry Power Station, Units 1*
6   *and 2.* NUREG-1437, Supplement 6, Washington, D.C.
7
8   U.S. Nuclear Regulatory Commission (NRC). 2004. *Draft Environmental Impact Statement*
9   *(EIS) for an Early Site Permit (ESP) at the North Anna ESP Site.* NUREG-1811, Draft Office of
10   Nuclear Reactor Regulation, Division of Regulatory Improvement Programs, Washington, D.C.
11

10.0 Conclusions and Recommendations

This chapter of the Supplement to the Draft Environmental Impact Statement was changed to reflect the higher power level and the proposed cooling system approach for Unit 3 in Revision 6 of the Environmental Report and presented in its entirety.

On September 25, 2003, the U.S. Nuclear Regulatory Commission (NRC) received an application from Dominion Nuclear North Anna, LLC (Dominion) for an early site permit (ESP) for a location adjacent to North Anna Power Station (NAPS) Units 1 and 2. The North Anna ESP site is located in Louisa County, Virginia, approximately 10 km (6 mi) northeast of the town of Mineral. Dominion submitted revisions to the Environmental Report (ER) on October 2, 2003, July 15, 2004, September 7, 2004, May 12, 2005, July 25, 2005, and April 13, 2006 (Dominion 2006a). On December 10, 2004, the staff prepared a Draft Environmental Impact Statement (Draft EIS) with its evaluation of the Dominion application through ER Revision 3 (NRC 2004). Any reference in this Supplement to the Draft Environmental Impact Statement (SDEIS) to the ER refers to Revision 6, including Dominion's responses to the staff's request for additional information on Revision 6 (Dominion 2006b), unless otherwise stated.

In Revision 6 to the North Anna ESP application, Dominion proposed (1) changing its approach for cooling the proposed Unit 3 reactor from the once-through cooling system (as described in previous versions of the ER) to a closed-cycle system and (2) increasing the maximum power level per unit from 4300 megawatts-thermal (MW(t)) to 4500 MW(t) for proposed Units 3 and 4 (referred to hereafter as Units 3 and 4). Under the revised cooling system approach, Unit 3 would use a closed-cycle, combination wet and dry cooling system. The proposed increase in power level corresponds to the revision of the maximum power of an economic simplified boiling water reactor (ESBWR), one of the reactor designs included in the plant parameter envelope (PPE) and evaluated in the Draft EIS.

The NRC staff determined that the changes to the proposed action were substantial; therefore, the staff decided to prepare a Supplement to its Draft EIS (referred to as the SDEIS) pursuant to Title 10 of the Code of Federal Regulations (CFR) Section 51.72. On May 16, 2006, following receipt of Dominion's ER Revision 6, the staff published a Notice of Intent to prepare a Supplement to the Draft EIS for the North Anna ESP application in the *Federal Register* (71 FR 28392). The scope of this SDEIS is limited to the environmental impacts associated with the change in the cooling system for Unit 3 and the increase in the power level for both units. The evaluation presented in this SDEIS replaces the evaluation of the impacts associated with the originally proposed once-through cooling for Unit 3 and modifies the analysis of impacts related to the power level increase. These revised evaluations, along with public comments received on the analysis presented in this SDEIS, will be incorporated into the Final EIS together with comments received concerning the Draft EIS and the staff's consideration of such comments.

1   An ESP is a Commission approval of a site or sites for one or more nuclear power facilities.
2   Issuance of an ESP is an action separate from the issuance of a construction permit (CP) or a
3   combined construction permit and operating license (combined license or COL) for such a
4   facility. An ESP application may refer to a reactor's or reactors' design parameters or a PPE,
5   which is a set of values of plant design parameters that an ESP applicant expects will bound the
6   design characteristics of the reactor or reactors that might be built at a selected site;
7   alternatively an ESP may refer to a detailed reactor design. An ESP is not a license to build a
8   nuclear power plant; rather, the application for an ESP initiates a process undertaken to assess
9   whether a proposed site is a suitable location for such a plant should the applicant decide to
10   pursue a CP or COL.

12   Section 102 of the National Environmental Policy Act of 1969 (NEPA) (42 USC 4321) directs
13   that Federal agencies prepare an EIS for major Federal actions that significantly affect the
14   quality of the human environment. The NRC has implemented Section 102 of NEPA in
15   10 CFR Part 51. Subpart A of 10 CFR Part 52 contains the NRC regulations related to ESPs.
16   In addition, as set forth in 10 CFR 52.18, the Commission has determined that an EIS will be
17   prepared during the review of an application for an ESP. The purpose of Dominion's proposed
18   action, issuance of the ESP, is to provide stability in the licensing process by addressing site
19   safety and environmental issues before the plants are built rather than after construction is
20   completed. Part 52 of Title 10 describes the ESP as a "partial construction permit." An
21   applicant for a CP or COL for a nuclear power plant or plants to be located at a site for which an
22   ESP has been issued can reference the ESP, and matters resolved in the ESP proceeding are
23   considered resolved in the subsequent proceeding. However, issuance of either a CP (and OL)
24   or COL to construct and operate a nuclear power plant is a major Federal action that requires its
25   own environmental review in accordance with 10 CFR Part 51.

27   The holder of an ESP, or an applicant for a CP or COL that references an ESP that includes a
28   site redress plan, may, in accordance with 10 CFR 52.25, perform the site preparation and
29   preliminary construction activities enumerated in 10 CFR 50.10(e)(1), provided that the final
30   ESP EIS concludes that the activities will not result in any significant adverse environmental
31   impacts that cannot be redressed. Dominion provided a site redress plan as part of its ESP
32   application. Pursuant to 10 CFR 52.17(a)(2), Dominion did not address the benefits of the
33   proposed action (e.g., the need for power). In accordance with 10 CFR 52.18, the EIS is
34   focused on the environmental effects of construction and operation of a reactor, or reactors, that
35   have characteristics that fall within the design parameters that would be specified in the ESP if it
36   is granted.

38   Three primary issues – site safety, environmental impacts, and emergency planning – must be
39   addressed in the ESP application. Likewise, in its review of the application, the NRC assesses
40   the applicant's proposal in relation to these issues and determines whether the application
41   meets the requirements of the Atomic Energy Act of 1954 and NRC regulations. Site safety and

1   emergency planning are addressed in the staff's safety evaluation report (NRC 2005). This
2   SDEIS addresses the environmental impacts related to the changes proposed in Revision 6 to
3   the ER.

5   To guide its assessment of environmental impacts of a proposed action or alternative actions,
6   the NRC has established a standard of significance for impacts using Council on Environmental
7   Quality (CEQ) guidance (40 CFR 1508.27). Using this approach, the NRC has established
8   three significance levels – SMALL, MODERATE, or LARGE – which are defined below:

10   SMALL – Environmental effects are not detectable or are so minor that they will neither
11   destabilize nor noticeably alter any important attribute of the resource.
12

13   MODERATE – Environmental effects are sufficient to alter noticeably, but not to
14   destabilize, important attributes of the resource.
15

16   LARGE – Environmental effects are clearly noticeable and are sufficient to destabilize
17   important attributes of the resource.

19   Mitigation measures were considered for each resource area and are presented in the
20   appropriate sections.

22   NEPA Section 102(2)(C)(ii),(iv)-(v) requires that an EIS include information on:

24   • any adverse environmental effects that cannot be avoided should the proposal be
25   implemented
26

27   • any irreversible and irretrievable commitments of resources that would be involved if the
28   proposed action is implemented
29

30   • the relationship between local short-term uses of the environment and the maintenance
31   and enhancement of long-term productivity.

33   The NEPA information is provided in Sections 10.1 through 10.3.

35   Activities permitted under an ESP with an approved site redress plan include preparation of the
36   site for construction of the facility, installation of temporary construction support facilities,
37   excavation for facility structures, construction of service facilities, and construction of certain
38   structures, systems, and components that do not prevent or mitigate the consequences of
39   postulated accidents. These activities are identified in the site redress plan. The following
40   discussion addresses the impacts of construction and operation of two units at the North Anna
41   ESP site and fulfils NEPA Section 102(2)(C)(ii),(iv)-(v). The construction impacts bound any
42   impacts of the site preparation and preliminary construction activities allowed under
43   10 CFR 52.25(a).

10.1 Unavoidable Adverse Environmental Impacts

Section 102(2)(C)(ii) of NEPA requires that an EIS include information on any adverse environmental effects that cannot be avoided should the proposal be implemented. Unavoidable adverse environmental impacts are those potential impacts of construction and operation of the proposed new units that cannot be avoided and for which no practical means of mitigation are available.

There would be no unavoidable adverse environmental impacts associated with the granting of the ESP with the exception of impacts associated with the site preparation and preliminary construction activities enumerated in 10 CFR 50.10(e)(1). The impacts associated with the site preparation and preliminary construction activities are bounded by the overall construction activities. However, there are unavoidable adverse environmental impacts associated with the construction and operation of two units at the North Anna ESP site which are described below.

If the ESP is granted, the ESP holder could, pursuant to 10 CFR 5.25, perform the following site preparation and preliminary construction activities consistent with the type enumerated in 10 CFR 50.10(e)(1):

- preparation of the site for construction of the facility (including such activities as clearing, grading, and construction of temporary access roads and borrow areas)

- installation of temporary construction support facilities (including such items as warehouse and shop facilities, utilities, concrete mixing plants, docking and unloading facilities, and construction support buildings)

- excavation for facility structures

- construction of service facilities (including such facilities as roadways, paving, railroad spurs, fencing, exterior utility and lighting systems, and sanitary sewage treatment facilities)

- construction of structures, systems, and components that do not prevent or mitigate the consequences of postulated accidents, that could cause undue risk to the health and safety of the public.

If the ESP is granted and any or all of the activities above are performed, but the ESP is not referenced in an application for a CP under 10 CFR Part 50 or a COL under 10 CFR Part 52 while the ESP remains valid, the ESP holder would be required to redress the site according to the site redress plan included in Part 4, Chapter 1 of the ESP application (Dominion 2006c). The staff reviewed the list of allowed site preparation and preliminary construction activities in the event that the ESP is granted and reviewed the full site redress plan submitted by Dominion. In accordance with 10 CFR 52.17(c), the application demonstrated that there is reasonable

1   assurance that redress carried out under the plan will achieve an environmentally stable and
2   aesthetically acceptable site suitable for whatever non-nuclear use may conform with local
3   zoning laws. Accordingly, in accordance with 10 CFR 52.25(a), the staff concludes that the
4   potential site preparation and preliminary construction activities described in Dominion's site
5   redress plan would not result in any significant adverse environmental impacts that could not be
6   redressed. As discussed in Section 1.5 of this SDEIS, the staff proposes to include a condition
7   prohibiting Dominion from conducting any pre-construction activity that would result in a
8   discharge into navigable waters without first submitting to the NRC a Virginia Water Protection
9   Permit or a determination by the Virginia Department of Environmental Quality (VDEQ) that no
10 certification is required.
11
12 ***Unavoidable Adverse Impacts During Construction***
13
14 Chapter 4 of the Draft EIS discusses the impacts from construction in detail. Chapter 4 of this
15 SDEIS provides summaries of each section, and examines the change in impacts as a result of
16 the changes proposed in ER Revision 6. The unavoidable adverse impacts related to
17 construction, including those from the revised cooling system and higher power level, are listed
18 in Table 10-1 and summarized below. The primary unavoidable adverse environmental impacts
19 during construction would be related to land use. All construction activities for Units 3 and 4,
20 including ground-disturbing activities, would occur within the existing NAPS site boundary.
21 According to Dominion, the area that would be affected on a long-term basis as a result of
22 permanent facilities is approximately 52 ha (128 ac); up to an additional 27.5 ha (67.9 ac) could
23 be disturbed on a short-term basis as a result of temporary activities and facilities and laydown
24 areas (Dominion 2006a).
25
26 The construction impacts on the terrestrial ecology of the site would be expected to be
27 short-term. Construction of two units would result in the removal of approximately 32 ha (80 ac)
28 of forested habitat within the site. The ESP site does not contain any old growth timber or
29 unique or sensitive plants or communities. Therefore, construction activities would not
30 noticeably reduce the local or regional diversity of plants or plant communities. There are no
31 important animal species or habitats on the ESP site. No areas designated by the U.S. Fish
32 and Wildlife Service as critical habitat for endangered or threatened species exist at or near the
33 site, nor are threatened or endangered plants or animals known to exist at the site. Therefore,
34 construction would be expected to have no impact on any threatened or endangered species or
35 other important species or habitats. Socioeconomic impacts of construction include an increase
36 in traffic. Atmospheric and meteorological impacts include fugitive dust from construction
37 activities that can be mitigated by the dust control plan. Radiological doses to construction
38 workers from the adjacent units are expected to be well below regulatory limits. Regarding
39 environmental justice, there are no unusual resource dependencies by low income or minority
40 groups and therefore no adverse unavoidable impacts.
41

1
2

Table 10-1. Unavoidable Adverse Environmental Impacts from Construction

3

Impact Category	Adverse Impacts Based on Dominion's Proposal	Actions to Mitigate Impacts	Unavoidable Adverse Impacts
Land use	Yes	Comply with requirements of applicable Federal, State, and local permits	52 ha (128 ac) disturbed on long-term basis, additional 27.5 ha (67.9 ac) would be disturbed on a short-term basis.
Hydrological and water use	Yes	Obtain a Clean Water Act 401 certification prior to site preparation activities; use best construction management practices;	Fill and grading operations at the North Anna ESP site would alter two ephemeral streams.
Ecological Terrestrial Aquatic	Yes Yes	(a) Use of construction best management practices, adherence to applicable permit conditions, and avoidance of sensitive areas. Where possible, reestablish habitat after construction. (b) Performing wetland surveys to determine Clean Water Act Section 404 applicability	(a) Removal of trees and vegetation and habitat. (b) Disturbance of intermittent streams, destruction of wetlands.
Socioeconomic	Yes	Implement traffic management plan	Increased traffic congestion
Radiological	Yes	Use of as low as reasonably achievable (ALARA) principles	Dose to construction workers
Atmospheric and meteorological	Yes	Implement dust control plan	Equipment emissions and fugitive dust from operation of earth-moving equipment are sources of air pollution.
Environmental justice	No	Not applicable	Not applicable

4
5
6
7
8
9
10
11
12
13
14
15

Unavoidable Adverse Impacts During Operation

16
17
18 Chapter 5 of the Draft EIS provides a detailed discussion of the impacts from operation.
19 Chapter 5 of this SDEIS provides a summary of the Draft EIS sections, and analyzes the
20 impacts of the changes presented in ER Revision 6. The unavoidable adverse impacts related
21 to operation are listed in Table 10-2 and summarized below.
22
23 Dominion changed its proposed cooling system from a once-through system for Unit 3, as
24 described in ER Revisions 3, 4, and 5, to the closed-cycle, combination wet and dry cooling
25 system described in Revision 6. This change reduces the previously predicted thermal impacts
26 and impingement and entrainment of Lake Anna's aquatic populations.
27
28 Socioeconomic impacts are primarily increased demand for services, with the increased tax
29 revenue to support the increase in services. The visual impact of lower water levels, and their

1   effect on shoreline exposure during intermittent severe drought, could temporarily impact the
2   area. Regarding environmental justice, there are no unusual resource dependencies by low
3   income or minority groups and therefore no adverse unavoidable impacts. Meteorological
4   impacts are expected to be negligible, although wet cooling towers would put more moisture
5   into the air in the form of a visible condensation plume. Pollutants emitted during operations are
6   considered insignificant. The unavoidable adverse impacts from operation for land use are
7   small and further mitigation is not warranted.
8
9   **Table 10-2**. Unavoidable Adverse Environmental Impacts from Operation
10

Impact Category	Adverse Impacts Based on Dominion's Proposal	Actions to Mitigate Impacts	Unavoidable Adverse Impacts
Land use	No	Local land management plans; comply with requirement of applicable Federal, State, and local permits.	Possible new housing and retail space added in vicinity because of potential growth.
Hydrological and water use	Yes	Comply with Commonwealth permit limits.	Occasional and temporary decrease in level of Lake Anna and reduction in available water released from dam into the North Anna River.
Ecological			
Terrestrial	No	None	None.
Aquatic	Yes	None	Proportion of resources subject to impingement and entrainment would be small.
Socioeconomic	Yes	Consider plume abatement measures.	Impacts to recreation because the level of Lake Anna would be lower during drought conditions. Periodic adverse visual aesthetic impact due to Unit 3 cooling tower plume.
Radiological	Yes	Use of ALARA principles	Dose to workers, the public, and biota.
Atmospheric and meteorological	No	None	None
Environmental justice	No	Not applicable	Not applicable

10.2 Irreversible and Irretrievable Commitments of Resources

Section 102(2)(C)(v) of NEPA requires that an EIS include information on any irreversible and irretrievable commitments of resources that would occur if the proposed action is implemented. The only irreversible and irretrievable commitments of resources that would be expended if the proposed action is implemented would be resources used by Dominion for site preparation activities. If not used during the duration of the ESP, any such resource commitments for site preparation activities would be used at the CP/COL stage or could potentially be used for other activities even if the ESP is issued but not referenced in a CP or a COL application.

Irretrievable commitments of resources during construction of the proposed new units generally would be similar to that of any major construction project. The actual commitment of construction resources (e.g., concrete, steel, and other building materials) would depend on the reactor design selected at the CP/COL stage. Hazardous materials such as asbestos would not be used, if possible. If materials such as asbestos were used, the use would be in accordance with safety regulations and practices. The actual estimate of construction materials would be performed at the CP/COL stage when the reactor design is selected.

The staff expects that the use of construction materials in the quantities associated with those expected for the two new units, while irretrievable, would be a small impact with respect to the availability of such resources.

The main resource that would be irretrievably committed during operation of two new nuclear unit would be uranium for the fuel and ultimately the offsite storage space for the spent fuel assemblies. The availability of uranium ore and existing stockpiles of highly enriched uranium in the United States and Russia that could be processed into fuel is sufficient, so the irreversible and irretrievable commitment would be of only small consequence.

10.3 Relationship Between Short-Term Uses and Long-Term Productivity of the Human Environment

Section 102(2)(C)(iv) of NEPA requires that an EIS include information on the relationship between local short-term uses of the environment and the maintenance and enhancement of long-term productivity. The only short-term use of the environment that could occur if the proposed action is granted would be site preparation and limited construction activities. Any such activities are unlikely to adversely affect the long-term productivity of the environment. The evaluation of the relationship between local short-term uses of the environment and the maintenance and enhancement of long-term productivity for the construction and operation of the two new units can only be performed by discussing the benefits of operating the units. The societal benefit is the production of electricity. In accordance with 10 CFR 52.18, an EIS for an

1 ESP need not include an assessment of the benefits of the proposed action. Therefore, an
2 assessment of the evaluation of the relationship between local short-term uses of the
3 environment and the maintenance and enhancement of long-term productivity for the
4 construction and operation of the two units would be performed at the CP/COL stage should the
5 NRC grant the ESP and an applicant references it in an application for a CP or COL. This issue
6 is, therefore, not resolved.
7

10.4 Cumulative Impacts

9
10 The staff considered the potential cumulative impacts resulting from construction and operation
11 of Units 3 and 4 in the context of past, present, and future actions at the North Anna ESP site in
12 Chapter 7 of the Draft EIS and this SDEIS, and summaries are provided in this SDEIS with
13 additional impact analysis regarding changes presented in ER Revision 6. For each impact
14 area, the staff determined that the potential cumulative impacts resulting from construction and
15 operation are SMALL, and mitigation is not warranted. The geographical area over which past,
16 present, and future actions could contribute to cumulative impacts is dependent on the type of
17 action considered. Several impact categories have the potential for MODERATE impacts, most
18 of which would occur under temporary circumstances or as the result of a larger than expected
19 concentration of construction workers settling near the North Anna ESP site. Some impact
20 issues were not resolved. The cumulative impacts for these issues would have to be addressed
21 in a future EIS, should an applicant for a CP or COL reference an ESP for the North Anna ESP
22 site.
23

10.5 Staff Conclusions and Recommendations

25
26 The staff's preliminary recommendation, in view of the environmental impacts described in the
27 Draft EIS, and the impacts reviewed in this SDEIS in relation to the changes presented in ER
28 Revision 6, is that the ESP for North Anna Units 3 and 4 should be issued. This
29 recommendation is based on (1) the ER submitted by Dominion, as revised; (2) consultation
30 with Federal, State, Tribal and local agencies; (3) the staff's independent review; (4) the
31 assessments summarized in the Draft EIS and this SDEIS, including the potential mitigation
32 measures identified in the ER and in both the Draft EIS and SDEIS. In addition, in making its
33 recommendation, the staff has concluded that the alternative sites considered are not obviously
34 superior to the proposed site. Finally, the staff concludes that the site preparation and
35 preliminary construction activities enumerated in 10 CFR 50.10(e)(1) would not result in any
36 significant adverse environmental impact that cannot be redressed.
37
38 A comparative summary showing the environmental impacts of constructing and operating two
39 new units at the North Anna ESP site or at any of the alternative sites is shown in Table 10-3.
40 The estimated environmental significance of the no-action alternative, or denial of the ESP
41 application, is also shown. Table 10-3 shows that the significance of the environmental impacts

1   of the proposed action is SMALL for all impact categories at all sites with the exception of
2   certain land use, ecology, water use and quality, socioeconomic and historic and cultural
3   resource impacts. The alternative sites may have adverse environmental effects in at least
4   some categories that reach MODERATE to LARGE significance. The staff concludes that none
5   of the alternative sites assessed is obviously superior to the North Anna ESP site.

6

7   The range of impacts estimated by the NRC staff for resolved issues is predicated on certain
8   assumptions; those are identified in each section. Should the Commission issue an ESP for the
9   North Anna ESP site, and it is referenced in an application for a CP or COL, the staff will verify
10   that the assumptions identified in this EIS remain applicable. In addition, certain issues are not
11   resolved because of a lack of information. An applicant for a CP or COL referencing an ESP for
12   the North Anna ESP site would need to provide the necessary information to resolve these
13   issues, if the proposed action ultimately would affect the resources associated with these
14   issues.

15

16   **Table 10-3.** Comparison of Environmental Impacts of Constructing and Operating Two Units
17   at the North Anna ESP Site and the Alternatives

18

Impact Category	Proposed Action ESP Permit at North Anna	No-Action Alternative Denial of ESP	Alternative Site Options Surry	Portsmouth	Savannah River
Land use	SMALL	SMALL	SMALL	SMALL	SMALL to MODERATE
Ecology	SMALL	SMALL	SMALL	SMALL	SMALL to MODERATE
Water use and quality	Unresolved, likely to be SMALL	SMALL SMALL	Unresolved, likely to be SMALL	Unresolved, likely to be SMALL	Unresolved, likely to be SMALL
Air Quality	SMALL	SMALL	SMALL	SMALL	SMALL
Waste	SMALL	SMALL	SMALL	SMALL	SMALL
Human health	SMALL	SMALL	SMALL	SMALL	SMALL
Socioeconomics	MODERATE ADVERSE to LARGE BENEFICIAL	SMALL	LARGE ADVERSE to LARGE BENEFICIAL	MODERATE ADVERSE to LARGE BENEFICIAL	SMALL ADVERSE to LARGE BENEFICIAL
Historic and cultural resources	SMALL	SMALL	MODERATE to LARGE	SMALL	SMALL
Environmental justice	SMALL	SMALL	SMALL	SMALL	SMALL

10.6 References

10 CFR Part 50. Code of Federal Regulations, Title 10 *Energy*, Part 50, "Domestic Licensing of Production and Utilization Facilities."

10 CFR Part 51. Code of Federal Regulations, Title 10, *Energy*, Part 51, "Environmental Protection Regulations for Domestic Licensing and Related Regulatory Functions."

10 CFR Part 52. Code of Federal Regulations, Title 10 *Energy*, Part 52, "Early Site Permits; Standard Design Certifications; and Combined Licenses for Nuclear Power Plants."

68 FR 65961. "Dominion Nuclear North Anna, LLC, North Anna Early Site Permit; Notice of Intent to Prepare an Environmental Impact Statement and Conduct Scoping Process." *Federal Register*, Vol. 68, No. 266, November 24, 2003.

Atomic Energy Act of 1954. 42 USC 2011, et seq.

Clean Water Act (CWA) (also referred to as the Federal Water Pollution Control Act). 33 USC 1251, et seq.

Dominion Nuclear North Anna, LLC (Dominion). 2006a. *North Anna Early Site Permit Application – Part 3 – Environmental Report*. Revision 6, Glen Allen, Virginia.

Dominion Nuclear North Anna, LLC (Dominion). 2006b. Letter from E. Grecheck (DNNA) to NRC dated May 24, 2006, submitting additional information in response to an NRC request dated May 10, 2006, (ML061510131).

Dominion Nuclear North Anna, LLC (Dominion). 2006c. *North Anna Early Site Permit Application – Part 4 – Programs and Plans*. Revision 6, Glen Allen, Virginia.

National Environmental Policy Act of 1969 (NEPA). 42 USC 4321 *et seq*.

U.S. Nuclear Regulatory Commission (NRC). 1987. *Standard Review Plans for the Review of Safety Analysis Reports for Nuclear Power Plants*. NUREG-0800, Washington, D.C.

U.S. Nuclear Regulatory Commission (NRC). 1996. *Generic Environmental Impact Statement for License Renewal of Nuclear Plants*. NUREG-1437, Volumes 1 and 2, Washington, D.C.

U.S. Nuclear Regulatory Commission (NRC). 2000. *Standard Review Plan for Environmental Review for Nuclear Power Plants*. NUREG-1555, Washington, D.C.

1   U.S. Nuclear Regulatory Commission (NRC). 2004a. *Draft Environmental Impact Statement for*
2   *an Early Site Permit (ESP) for the North Anna ESP Site.* NUREG-1811, Washington, D.C.
3
4   U.S. Nuclear Regulatory Commission (NRC). 2005. Safety Evaluation Report for an Early Site
5   Permit (ESP) at the North Anna ESP Site. September 2005.

Appendix A

**Contributors to the Environmental Impact Statement
Related to Dominion Nuclear North Anna, LLC's Application for an
Early Site Permit at North Anna Nuclear Plant Site**

Appendix A

Contributors to the Environmental Impact Statement Related to Dominion Nuclear North Anna, LLC's Application for an Early Site Permit at North Anna Nuclear Plant Site

Additional staff contributors since the Draft Environmental Impact Statement have been added to this list.

The overall responsibility for the preparation of this environmental impact statement was assigned to the Office of Nuclear Reactor Regulation, U.S. Nuclear Regulatory Commission (NRC). The statement was prepared by members of the Offices of Nuclear Reactor Regulation with assistance from other NRC organizations and Pacific Northwest National Laboratory.

Name	Affiliation	Function or Expertise
NUCLEAR REGULATORY COMMISSION		
Andrew Kugler	Nuclear Reactor Regulation	Section Chief/Project Manager
John Tappert	Nuclear Reactor Regulation	Section Chief
Jack Cushing	Nuclear Reactor Regulation	Project Manager
Stacey Imboden	Nuclear Reactor Regulation	Project Management Support
Alicia Williamson	Nuclear Reactor Regulation	Project Management Support
Samuel Quiones Hernandez	Nuclear Reactor Regulation	Project Management Support
Barry Zalcman	Nuclear Reactor Regulation	Technical Monitor
Tom Kenyon	Nuclear Reactor Regulation	Project Management
James Wilson	Nuclear Reactor Regulation	Biologist
Michael Masnik	Nuclear Reactor Regulation	Biologist
Harriet Nash	Nuclear Reactor Regulation	Biologist
Brad Harvey	Nuclear Reactor Regulation	Meteorology
Rich Emch	Nuclear Reactor Regulation	Radiological Impacts
Charles Hinson	Nuclear Reactor Regulation	Radiological Impacts
Steve Klementowicz	Nuclear Reactor Regulation	Radiological Impacts
Audrey Hayes	Nuclear Reactor Regulation	Radiological Impacts
Jay Lee	Nuclear Reactor Regulation	Design Basis and Severe Accidents
Robert Palla	Nuclear Reactor Regulation	Severe Accidents
Amy Snyder	Nuclear Material Safety and Safeguards	Fuel Cycle Impacts
James Park	Nuclear Material Safety and Safeguards	Fuel Cycle Impacts
Cynthia Barr	Nuclear Material Safety and Safeguards	Fuel Cycle Impacts
Nina Barnett	Nuclear Reactor Regulation	Administrative Support
Yvonne Edmonds	Nuclear Reactor Regulation	Administrative Support
Jennifer Davis	Nuclear Reactor Regulation	Cultural Resources

Appendix A

	Name	Affiliation	Function or Expertise
1		PACIFIC NORTHWEST NATIONAL LABORATORY[a]	
2	Mary Ann Parkhurst		Task Leader
3	Beverly Miller		Deputy Task Leader
4	Kimberly Leigh		Deputy Task Leader
5	William Sandusky		Air Quality
6	John Jaksch		Socioeconomics
7	Mike Scott		Socioeconomics
8	Duane Neitzel		Aquatic Ecology
9	Jeffrey Ward		Aquatic Ecology
10	Mike Sackschewsky		Terrestrial Ecology
11	Greg Stoetzel		Radiation Protection
12	Paul Nickens		Cultural Resources
13	Paul Hendrickson		Land Use, Related Federal Programs, Alternatives
14	Lance Vail		Water Use, Hydrology
15	Chris Cook		Water Use, Hydrology
16	Stuart Saslow		Water Use, Hydrology
17	Eva Hickey		Decomissioning
18	Van Ramsdell		Design Basis and Severe Accidents
19	Dennis Strenge		Severe Accidents
20	Maha Mahasenan		Transportation
21	Philip Daling		Transportation
22	Michael Smith		Technical Review
23	Cary Counts		Technical Editing
24	Barbara Wilson		Publications Assistant
25	Debbie Schulz		Document Production
26	Jean Cheyney		Document Production
27	Mike Parker		Document Production
28	Susan Tackett		Document Production
29	Trina Russell		Document Production
30	Rose Urbina		Document Production
31	Seleste Williams		Document Production

32   (a)   Pacific Northwest National Laboratory is operated for the U.S. Department of Energy by Battelle Memorial Institute.

33

Appendix B

Organizations Contacted

Appendix B

Organizations Contacted

This appendix is not affected by the changes presented in the ER Revision 6. No additional organizations were contacted for the analysis of the changes to the Unit 3 cooling system or the increase in maximum power level for the Plant Parameter Envelope. This information is included here for reference purposes.

During the course of the staff's independent review of potential environmental impacts from siting two new nuclear units at the North Anna site, the following Federal, State, regional, Tribal and local agencies were contacted:

Lake Anna State Park, Spotsylvania, Virginia

Louisa County Historical Society, Louisa, Virginia

Virginia Department of Conservation and Recreation, Richmond, Virginia

Virginia Department of Historic Resources, Richmond, Virginia

Chickahominy Indian Tribe, Providence Forge, Virginia

Chickahominy Indians – Eastern Division, Providence Forge, Virginia

Mattaponi Indian Tribe, West Point, Virginia

Monacan Indian Nation, Madison Heights, Virginia

Nansemond Indian Tribe, Suffolk, Virginia

Pamunkey Indian Tribe, King William, Virginia

Rappahannock Tribe, Indian Neck, Virginia

Upper Mattaponi Indian Tribe, Mechanicsville, Virginia

Virginia Council on Indians, Richmond, Virginia

U.S. Army Corps of Engineers

South Carolina Field Office, U.S. Fish and Wildlife Service, Charleston, South Carolina

Ohio Field Office, U.S. Fish and Wildlife Service, Reynoldsburg, Ohio

Appendix B

1 Virginia Department of Environmental Quality, Richmond, Virginia
2
3 Department of Conservation and Recreation, Richmond, Virginia
4
5 Virginia Department of Game and Inland Fisheries, Richmond, Virginia
6
7 Wildlife Diversity Division, Virginia Department of Game and Inland Fisheries, Richmond,
8 Virginia
9
10 Department of Mines, Minerals, and Energy, Richmond, Virginia
11
12 Marine Resources Commission, Newport News, Virginia
13
14 Virginia Department of Transportation, Richmond, Virginia
15
16 Department of Agriculture and Consumer Services, Richmond, Virginia
17
18 Chesapeake Bay Field Office U.S. Fish and Wildlife Service, Annapolis, Maryland
19
20 Budget Director, Spotsylvania County, Spotsylvania, Virginia
21
22 Finance Director, Louisa County, Louisa, Virginia
23
24 Treasurer, Orange County, Orange, Virginia
25
26 Reservoir Coordinator, Nuclear Site Services, Dominion Generation, North Anna Site
27
28 Commissioner of Revenue, Louisa County, Louisa, Virginia
29
30 Assessor, Louisa County, Louisa, Virginia
31
32 Director of the Department of Community Development, Louisa County, Louisa, Virginia
33
34 Director of the Planning Division, Louisa County, Louisa, Virginia
35
36 Director Department of Planning, Spotsylvania County, Spotsylvania, Virginia
37
38 Customer Services Supervisor, Department of Public Utilities Henrico County, Virginia
39
40 Director of Economic Development, Spotsylvania County, Spotsylvania, Virginia
41
42 President of Fredericksburg Regional Alliance, Fredericksburg, Virginia
43

1   Realtor, Century 21, Fredericksburg, Virginia
2
3   Owner/Broker Century 21, Fredericksburg, Virginia
4
5   Rappahannock Area Development Commission, Fredericksburg, Virginia
6
7   Louisa County Farm Service Agency, Louisa, Virginia
8
9   Administrative Assistant for School Admissions, Spotsylvania Public Schools, Spotsylvania,
10   Virginia
11
12   School Superintendent, Louisa County Public Schools, Louisa, Virginia
13
14   School Superintendent, Orange County Public Schools, Orange, Virginia
15
16   County Administrator, Louisa County, Louisa, Virginia
17
18   Director Office of Economic Development, Orange County, Orange, Virginia
19
20   Director Planning and Zoning, Orange County, Orange, Virginia
21
22   Director of Economic Development, Louisa County, Louisa, Virginia
23
24   Louisa Town Manager, Louisa, Virginia
25
26   Real Estate Agent, Century 21, Mineral, Virginia
27
28   Director of Social Services, Orange County, Virginia
29
30   Director of Social Services, Louisa County, Virginia
31
32   County Administrator, Orange County, Virginia
33
34   Town Manager, Orange, Virginia
35
36   Director of Public Works, Orange, Virginia
37
38   Managing Broker, Century 21, Orange, Virginia
39
40   Branch Manager, Virginia Community Bank, Louisa, Virginia
41
42   Town Manager, Mineral, Virginia
43
44   Interim County Manager, Spotsylvania County, Spotsylvania, Virginia
45
46   Deputy Superintendent, Colonial National Historic Park, National Park Service

Appendix C

Chronology of NRC Staff Environmental Review Correspondence Related to Dominion Nuclear North Anna, LLC's Application for Early Site Permit at North Anna Nuclear Plant Site

Appendix C

Chronology of NRC Staff Environmental Review Correspondence Related to Dominion Nuclear North Anna, LLC's Application for Early Site Permit at North Anna Nuclear Plant Site

This appendix contains the correspondence since the issuance of the Draft Environmental Impact Statement.

This appendix contains a chronological listing of correspondence between the U.S. Nuclear Regulatory Commission (NRC) and Dominion Nuclear North Anna, LLC (Dominion) and other correspondence related to the NRC staff's environmental review, under 10 CFR Part 51, for Dominion's application for an early site permit at the North Anna Nuclear Plant site since just prior to the issuance of the Draft Environmental Impact Statement (EIS) in December 2004. Previous correspondence is contained in Appendix C of NUREG 1811, *Draft Environmental Impact Statement for an Early Site Permit (ESP) at the North Anna ESP Site"* (Accession No. ML043380308). All documents, with the exception of those containing proprietary information, have been placed in the Commission's Public Document Room, at One White Flint North, 11555 Rockville Pike (first floor), Rockville, MD, and are available electronically from the Public Electronic Reading Room found on the Internet at the following web address: http://www.nrc.gov/reading-rm.html. From this site, the public can gain access to the NRC's Agencywide Document Access and Management Systems (ADAMS), which provides text and image files of NRC's public documents in the Publicly Available Records (PARS) component of ADAMS. The ADAMS accession numbers for each document are included below.

November 30, 2004	NUREG 1811, *Draft Environmental Impact Statement for an Early Site Permit (ESP) at the North Anna ESP Site* (Accession No. ML043380308)
December 2, 2004	Letter to EPA transmitting NUREG 1811, *Draft Environmental Impact Statement for an Early Site Permit (ESP) at the North Anna ESP Site* (Accession No. ML043370446)
December 2, 2004	Letter to Dominion transmitting Federal Register Notice of Availability of the *Draft Environmental Impact Statement for an Early Site Permit (ESP) at the North Anna ESP Site* (Accession No. ML043370460)
December 27, 2004	Meeting Notice for meeting on the Draft Environmental Impact Statement (Accession No. ML043650007)

Appendix C

January 26, 2005	Meeting Notice of rescheduled meeting on the Draft Environmental Impact Statement (Accession No. ML05027019)	
January 31, 2005	Biological Assessment for the Early Site Permit (ESP) of the North Anna ESP Site and a Request for Informal Consultation (Accession No. ML050320461)	
January 31, 2005	Trip report for the January 6, 2005, drop in visit with the County Commissioners of Spotsylvania, Orange, and Louisa Counties (Accession No. ML050340579)	
February 23, 2005	EPA letter requesting an extension of the comment period (Accession No. ML050610265)	
March 17, 2005	NRC response to EPA request for extension of comment (Accession No. ML050500497)	
March 18, 2005	Supplemental Request for Additional Information (RAI) (Accession No. ML050840226)	
March 20, 2005	Meeting summary for public meeting held to on February 17, 2005, in Mineral, Virginia, to receive comments on the Draft Environmental Impact Statement (Accession No. ML050880304)	
March 22, 2005	Trip report for the January 19, 2005, drop-in visit with the Commonwealth of Virginia (Accession No. ML050810272)	
March 31, 2005	Summary of a telephone call between NRC and Dominion concerning the RAI pertaining to the North Anna ESP application (Accession No. ML050920010)	
April 12, 2005	Dominion's response to the March 18, 2005, RAI number 4, requesting documentation of Dominion's commitment to the Commonwealth of Virginia regarding the striped bass (Accession No. ML0501090376)	
April 13, 2005	E-mail from Jack Cushing (NRC) to Ellie Irons, Commonwealth of Virginia, Department of Environmental Quality, requesting clarification of commitment between the Commonwealth and Dominion regarding the striped bass (Accession No. ML051040399)	

1	April 13, 2005	Dominion's response to the March 18, 2005, RAI numbers 1, 2, and 3
2		(Accession No. ML051100321)
3		
4	April 21, 2005	E-mail from the Commonwealth of Virginia (Ellie Irons) clarifying
5		mitigation for the striped bass (Accession No. ML051120483)
6		
7	May 12, 2005	Revision 4 to North Anna ESP application (Accession No. ML051450310)
8		
9	May 20, 2005	Letter from the U.S. Fish and Wildlife Service Chesapeake Bay Field
10		Office concurring with the NRC's biological assessment
11		(Accession No. ML051600263)
12		
13	June 14, 2005	Summary of Telephone Conference with the Virginia Department of
14		Historic Resources Regarding the North Anna ESP Review
15		(Accession No. ML05166060)
16		
17	June 16, 2005	Memo to Andrew Kugler, NRC, regarding report containing comments
18		received pertaining to the Draft Environmental Impact Statement for the
19		North Anna ESP application (Accession No. ML051720560)
20		
21	June 30, 2005	Letter to U. S. Army Corps of Engineers transmitting NUREG 1811, *Draft*
22		*Environmental Impact Statement for an Early Site Permit (ESP) at the*
23		*North Anna ESP Site* (Accession No. ML051880003)
24		
25	July 7, 2005	Letter from Virginia Department of Environmental Quality requesting
26		comment and response document pertaining to the North Anna ESP Draft
27		Environmental Impact Statement (Accession No. ML052010112)
28		
29	July 15, 2005	Letter from U.S. Army Corps of Engineering Regarding NUREG 1811,
30		*Draft Environmental Impact Statement for the North Anna Early Site*
31		*Permit (ESP) at the North Anna ESP Site* (Accession No. ML052020342)
32		
33	July 15, 2005	Trip Report of a tour of Doswell Limited Partnership Combined Cycle
34		Facility (Accession No. ML052170374)
35		
36	July 19, 2005	E-mail from NRC to the Virginia Department of Environmental Quality
37		regarding request for comment and response document pertaining to
38		North Anna ESP Draft Environmental Impact Statement
39		(Accession No. ML052010108)
40		

1	July 20, 2005	Letter from the NRC to Mr. David Christian, Dominion, requesting
2		additional information regarding compliance with Section 307 of the
3		Coastal Zone Management Act and Section 401 of the Federal Water
4		Pollution Control Act (Accession No. ML052010524)
5		
6	July 25, 2005	Transmittal of Final Safety Evaluation Report review items and Revision 5
7		to the North Anna ESP application (Accession No. ML052150226)
8		
9	August 16, 2005	Letter transmitting revised schedule (Accession No. ML051520461)
10		
11	September 8, 2005	Supplemental RAI regarding the environmental portion of the ESP
12		application for the North Anna Site (Accession No. ML052520272)
13		
14	September 22, 2005	Dominion's response to the supplemental RAI
15		(Accession No. ML052660062)
16		
17	September 27, 2005	Letter from the NRC to Dr. Ethel Eaton, Virginia Department of Historic
18		Resources, regarding the North Anna ESP Review (Accession
19		No. ML052730103)
20		
21	October 6, 2005	Dominion's response to the supplemental RAI dated July 20, 2005
22		(Accession No. ML052790657).
23		
24	October 24, 2005	Letter from Dominion to NRC regarding North Anna ESP application
25		planned revision to the Unit 3 cooling water approach (Accession
26		No. ML052980117)
27		
28	October 25, 2005	Letter from Mr. Brooks, Deputy Superintendent, National Park Service, to
29		NRC, providing comments on the National Park Service's concern
30		regarding the alternative Surry ESP site potential impact on the viewshed
31		(Accession No. ML053080128)
32		
33	November 2, 2005	Letter from NRC to Dominion responding to Dominion's notification of the
34		modification of the cooling system for the proposed Unit 3 at the North
35		Anna ESP Site (Accession No. ML053000566)
36		
37	November 3, 2005	Letter from the Virginia Department of Historic Resources regarding
38		consultation under Section 106 of the National Historic Preservation Act
39		(Accession No. ML0531301730)
40		

1 2 3	December 5, 2005	Letter from NRC to Dominion regarding a revision to the North Anna ESP schedule (Accession No. ML0532100541)
4 5 6 7	January 13, 2006	Dominion North Anna Early Site Permit Application Supplement to address a modified approach to Unit 3 cooling and to ensure the plant parameter envelope remains bounding (Accession No. ML060250396)
8 9 10	February 10, 2006	Letter from NRC to Dominion regarding the North Anna ESP application review schedule (Accession No. ML060390208).
11 12 13 14	March 2, 2006	Letter from NRC to Dominion regarding information needs in the revision to the ESP application in regards to the change in cooling system and the increase in power level (Accession No. ML060610065)
15 16 17	March 13, 2006	Letter from NRC to Dominion regarding possible bald eagle nest (Accession No. ML060650396)
18 19 20	April 3, 2005	North Anna ESP application, response to NRC Question 10.q - water budget analysis spreadsheets (Accession No. ML061040606)
21 22 23 24	April 11, 2006	Meeting summary of the March 10, 2006, meeting with Dominion to discuss the supplement to the North Anna ESP application (Accession No. ML060860305)
25 26 27	April 13, 2006	North Anna ESP application response to NRC questions and Revision 6 to the plant application (Accession No. ML061180180)
28 29	May 4, 2006	North Anna review schedule letter (Accession No. ML061230005)
30 31 32 33	May 5, 2006	Press release regarding review schedule and informing public of the intent to prepare a supplement to the Draft Environmental Impact Statement (Accession No. ML061250437)
34 35 36 37	May 10, 2006	Letter transmitting Federal Register Notice of Intent to Prepare a supplement to the Draft Environmental Impact Statement (Accession Nos. ML061240025 and ML061240029)
38 39	May 10, 2006	Letter to Dominion transmitting RAIs (Accession No. ML061290142)
40 41	May 12, 2006	Summary of May 3-4, 2006 site audit to support the review of the North Anna ESP application (Accession No. ML061320447)

1	May 24, 2006	North Anna ESP application, response to NRC May 10, 2006, RAI
2		May 12, 2006; site audit summary report comments; and NRC site audit
3		follow-up questions (Accession No. ML061510131)
4		
5	June 16, 2006	Letter from Virginia Department of Environmental Quality regarding
6		401 certification (Accession No. ML061720278)
7		
8	June 21, 2006	Letter Dominion Nuclear North Anna, LLC to the NRC transmitting the
9		North Anna Early Site Permit Application Response to NRC Questions
10		and Revision 7 to the North Anna ESP Application
11		
12	June 28, 2006	Letter to the U.S. Fish and Wildlife Service regarding eagle nests near
13		Lake Anna (Accession No. ML061510149)
14		
15		

Appendix D

Scoping Meeting Comments and Responses

1
2   **Appendix D**
3
4   **Scoping Meeting Comments and Responses Site**
5
6
7   *This appendix was intentionally left blank in this Supplement to the Draft Environmental Impact*
8   *Statement. It is not affected by the changes presented in ER Revision 6. It was shown in its*
9   *entirety in the Draft Environmental Impact Statement and it will appear in the Final*
10   *Environmental Impact Statement.*

Appendix E

Draft Environmental Impact Statement
Comments and Responses

Appendix E

Draft Environmental Impact Statement
Comments and Responses

This appendix was intentionally left blank in the Draft Environmental Impact Statement (EIS) and in this Supplement to the Draft EIS. In the Final EIS, Appendix E will include written comments and responses received on the Draft EIS and those received at the public meeting held in Mineral, Virginia on February 17, 2005. In addition, comments received on this Supplement to the Draft EIS will be added with staff responses.

Appendix F

Key Correspondence

Appendix F

Dominion Nuclear North Anna LLC's
Key Early Site Permit Consultation Correspondence

Appendix F has been changed from the Draft Environmental Impact Statement (Draft EIS), and contains correspondence received starting after the publication of the Draft Environmental Impact Statement for an Early Site Permit (ESP) at the North Anna ESP Site from December 2004 until the present time.

Correspondence received during the evaluation process of the early site permit (ESP) application for Dominion Nuclear North Anna LLC (Dominion) for the proposed North Anna site is identified in Table F-1. Copies of the correspondence are included at the end of this table.

Table F-1. Key Consultation Correspondence for the Dominion ESP Supplemental EIS

Source	Recipient	Date of Letter/E-mail
United States Nuclear Regulatory Commission (Pao-Tsin Kuo)	U.S. Fish and Wildlife Service (Mr. David Sutherland)	January 31, 2005
United States Nuclear Regulatory Commission (Pao-Tsin Kuo)	U.S. Army Corps of Engineers (Mr. Regena Bronson)	June 30, 2005
Department of the Army Corps of Engineers (Bruce F. Williams)	United States Nuclear Regulatory Commission (Jack Cushing)	July 15, 2005
United States Nuclear Regulatory Commission (Pao-Tsin Kuo)	Virginia Department of Historic Resources (Dr. Ethel Eaton)	September 27, 2005
United States Department of the Interior, National Park Service (Skip Brooks)	United States Nuclear Regulatory Commission (Jack Cushing)	October 25, 2005
Virginia Department of Historic Resources (Roger W. Kirchen)	United States Nuclear Regulatory Commission (Pao-Tsin Kuo)	November 3, 2005
Virginia Department of Environmental Quality (Jeffery A. Steers)	United States Nuclear Regulatory Commission (Jack Cushing)	June 16, 2006

January 31, 2005

Mr. David Sutherland
Chesapeake Bay Field Office
U.S. Fish and Wildlife Service
177 Admiral Cochrane Drive
Annapolis, MD 21401

SUBJECT: BIOLOGICAL ASSESSMENT FOR THE EARLY SITE PERMIT (ESP) OF THE
           NORTH ANNA ESP SITE AND A REQUEST FOR INFORMAL CONSULTATION

Dear Mr. Sutherland:

The U.S. Nuclear Regulatory Commission (NRC) has prepared the enclosed biological
assessment (BA) to evaluate whether the proposed action of the North Anna ESP would have
adverse effects on listed species. The North Anna ESP site is located within the North Anna
Power Station (NAPS) site adjacent to Lake Anna near Mineral, Virginia. The proposed Federal
action is the issuance, under provisions of Title 10 of the *Code of Federal Regulations* Part 52
(10 CFR Part 52), of an ESP for the North Anna ESP site for postulated additional nuclear
power facilities, and to conduct site preparation and limited construction activities. The site
preparation and limited construction activities allowed by 10 CFR 52.25 include clearing,
grading, and constructing non-safety-related facilities. The proposed action does not include
approval to construct and operate new units; therefore, the BA does not analyze environmental
impacts that could result from construction and operation of two new nuclear units at the North
Anna ESP site. Impacts associated with actual facility construction and operation will be
assessed during the NRC staff's review of an application for a combined license or construction
permit, should the applicant choose to go forward with the project.

The existing transmission system at the NAPS is sufficient to transmit all power generated by
existing and proposed nuclear units at NAPS. The NRC's recent analysis of the existing
transmission system at NAPS (NRC 2002) concluded that continued operation would not impact
threatened or endangered species. Because no changes to transmission lines or rights-of-way
are anticipated, this BA does not consider them for further analysis.

By letter dated December 21, 2003, (NRC 2003b), the NRC requested the Federally listed
threatened or endangered species that may be in the vicinity of NAPS and its associated
transmission lines. In a letter dated October 25, 2004, (FWS 2004a) the U.S. Fish and Wildlife
Service (FWS) provided the Federally listed threatened or endangered species. The FWS
identified the following: one endangered species, dwarf wedgemussel (*Alasmidonta heterdon*);
and four threatened species, bald eagle (*Haliaeetus leucocephalus*), small whorled pogonia
(*Isotria medeoloides*), sensitive joint-vetch (*Aeschynomene virginica*), and swamp pink
(*Helonias bullata*). For documentation purposes, the NRC has addressed the potential impact
of the North Anna ESP site on these five species in the enclosed BA.

The NRC has determined that the proposed action would not affect the dwarf wedgemussel
because there is no suitable habitat for the dwarf wedgemussel on the North Anna ESP site.

D. Sutherland   -2-

Because bald eagles have been observed in the vicinity of the project site, the NRC determined that the proposed action may affect, but is not likely to adversely affect, the bald eagle. The NRC concluded that the proposed action would not affect the small whorled pogonia, sensitive joint-vetch, and swamp pink because no known habitats exist for these protected plant species on the North Anna ESP site. Finally, no designated critical habitat exists for any of the five listed species.

We are placing this BA in our project files and are requesting your concurrence with our determination. In reaching our conclusion, the NRC staff relied on information provided by the applicant, on research performed by NRC staff, and information from FWS (i.e., current listings of species provided by the FWS, Gloucester, Virginia Field Office).

If you have any questions regarding this BA or the staff's request, please contact Mr. Jack Cushing, Environmental Project Manager, at 301-415-1424, or by e-mail at jxc9@nrc.gov.

Sincerely,
/RA/

Pao-Tsin Kuo, Program Director
License Renewal and Environmental Impacts Program
Division of Regulatory Improvement Programs
Office of Nuclear Reactor Regulation

Docket No.: 52-008

Enclosure: As stated

cc w/encl.: See next page

ENCLOSURE
BIOLOGICAL ASSESSMENT

BIOLOGICAL ASSESSMENT

North Anna
Early Site Permit Application

Louisa County, Virginia

Docket Number 52-008

January 2005

1.0 Introduction

On September 25, 2003[1], the U.S. Nuclear Regulatory Commission (NRC) received an application from Dominion Nuclear North Anna, LLC (Dominion) for an early site permit (ESP) for an ESP site (the North Anna ESP site) located within the existing North Anna Power Station (NAPS) site near the town of Mineral, in Louisa County, Virginia (Figure 1). Under the NRC regulations in Title 10 of the Code of Federal Regulations (CFR) Part 52 and in accordance with the applicable provisions of 10 CFR Part 51, which are the NRC regulations implementing the National Environmental Policy Act of 1969 (NEPA), the NRC is required to prepare an environmental impact statement (EIS) as part of its review of an ESP application. The NRC staff published in the Federal Register a Notice of Intent (68 FR 65961) to conduct scoping, prepare an EIS, and publish a draft EIS for public comment. The comment period for the draft EIS ends on March 1, 2005. The draft EIS is available on the NRC website at www.nrc.gov/reading-rm/doc-collections/nuregs/staff/sr1811/ index.html. The final EIS will be issued after considering public comments on the draft. A separate safety evaluation report will also be prepared in accordance with 10 CFR Part 52.

The North Anna ESP site proposed by Dominion is located in Louisa County in central Virginia, near the town of Mineral. It is completely within the confines of the current NAPS site, which is located on a peninsula on the southern shore of Lake Anna, approximately eight kilometers (km) (five miles [mi]) upstream of the North Anna Dam. Lake Anna is approximately 27 km (17 mi) long, with 435 km (272 mi) of shoreline. The lake was created in 1971 by the construction of a dam on the main stem of the North Anna River. Virginia Electric and Power Company (Virginia Power), a subsidiary of Dominion Resources, Inc., owns the land above and below the lake surface and around the lake up to the expected high-water mark.

As part of the environmental review process, the NRC staff sent letters to staff at the United States Fish and Wildlife Service (FWS) and National Oceanic and Atmospheric Administration (NOAA) Fisheries (NRC 2003a,b) requesting lists of threatened and endangered species that potentially could be affected by the construction and operation of new power plants at NAPS. Specifically, the staff requested a list of species and information on protected, proposed, and candidate species, and critical habitat that may be in the vicinity of North Anna.

In a letter dated January 6, 2004 (NOAA 2004), NOAA Fisheries stated that "no federally listed or proposed threatened or endangered species under the jurisdiction of NOAA Fisheries are known to exist in the vicinity of the North Anna Power Station." The FWS replied by letter dated October 25, 2004 (FWS 2004a) with attached tables that identify two animal and three plant species listed by the Endangered Species Act (ESA) that occur or may occur in the counties adjacent to the NAPS. These species are the dwarf wedgemussel (*Alasmidonta heterodon*), bald eagle (*Haliaeetus leucocephalus*), small whorled pogonia (*Isotria medeoloides*), sensitive joint-vetch (*Aeschynomene virginica*), and swamp pink (*Helonias bullata*).

---

[1] The September 25, 2003, Environmental Report (ER) for this application was revised by letters dated October 2, 2003 (Revision 1), July 15, 2004 (Revision 2), and September 7, 2004 (Revision 3). Any reference in this Biological Assessment (BA) to the ER refers to Revision 3 (Dominion 2004), unless otherwise stated.

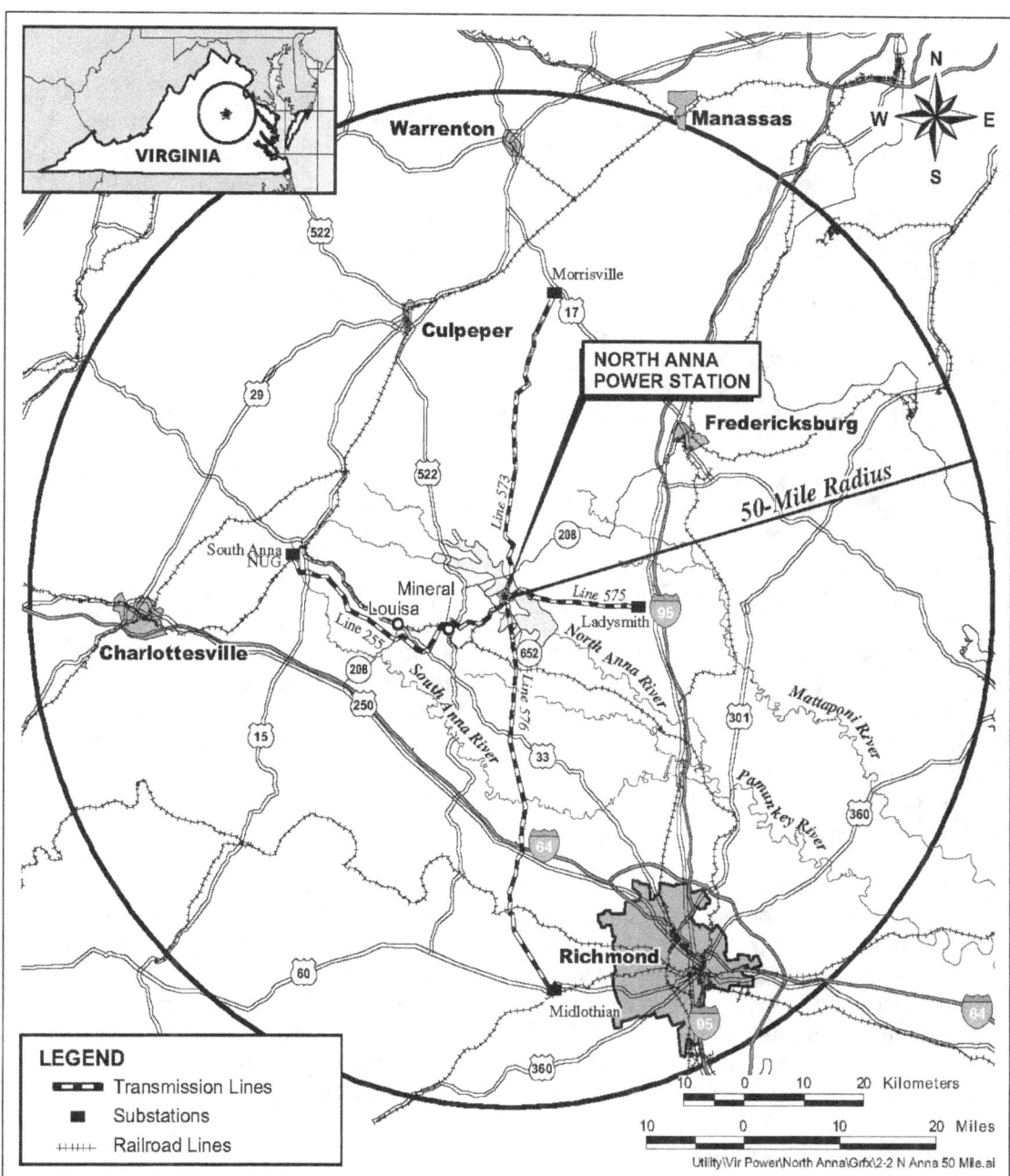

Figure 1. Location of North Anna ESP Site, 80-km (50-mi) Region

2

2.0 Project Description

The proposed Federal action is the issuance, under the provisions of 10 CFR Part 52, of an ESP for the North Anna ESP site for additional nuclear power facilities, and to conduct site preparation and limited construction activities identified in the application. The proposed action does not include approval to construct and operate new units but rather allows limited construction associated with site preparation activities. The complete construction and operation of new units are not presently proposed; therefore, this BA does not analyze the environmental impacts that could result from the actual construction and operation of two new nuclear units at the North Anna ESP site. Site preparation impacts are analyzed to determine whether activities proposed under the site redress plan might impact threatened and endangered species that occur in the vicinity of the NAPS.

No specific plant design has been selected by Dominion for the ESP site; instead, a set of bounding plant parameters has been specified to envelope future site development. This plant parameter envelope is based on the addition of power generation from two distinct units, to be designated as North Anna Units 3 and 4. Cooling water for Unit 3, the first of the proposed new units, would be provided by Lake Anna. Unit 4 would use dry cooling towers.

In this BA, the proposed ESP site is evaluated only for those activities related to the site preparation activities and the limited construction activities allowed by 10 CFR 52.25. The site redress plan provides for redress of impacts associated with site preparation and limited construction activities, if the applicant ultimately decides not to pursue construction of one or more nuclear units after the permitted activities have occurred. The activities permitted under 10 CFR 52.25 would allow for these site preparation and limited construction activities such as clearing and grading, and the construction of non-safety related facilities, which could include intake and discharge structures, cooling towers, turbine buildings, and non-safety related support facilities.

Dominion evaluated the existing transmission system that connects the NAPS site with the regional transmission grid, and determined that the existing transmission lines are sufficient to transmit all of the power generated by the existing and the postualated new nuclear units at the NAPS site. Therefore, no changes to the existing transmission system are proposed. The NRC examined the potential impacts of continued operation of the NAPS transmission lines in connection with the license renewal for NAPS Units 1 and 2 (NRC 2002) and determined that there would be no effect to threatened or endangered species. Because no changes to the lines or rights-of-way are anticipated, the transmission lines are not considered in this BA.

3.0 Potential Environmental Impacts

Site preparation activities may result in the removal of approximately 32 hectares (ha) (80 acres [ac]) of forested habitats, as well as grading of areas previously disturbed during construction of the existing NAPS units. In addition to direct habitat loss, there would likely be a temporary increase in ambient noise levels typical of land development and construction activities. Construction of intake and discharge structures would impact small portions of the Lake Anna shoreline.

3

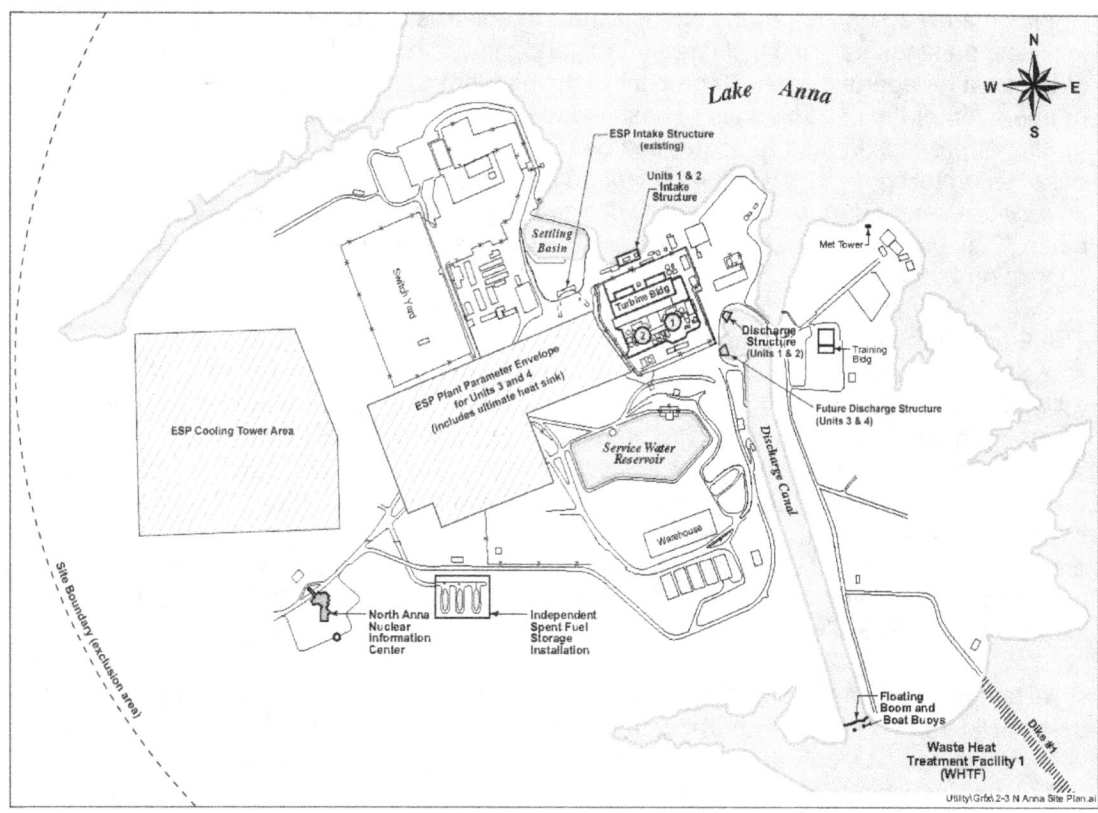

Figure 2. North Anna ESP Site Boundaries within the Existing NAPS Site

Much of the proposed North Anna ESP site construction area consists of dirt roads, cleared areas, parking lots, buildings, and early succession habitats (Figure 2). The western portion of the current and proposed laydown area, located northeast of the current switchyard, can be classified as "old-field" habitat. None of the current or proposed laydown area is forested. The area proposed for temporary offices, located east of the switchyard, is an existing office complex; thus, undisturbed habitats would not be impacted. The proposed cooling tower site consists primarily of forested habitat.

4.0 Description of the Project Area

4.1 Terrestrial Biological Communities of the North Anna Site

The ESP site is located within the Piedmont Physiographic Province as described by Omernik (1987). Although forests in the Piedmont Province are nominally characterized by oak-hickory-pine forest (Woods et al. 1999), this portion of north-central Virginia has been settled since the colonial era and, therefore, no longer contains virgin forests. Vegetative cover surrounding the ESP site is an irregular patchwork of row crops, pastures, pine plantations, abandoned (old) fields, and second-growth forests of hardwoods and mixed pine-hardwoods (Dominion 2004).

4

Approximately 30 percent of the North Anna site consists of power generation and maintenance facilities, parking lots, roads, cleared areas, and mowed grass. Hardwood forests and planted pines exist on approximately 70 percent of the site that has not been cleared for the construction or operation of the existing units. These wooded areas are remnants of forests that were used for timber production prior to acquisition by Virginia Power and are dominated by a variety of oaks (*Quercus* spp.), yellow poplar (*Liriodendron tulipifera*), sweet gum (*Liquidambar styraciflua*), and red maple (*Acer rubrum*) trees. Scattered loblolly pines (*Pinus taeda*), Virginia pines *(P. virginiana)*, and short-leaf pines *(P. echinata)* exist in some wooded areas (Dominion 2004).

The Piedmont region of Virginia is characterized as an irregular plain with low, rounded ridges and shallow ravines (Woods et al. 1999). There are no steep ridges on the ESP site. The rolling terrain at the site extends down slope to the waters of Lake Anna, resulting in essentially no marsh habitat along the shoreline at the site. Hydrophytic vegetation, such as cattail (*Typha* spp.) and rushes (*Juncus* spp.), are typically absent or extend only to approximately 0.3 meters (m) to 1 m (one to three feet [ft]) beyond the shoreline (Dominion 2004). Two intermittent streams flowing north into an unnamed arm of Lake Anna, just northwest of the power-block area, bisect the area where cooling towers are proposed to be located. A narrow band of wetlands is associated with each of these streams. A small (<.5 ha [one ac]) isolated wetland is located within the ESP site.

Wildlife species found in the forested portions of the North Anna site are those typically found in upland Piedmont forests of north-central Virginia. Frequently observed mammals, such as the white-tailed deer (*Odocoileus virginianus*), raccoon (*Procyon lotor*), opossum (*Didelphis virginiana*), gray squirrel (*Sciurus carolinensis*), and gray fox (*Urocyon cinereoagenteus*), exist at the site, as do smaller mammals such as moles (Talpidae), shrews (Soricidae), and a variety of mice (Muridae) and voles (*Microtus* spp.). Woodchucks (*Marmota monax*) live in the grassy areas near forest edges at the site, and beavers (*Castor canadensis*) occur in Lake Anna and its tributaries. Various birds and herpifauna (e.g., snakes, turtles, lizards, and toads) live in the uplands and along the edge of Lake Anna (Dominion 2004).

Virginia Power has cooperated with the National Audubon Society in conducting periodic "Christmas Bird Counts" during December or January. Common bird species recorded in upland areas on and near the North Anna site during these surveys include the American crow (*Corvus brachyrhynchos*), blue jay (*Cyanocitta cristata*), Carolina chickadee (*Poecile carolinensis*), mourning dove (*Zenaida macroura*), black vulture (*Coragyps atratus*), turkey vulture (*Cathartes aura*), European starling (*Sturnus vulgaris*), song sparrow (*Melospiza melodia*), white-throated sparrow (*Zonotrichia albicollis*), dark-eyed junco (*Junco hyemalis*), northern cardinal (*Cardinalis cardinalis*), house finch (*Carpodacus mexicanus*), tufted titmouse (*Baeolophus bicolor*), red-bellied woodpecker (*Melanerpes carolinus*), downy woodpecker (*Picoides pubescens*), and northern flicker (*Colaptes auratus*) (Audubon Society 2004). Species known to nest within forested areas at the North Anna site, along forested edges, and in open areas (for example, northern cardinal, Carolina chickadee, blue jay) are those that commonly nest in upland Virginia habitats. Virginia Power has placed bluebird nest boxes in suitable habitats at the North Anna site and has constructed roofed structures for swallows in some locations. Eastern bluebirds (*Sialia sialis*) annually use the nest boxes, and barn swallows (*Hirundo rustica*) nest beneath the roofed structures (Dominion 2004).

5

Several species of residential and migratory wading birds and waterfowl use Lake Anna. Numerous gulls, ducks, and geese were noted during Christmas Bird Counts (Audubon Society 2004), as were great blue herons (*Ardea herodias*). Virginia Power biologists have documented breeding at Lake Anna by mallards (*Anas platyrhynchos*), wood ducks (*Aix sponsa*), and Canada geese (*Branta canadensis*) (VEPCo 1986). Virginia Power, in association with the Louisa County Chapter of Ducks Unlimited, has placed wood duck nest boxes on Lake Anna, and wood ducks have used several of these nest boxes (VEPCo 1986). Belted kingfishers (*Ceryle alcyon*), great blue herons, and green-backed herons (*Butorides virescens*) are present at Lake Anna throughout the year, and belted kingfishers and green-backed heron presumably nest on or near the Lake Anna shoreline. There are no known great blue herons rookeries at Lake Anna (Dominion 2004). Waterfowl are typically most abundant at Lake Anna during the winter. Lake Anna provides important habitat for migratory waterfowl on the Atlantic flyway, especially during extremely cold winters when the elevated water temperature from station operation maintains a large ice-free body of water. The most common ducks observed during winter are mallard, American black duck (*Anas rubripes*), bufflehead (*Bucephala albeola*), and greater scaup (*Aythya marila*). The Canada goose, American coot (*Fulica americana*), ringed-billed gull (*Larus delawarensis*), and herring gull (*L. argentatus*) are also abundant on Lake Anna during the winter (Audubon Society 2004; VEPCo 1986).

4.2 Aquatic Biological Communities of the North Anna Site

The aquatic resources in the vicinity of the North Anna ESP site, the Waste Heat Treatment Facility (WHTF), and the North Anna River, are associated with Lake Anna (VEPCo 2001). Lake Anna was created to serve as the cooling water source for NAPS. The lake was formed during 1971 by erecting a dam on the main stem of the North Anna River, just upstream of the confluence of the North Anna River and Northeast Creek.

Lake Anna is typical of many shallow reservoirs found in the southern and mid-Atlantic states. Since impoundment, Lake Anna has gone through the typical ecological succession of reservoirs. The initial biotic community was highly productive because initial nutrient levels were high. Productivity subsequently decreased and ultimately stabilized (Paterson and Fernando 1970; Voshell and Simmons 1978). Aquatic communities in Lake Anna experienced gradual post-impoundment changes from riverine to lake communities. Some of these communities had stabilized in Lake Anna by 1975 (VEPCo 1986), and all have been relatively stable since 1985 (VEPCo 1986; VEPCo 2002).

Lake Anna contains numerous phytoplankton, zooplankton, and benthic macroinvertebrate communities. Seventy-seven genera of phytoplankton have been identified, and diatoms, green algae, blue-green algae (Cyanobacteria), and cryptomonads are the dominant forms. The zooplankton are dominated by small-bodied forms (rotifers and copepods). This has been attributed to selective predation upon larger-bodied zooplankton by landlocked schooling clupeids such as various shad species (Brooks and Dodson 1965). A total of 124 benthic taxa have been identified from Lake Anna (VEPCo 1986). Three bivalve species were collected in the North Anna basin prior to impoundment: *Elliptio complanatus*, *E. productus*, and *Sphaerium striatum* (AEC 1973).

In more recent years, the introduced Asiatic clam (*Corbicula* spp.) has dominated collections from both Lake Anna and the lower North Anna River. The Asiatic clam has spread rapidly throughout the United States since its first discovery in 1938 (VEPCo 1986). Its populations

6

expand rapidly when they invade a new habitat, and densities stabilize as the species reach carrying capacity of the habitat. Asiatic clams are present throughout Lake Anna with the greatest population densities found at mid-lake (VEPCo 1989). After its initial invasion of Lake Anna, densities increased sharply from 1979 to 1981. Populations remained relatively stable between 1984 and 1988 (VEPCo 1989). Virginia Power received approval from VDEQ to discontinue Asiatic clam sampling in 1989. The zebra mussel (*Dreissena polymorpha*) has not been observed in Lake Anna.

Small numbers of unionid mussels (*Elliptio* spp.) and fingernail clams (Sphaeriidae) have also been collected. Acid drainage and sediment from the Contrary Creek mine site historically depressed freshwater mussel populations downstream from the Contrary Creek-North Anna River confluence; the first major mussel beds prior to the inpoundment of Lake Anna did not occur until 100 m (328 ft) downstream of the confluence of the North and South Anna Rivers (Reed and Simmons 1972). There are indications that mussel populations (*Elliptio* spp.) are recovering in the lower North Anna River (VEPCo 1986).

Thirty-nine species of fish (representing 12 families) have been identified in Lake Anna (VEPCo 1986). Species include those historically found in the North Anna River, those that had been in local farm ponds inundated by the new reservoir, and species introduced by the Virginia Department of Game and Inland Fisheries (VDGIF).

Recreational species include largemouth bass (*Micropterus salmoides*), smallmouth bass (*M. dolomieu*), striped bass (*Morone saxatilis*), walleye (*Stizostedion vitreum*), bluegill (*Lepomis macrochirus*), yellow perch (*Perca flavescens*), black crappie (*Pomoxis nigromaculatus*), white perch (*M. americana*), pumpkinseed (*L. gibbosus*), redear sunfish (*L. microlophus*), redbreast sunfish (*L. auritus*), channel catfish (*Ictalurus punctatus*), and white catfish (*Ameiurus catus*). Forage species include threadfin shad (*Dorosoma petenense*) and gizzard shad (*D. cepedianum*). Striped bass and walleye are stocked annually by VDGIF. In 1994, sterile triploid herbivorous grass carp (*Ctenopharyngodon idella*) was stocked by Virginia Power to control the growth of the nuisance submerged aquatic plant hydrilla (*Hydrilla verticillata*) with the approval of the VDGIF.

Before the North Anna River was impounded, the fish community of the river downstream of the Contrary Creek inflow was dominated by pollution-tolerant species. In the years following impoundment (and reclamation of the Contrary Creek mine site), there was a steady increase in measures of abundance and diversity of fish. During 1984 to 1985, 38 species from ten families were found in the North Anna River, compared to 25 species from eight families in the control stream, the South Anna River (VEPCo 1986). When species from the North Anna Reservoir were subtracted from the North Anna River totals, the two fish communities (North and South Anna River communities) showed striking similarities, indicating that the operation of the existing units had little or no effect on fish populations downstream from the dam.

The WHTF is the body of water into which waste heat from the existing units is discharged via the discharge canal. It is physically separated from the rest of Lake Anna by a series of dikes. The same aquatic communities occur in the WHTF that occur in the main reservoir. Fish can swim from the main reservoir into the WHTF and back. However, fish are not stocked in the WHTF, and angler access to this fishery is restricted to the land owners along this part of the shoreline.

7

There is no commercial fishing in Lake Anna or the North Anna River. There are no runs of anadromous fish in the North Anna River. The North Anna River is a tributary of the Pamunkey River, which has an annual run of American shad, but these shad do not move into the North Anna River (Jenkins and Burkhead 1994; Bilkovic et al. 2002). The Pamunkey Fish Hatchery in King William County, Virginia, is approximately 121 km (75 mi) downstream of the North Anna Dam. Shad reared at this facility are normally stocked in the Pamunkey River and the James River as fry. Young American eels (*Anguilla rostrata*) are found in the North Anna River, but are not sought by commercial fishermen. The American eel is a catadromous species, meaning that these fish begin their lives in the open ocean and migrate into coastal rivers where they spend much of their lives in fresh water (Rohde et al. 1994). Upon reaching sexual maturity, at age five to seven years, the eels migrate back to the ocean where they spawn and die. Eels in the North Anna River are juveniles, also known as "yellow eels."

The lower North Anna River downstream from the North Anna Dam is small, approximately 23 to 46 m (75 to 150 ft) wide, but supports a diverse assemblage of stream fishes. It is a popular fishing spot. Unless stream flow is unusually high, powerboats are impractical. Most anglers fish from shore or from canoes and kayaks. Recreational fishermen generally seek largemouth and smallmouth bass or redbreast sunfish. Bluegill and redear sunfish are present as well, but receive less attention from anglers.

5.0 List of Federally Threatened and Endangered Species

This section describes the threatened and endangered animal and plant species that potentially exist at or near the proposed ESP site. The FWS provided a list of species in the counties of interest (FWS 2004a) and also maintains current lists of threatened or endangered species on its website (FWS 2004b). The Virginia Department of Game and Inland Fisheries (VDGIF) (VDGIF 2004) and Virginia Department of Conservation and Recreation (VDCR 2004) also maintain lists of State-protected species on their websites. Species potentially occurring near the proposed North Anna ESP site that are listed as threatened or endangered by the FWS are listed in Table 1.

Table 1. Federally Threatened or Endangered Species Known or Potentially Occurring Near the Proposed North Anna ESP Site .

Scientific Name	Species	Counties	Status*
Invertebrates			
Alasmidonta heterdon	dwarf wedgemussel	Louisa, Orange, Hanover	FE
Birds			
Haliaeetus leucocephalus	bald eagle	Louisa, Orange, Caroline, Spotsylvania, Hanover	FT
Vascular Plants			
Isotria medeoloides	small whorled pogonia	Spotsylvania, Hanover, Caroline	FT
Aeschynomene virginica	sensitive joint-vetch	Hanover, Caroline	FT
Helonias bullata	swamp pink	Spotsylvania, Hanover, Caroline	FT

Status*: FE = Federally endangered, FT = Federally threatened
Sources: FWS 2004a, 2004b, VDCR 2004, VDGIF 2004

6.0 Description of Species and Habitats

In this section, each of the species listed in Table 1 is described, including its habitat requirements, status, and distribution in relation to the proposed project.

Dwarf Wedgemussel

The dwarf wedgemussel (*Alismidonta heterodon*) occurs sporadically in Atlantic coast rivers from Canada to North Carolina (FWS 1993). It is a small freshwater mussel (< 55 millimeters [2.17 inches]) long and roughly trapezoidal in shape. The outside of the shell is brown or yellowish-brown, with greenish rays visible in young or pale-colored specimens. The interior of the shell is bluish or silvery white and is iridescent in the posterior part of the shell. The hinge teeth are small but distinct. This species is unique in that it has two lateral teeth in the right valve and one in the left; other species have two lateral teeth in the left valve and one in the right (Environment Canada 2004).

The mussel is found in small streams to medium-sized rivers with slow to moderate current and fine sediment, sand, or gravel substrates. It appears to have poor tolerance for suspended silt. Stream-side vegetation seems to be required. The mussel releases a parasitic larvae, but the host fish species for the larvae is not known. The maximum life span of the mussel is believed to be 12-18 years. The mussel is vulnerable to pesticide and metal contamination, and to low oxygen levels. Impoundment of rivers has been a major negative factor on continued persistence of this species throughout its range, possibly due to dams blocking movements of host fish species (Environment Canada 2004).

The dwarf wedgemussel is known to occur in the South Anna River in Louisa County, VA (FWS 1993), but it has not been reported in the North Anna River or its tributaries. There are no rivers

9

or streams on the proposed North Anna ESP site that are suitable habitat for the dwarf wedgemussel.

Bald Eagle

Bald eagles (*Haliaeetus leucocephalus*) in Virginia are most common along the Chesapeake Bay, and along the lower reaches of several of the larger river systems such as the Potomac, Rappahannock, York, and James Rivers (VDGIF 2004, Watts and Byrd 2003). Most nest sites are found in the midst of large wooded areas adjacent to marshes or bodies of water, or in isolated trees located in marshes, on farmland, or in logged over areas where scattered seed trees remain (VDGIF 2004). Most eagle nests are less than 1.6 km (one mi) from feeding areas, although some can be as much as 3.2 km (two mi) from primary food sources. Wintering roost sites typically have the same characteristics as nest sites (VDGIF 2004). Bald eagle habitat usually occurs in undeveloped areas with little human activity. Bald eagles are primarily fish eaters but will prey upon mammals and birds when necessary, and they will eat carrion.

Bald eagles are occasionally observed along Lake Anna (six were observed during the 2003 Christmas Bird Count) (Audubon Society 2004). However, there are no known eagle nests near the proposed ESP site (NRC 2002). The VDGIF database indicates that one nest was located approximately eight km (five mi) downstream from Lake Anna Dam in 2000, but later surveys indicate this nest was not in use in 2003 (Watts and Byrd 2003). Dominion biologists indicated that there is a bald eagle nest near the north end of Lake Anna, approximately 16 km (10 mi) upstream of the existing units (Dominion 2004). Although the VDGIF information service does not include records of bald eagle nests on Lake Anna upstream from the NAPS, Watts and Byrd (2003) found that there was an occupied territory, but not an active nest, within the Mineral United States Geological Survey quadrangle in 2003. The Mineral quad is located west of the North Anna Power Station and includes the upper reaches of Lake Anna.

Small Whorled Pogonia

The small whorled pogonia (*Isotria medeoloides*) generally grows in open, dry, deciduous woods with acidic, sandy, loamy soil with low nutrient content. Suitable habitat for this species is limited on the NAPS site. It is not known to occur at the proposed North Anna ESP site (Dominion 2004; NRC 2002) and has not been reported in Louisa County. It has been reported to occur in the adjacent Spotsylvania and Caroline Counties, and potentially occurs in Hanover County (FWS 2004a, VDCR 2004).

Sensitive Joint-Vetch

The sensitive joint-vetch (*Aeschynomene virginica*) occurs in fresh to slightly-brackish tidal river systems in the intertidal area where the plants are flooded twice daily. Lake Anna and the North Anna River are not tidally influenced, and therefore, no habitat for the sensitive joint-vetch occurs at the proposed ESP site. The species is thought to potentially occur in Caroline and Hanover Counties (FWA 2004a) because suitable habitat is located within these counties, and the sensitive joint-vetch is known to occur in adjacent counties. However, any potential habitat would be located at least 48 km (30 mi) from the proposed North Anna ESP site.

Swamp Pink

The swamp pink (*Helonias bullata*), occurs in a variety of wetland habitats such as bogs, spring seeps, stream edges, and wet meadows. Sites are typically saturated year-round, but are rarely flooded. Soils are usually neutral to acidic. There is very little saturated ground or wetlands on the proposed North Anna ESP site; therefore, it is unlikely that there is suitable habitat within the affected area. The swamp pink is not known to occur at the North Anna site (Dominion 2004; NRC 2002) and has not been reported in Louisa County. It has been reported in Caroline County and is considered as potentially occurring in Hanover and Spotsylvania Counties (FWS 2004a).

7.0 Evaluation of Potential Impacts

Site preparation and limited construction activities would result in the removal of up to approximately 32 ha (80 ac) of forested habitat within the site. The ESP site does not contain any old growth timber, unique or sensitive plants, or unique or sensitive plant communities. Therefore, construction activities would not noticeably reduce the local or regional diversity of plants or plant communities. There are no areas designated by the FWS as critical habitat for endangered or threatened species at or near the site. No threatened or endangered plant species have been reported near the North Anna ESP site or in Louisa County, and no suitable habitat for threatened or endangered plant species is known to exist on the North Anna ESP site.

Movement of construction workers, materials, and equipment, and the operation of construction equipment (e.g., earth-moving equipment, portable generators, pile drivers, pneumatic equipment, and hand tools) would generate noise. Noise from human activities can affect wildlife by inducing physiological changes, nest or habitat abandonment, and behavioral modifications, or it may disrupt communications required for breeding or defense (Larkin 1996). However, it is not unusual for wildlife to adapt to noise from human activities (Larkin 1996). Although short-term noise levels from construction activities could be as high as approximately 110 decibels (e.g., impulse noise during pile-driving activities), these noise levels would not extend far beyond the boundaries of the ESP site. At a distance of 120 m (400 ft) from the construction site, noise levels from these activities would range from approximately 60 to 80 decibels. These noise levels are below the 80-to-85-decibel threshold at which birds and small mammals are startled or frightened (Golden et al. 1980). Thus, noise from construction activities would not be likely to disturb wildlife beyond 120 m (400 ft) from the construction site. Additionally, construction would occur adjacent to the existing operating Units 1 and 2, where wildlife has presumably become accustomed to typical, existing operating facility noise levels of approximately 50 to 60 decibels at the NAPS security fence (Dominion 2004).

There are no small streams to medium-sized rivers with slow-to-moderate current and fine sediment, sand, or gravel substrates on the ESP site. Two intermittent streams exist on the North Anna ESP site (Dominion 2004); however, they are not expected to support a population of dwarf wedgemussels. Besides being intermittent streams, they do not support fish populations that are essential to the life cycle of the dwarf wedgemussel. Proposed activities authorized under 10 CFR 52.25 would not adversely affect the North Anna River.

The 32 ha (80 ac) of forested habitat removed during construction presumably could be used by bald eagles for perching, roosting, or nesting. Eagles are occasionally observed in the vicinity of NAPS, but there is no indication that the proposed project site is regularly utilized by bald eagles. The nearest known bald eagle territory is believed to be approximately 16 km (ten mi) from site

11

preparation and construction activities at the proposed ESP site. The *Bald Eagle Protection Guidelines for Virginia* (USFWS and VGDIF 2000) recommends a buffer of 400 m (0.25 mi), in which construction activities should be limited. Although bald eagles may occasionally be observed near the plant, no nesting or roosting activity has ever been observed within an area that could be affected by construction or operational noise. No avian collisions with existing structures at the NAPS site have been noted (Dominion 2004); therefore, such collisions during the site preparation and construction phase would be unlikely.

8.0 Management Actions Related to the Species

To minimize construction-related impacts to wildlife, Dominion has stated that it would adhere to State permit conditions that may restrict the timing of certain construction activities (Dominion 2004). Dominion maintains a migratory bird protection program, including protection of nests and reporting bird (especially raptor) strikes and other events (Dominion 2001).

A few small wetland areas and two intermittent streams exist on the North Anna ESP site (Dominion 2004). Watercourses and wetlands would be avoided to the extent possible during any construction. Dominion has stated (Dominion 2004) that any work that has the potential to impact a wetland would be performed in accordance with applicable laws, regulations, permits, and authorizations. Wetland delineations and surveys would be conducted prior to commencement of construction activities. The Army Corps of Engineers has jurisdiction over wetlands under Section 404 of the Clean Water Act . If the areas are determined to be wetlands under the Clean Water Act, disturbance of the areas would either be avoided or other appropriate mitigation actions would be implemented as required by any applicable permits and regulations (Dominion 2004).

9.0 Conclusions

The proposed action is the issuance of an ESP for two additional nuclear power units at the North Anna ESP site. This BA has considered the potential impacts of site preparation and limited construction activities at the proposed site on species listed as threatened or endangered under the ESA, species proposed for such status, species considered candidates for listing under the ESA, or designated critical habitats for such listed species.

There is no habitat for the dwarf wedgemussel on the North Anna ESP site, and the proposed site preparation activities would not have an effect on, or occur near, the North Anna River or any other potential habitat areas. Therefore, the staff concludes that the proposed action would have no effect on the dwarf wedgemussel.

Because bald eagles have been observed in the vicinity of the North Anna ESP site, but have never been observed to nest or roost in the vicinity, the staff has concluded that the proposed action may affect, but are not likely to adversely affect bald eagles.

It is very unlikely that three protected plant species, small-whorled pogonia, sensitive joint-vetch, and swamp pink, may occur at the NAPS site. These species have never been reported in Louisa county, and there is no known habitat for these species on the North Anna ESP site. Therefore, the staff concludes that the proposed action would have no effect on the small-whorled pogonia, sensitive joint-vetch, and swamp pink.

10.0 References

10 CFR Part 51. Code of the *Federal Regulations*, Title 10, *Energy,* Part 51, " Environmental Protection Regulations for Domestic Licensing and Related Regulatory Functions."

10 CFR Part 52. Code of Federal Regulations, Title 10, *Energy,* Part 52, " Early Site Permits; Standard Design Certifications; and Combined Licenses for Nuclear Power Plants."

Audubon Society. 2004. "103rd Christmas Bird Count. Lake Anna Virginia, United States." http://audubon.birdsource.org/CBCOutput/review.html?speciesByState=false&year=103&circle=S 1022926. Accessed 3/1/2004.

Bilkovic, D. M., C. H. Hershner, and J. E. Olney. 2002. "Macroscale Assessment of American Shad Spawning and Nursery Habitat in the Mattaponi and Pamunkey Rivers, Virginia. *North American Journal of Fisheries Management* 22:1176-1192; 2002."

Brooks, J.L. and S.I. Dodson. 1965. "Predation, Body Size, and Composition of Plankton." *Science*, 150:28-35.

Dominion Nuclear North Anna, LLC (Dominion) 2001. "Migratory Birds." *Water/Waste Environmental Protection Manual.* Chapter 13. Glen Allen, Virginia.

Dominion Nuclear North Anna, LLC (Dominion). 2004. *North Anna Early Site Permit Applications – Part 3 - Environmental Report.* Revision 3, Glen Allen, Virginia.

Endangered Species Act of 1973. 16 USC 1531, et seq.

Environment Canada. 2004. http://www.speciesatrisk.gc.ca/search/speciesDetails_e.cfm?SpeciesID=591#description. Accessed 11/4/2004.

Golden, J., R. P. Ouellette, S. Saari, and P. N. Cheremisinoff. 1980. Environmental Impact Data Book. Ann Arbor Science Publishers, Inc. Ann Arbor, Michigan.

Jenkins, R. E. and N. M. Burkhead. 1994. *Freshwater Fishes of Virginia.* American Fisheries Society, Bethesda, Maryland.

Larkin, R. P. 1996. Effects of Military Noise on Wildlife: A Literature Review. USACERL Technical Report 96/21. U.S. Army Construction Engineering Research Lab. Champaign, Illinois. http://nhsbig.inhs.uiuc.edu/bioacoustics/ noise_and_wildlife.txt. Accessed 9/21/2004.

National Environmental Policy Act of 1969 (NEPA). 42 USC 4321, et seq.

National Oceanic and Atmospheric Administration (NOAA). 2004. Letter from Mary Colligan (NOAA) to Pao-Tsin Kuo, NRC.Date January6, 2004.

Omernik, J. 1987. "Ecoregions of the conterminous United States. Map (Scale 1:7500000)." *Annals of the Association of American Geographers* 77(1):118-125.

Paterson, C.G. and C.H. Fernando. 1970. Benthic Fauna Colonization of a New Reservoir with Particular Reference to the Chironomidae. *J. Fish. Res. Bd.*, *Canada* 27:213-232.

Reed, J.C. and G.M. Simmons. 1972. *An Ecological Investigation of the Lower North Anna and Upper Pamunkey River System*. Prepared for Virginia Electric and Power Company, Richmond, Virginia.

Rohde, F. C., R. G. Arndt, D. G. Lindquist, and J. F. Parnell. 1994. *Freshwater Fishes of the Carolinas, Virginia, Maryland, and Delaware*. The University of North Carolina Press, Chapel Hill, North Carolina.

U.S. Atomic Energy Commission (AEC). 1973. *Final Environmental Statement Related to the Continuation of Construction and the Operation of Units 1 and 2 and Construction of Units 3 and 4, North Anna Power Station*. Washington, D.C.

U.S. Fish and Wildlife Service and Virginia Department of Game and Inland Fisheries (FWS and VDGIF). 2000. Bald Eagle Protection Guidelines for Virginia. http://www.dgif.state.va.us/wildlife/publications/EagleGuidelines.pdf. Accessed 3/23/2004.

U.S. Fish and Wildlife Service (FWS). 1993. Dwarf wedge mussel recovery plan. Hadley, Massachusetts.

U.S. Fish and Wildlife Service (FWS) 2004a. Letter From Karen Mayne, FWS to the NRC, dated October 25, 2004. " NRC's North Anna and Surry Power Stations", Project number 9064.

U.S. Fish and Wildlife Service (FWS). 2004b. "Threatened and Endangered Species System (TESS) Listings by State and Territory as of 03/01/2004. Virginia." http://ecos.fws.gov/tess_public/TESSWebpageUsaLists?state=VA . Accessed 3/1/2004.

U.S. Nuclear Regulatory Commission (NRC). 2002. *Generic Environmental Impact Statement for License Renewal of Nuclear Plants, Supplement 7, Regarding North Anna Power Station, Units 1 and 2*, NUREG-1437, Office of Nuclear Reactor Regulation, Division of Regulatory Improvement Programs, Washington, D.C.

U.S. Nuclear Regulatory Commission (NRC). 2003a. Letter from P.T. Kuo, U.S. Nuclear Regulatory Commission, to M. Colligan, NOAA Fisheries, Application for an Early Site Permit for the North Anna Power Station Site.

U.S. Nuclear Regulatory Commission (NRC). 2003b. Letter from P.T. Kuo, U.S. Nuclear Regulatory Commission, to J. Wolflin, U.S. Fish and Wildlife Service, Application for an Early Site Permit for the North Anna Power Station Site.

Virginia Department of Conservation & Recreation (VDCR). 2004. "Virginia's Natural Communities, Rare, Threatened and Endangered Animals and Plants." www.dcr.state.va.us/dnh/nhrinfo.htm. Accessed 3/25/2004.

Virginia Department of Game and Inland Fisheries (VDGIF). 2004. "Geographic Search. Fish and Wildlife Information Service." Available at vafwis.org/perl/vafwis.pl/vafwis.login. (Note: This is a protected website that is accessible only through VDGIF authorization.)

14

Virginia Electric and Power Company (VEPCo). 1986. *Section 316(a) Demonstration for North Anna Power Station: Environmental Studies of Lake Anna and the Lower North Anna River.* Virginia Power Corporate Technical Assessment. Water Quality Department. Richmond, Virginia.

Virginia Electric and Power Company (VEPCo). 1989. *Environmental Study of Lake Anna and the Lower North Anna River.* Annual Report for 1988. Richmond, Virginia.

Virginia Electric Power Company (VEPCo). 2001. *Application for License Renewal for North Anna Power Station, Units 1 and 2, Appendix E, Environmental Report - Operating License Renewal Stage.* Richmond, Virginia.

Virginia Electric and Power Company (VEPCo). 2002. *Environmental Study of Lake Anna and the Lower North Anna River.* Annual Report for 2001 including Summary for 1998-2000, Richmond, Virginia.

Voshell, J.R., and G.M. Simmons, Jr. 1978. "The Odonota of a New Reservoir in the South-Eastern United States," *Odonatologica* 7(1):67-76.

Watts, B.D. and M. A. Byrd. 2003. Virginia bald eagle nest and productivity survey: Year 2003 report. Center for Conservation Biology Technical Report Series, CCBTR-03-03. College of William and Mary, Williamsburg, VA. 26 pp.

Woods, A.J., J.M. Omernik, and D.D. Brown. 1999. "Level III Ecoregions of Delaware, Maryland, Pennsylvania, Virginia, and West Virginia." National Health and Environmental Effects Research Laboratory. US EPA. Corvallis, Oregon. http://www.epa.gov/wed/pages/ecoregions/reg3_eco.htm. Accessed 3/1/2004.

June 30, 2005

Ms. Regena Bronson
U.S. Army Corps of Engineers
Potomac Virginia Field Office
P.O. Box 1704
Leonardtown, MD 20650

SUBJECT: NORTH ANNA EARLY SITE PERMIT REVIEW (TAC NO. MC1128)

Dear Ms. Bronson:

The U.S. Nuclear Regulatory Commission (NRC) staff is reviewing an application submitted by Dominion Nuclear North Anna, LLC (Dominion) for an early site permit (ESP). The proposed action requested in Dominion's application is for the NRC to: (1) approve a site within the existing North Anna Power Station (NAPS) boundaries as suitable for the construction and operation of one or more new nuclear power generating facilities; and (2) issue an ESP for the proposed site located at NAPS. An ESP does not authorize construction or operation of a nuclear power plant. Rather, the ESP application and review process makes it possible to evaluate and resolve certain safety and environmental issues related to siting before the applicant makes large commitments of resources. If the ESP is approved, the applicant can "bank" the site for up to 20 years for future reactor siting. To construct or operate a nuclear power plant, an ESP holder must obtain a construction permit and an operating license, or a combined license.

As part of its environmental review of Dominion's ESP application, the NRC prepared a draft environmental impact statement (DEIS) in accordance with 10 CFR 52.18. The DEIS includes the NRC staff's analysis of the environmental impacts of constructing and operating two nuclear units at the North Anna ESP site, or at alternative sites. It also includes the staff's preliminary recommendation to the Commission regarding the proposed action. In addition, as described in the DEIS, if the ESP includes a site redress plan, the ESP holder can conduct certain site preparation and preliminary construction activities allowed by Title 10 of the *Code of Federal Regulations* Section 50.10(e)(1) (10 CFR 50.10 (e)(1)), provided the final EIS concludes that such activities will not result in any significant environmental impact that cannot redressed. Dominion has included a site redress plan in its application. If the ESP is approved, Dominion would be allowed to conduct site preparation and preliminary construction activities pursuant to 10 CFR 52.25 and 10 CFR 50.10(e)(1), subject to receipt of any other necessary Federal, State, and/or local approvals. Dominion has stated that it does not plan to conduct such activities at this time. However, these activities, if performed, could include dredging and other activities potentially subject to Clean Water Act requirements. The environmental impacts of these activities are discussed in the DEIS.

R. Bronson   -2-

Pursuant to the "Memorandum of Understanding Between the Corps of Engineers, United States Army, and the United States Nuclear Regulatory Commission for Regulation of Nuclear Power Plants" (40 FR 37110 (dated August 25, 1975)), we request that the Army Corps of Engineers review and provide to the NRC any comments on the DEIS.

Enclosed is a copy of NUREG-1811 "The Draft Environmental Impact Statement for an Early Site Permit (ESP) at the North Anna ESP Site." We request your comments no later than August 12, 2005. Enclosed to aid in your review is a CD containing Dominion's application for an ESP. If you have any questions concerning the ESP application or other aspects of this project, please contact Mr. Jack Cushing, Senior Environmental Project Manager, at 301-415-1424 or by e-mail at JXC9@nrc.gov.

                              Sincerely,

                              /RA/

                              Pao-Tsin Kuo, Program Director
                              License Renewal and Environmental Impacts Program
                              Division of Regulatory Improvement Programs
                              Office of Nuclear Reactor Regulation

Docket No.: 52-008

Enclosure: As stated

cc wo/encl.: See next page

DEPARTMENT OF THE ARMY
NORFOLK DISTRICT, CORPS OF ENGINEERS
FORT NORFOLK, 803 FRONT STREET
NORFOLK, VIRGINIA 23510-1096

REPLY TO
ATTENTION OF:

July 15, 2005

Northern Virginia Regulatory Section
(Lake Anna)

US Nuclear Regulatory Commission
OWFN 11 F-1
Attn: Mr. Jack Cushing
Washington, DC 20555-0001

Dear Mr. Cushing:

This is in reference to your request for Corps' comments on the "Draft Environmental Impact Statement for an Early Site Permit (ESP) at the North Anna ESP Site" project in Caroline County.

Based on the ESP draft submitted by you, fill may be proposed in waters of the United States regulated under Section 404 of the Clean Water Act (33 U.S.C. 1344) and may require a Department of the Army Permit. However, before we could make such a decision, the following additional information is required for review:

> 1. A USGS topo map depicting the location and boundaries of the project site for the proposed Units 3 and 4.
>
> 2. If the project would impact wetlands, a wetland delineation utilizing the Corps 1987 Delineation Manual would be required to be submitted for our review and confirmation. The delineation must include sample locations on a plan and correspond to your data sheets.
>
> 3. The entire proposed plan of development with a depiction of all work that is subject to regulation under Section 404 of the Clean Water Act (i.e. intake and outfalls structures within jurisdictional waters and/or wetlands).
>
> 4. A survey for the federally listed threatened species; swamp pink (Helonias bullata), small whorled pogonia (Isotria medeoloides), and sensitive joint vetch (Aeschynomene virginica.
>
> 5. Identification of any archaeological, cultural, and historic properties that may exist on the subject site within the Corps' permit area.
>
> 6. Evidence that discharges of dredged or fill material into waters of the United States are avoided or minimized to the maximum extent practicable at the project site.

Any work in these areas may require authorization by state and local agencies. Thank you for providing us the opportunity to provide early comment on the project.

Should you have questions, please call Ms. Regena Bronson at 301.475.2720 in our Potomac Field Office.

Sincerely,

Bruce F. Williams
Chief, Northern Virginia
Regulatory Section

September 27, 2005

Dr. Ethel Eaton, Manager
Office of Review and Compliance
Virginia Department of Historic Resources
2801 Kensington Avenue
Richmond, VA 23221

SUBJECT: NORTH ANNA EARLY SITE PERMIT REVIEW (TAC NO. MC1128)

Dear Dr. Eaton:

This letter responds to your request for a programmatic agreement with the U.S. Nuclear
Regulatory Commission (NRC) under the National Historic Preservation Act (NHPA) raised
during our teleconference conducted on May 23, 2005, with members of your staff and
Dominion Nuclear North Anna, LLC (Dominion). The Virginia Department of Historic Resources
(VDHR) request for a programmatic agreement (PA) relates to Dominion's application for an
early site permit (ESP) at the North Anna site in Louisa County, Virginia.

The NRC stated the actions that it expected Dominion to take based on representations made
in Dominion's environmental report (ER), which is reflected in the NRC's draft environmental
impact statement (DEIS). Specifically, Dominion stated in its ER:

> "Prior to any activities that would disturb existing ground conditions, Dominion
> would assess the need, in coordination with VDHR, to undertake subsurface
> investigations for the identification of potentially significant historic or cultural
> resources in the area(s) to be disturbed. The investigations would be conducted
> in accordance with professional archeological practices and recommendations
> as developed in coordination with VDHR. Additionally, Dominion would
> implement the necessary administrative steps to make proper notifications in the
> event of any unanticipated discovery (including human remains). These steps
> would include stop-work, assessment, and notification protocol." [ER Revision 5
> Section 4.1.3, Page 3-4-6].

The above statement regarding coordination by Dominion with VDHR before ground disturbing
activities was relied on and is reflected in DEIS Section 4.6.

As set forth in our November 21, 2003, letter to you, the NRC staff is using the National
Environmental Policy Act (NEPA) process to comply with the obligations imposed under
§ 106 of the NHPA in accordance with the provisions in 36 CFR 800.8. Consistent with our
November 21 letter, the NRC has described in the EIS analyses of potential impacts to
historical and cultural resources and measures in place at the ESP site that would be expected
to avoid, minimize or mitigate any adverse effects on historic properties. The NRC staff also
forwarded the draft EIS to you for your review and comment. Accordingly, the NRC staff does
not believe a PA is warranted.

E. Eaton   -2-

If you have any questions concerning the ESP application or other aspects of this project, please contact Mr. Jack Cushing, Senior Environmental Project Manager, at 301-415-1424 or by e-mail at JXC9@nrc.gov.

                                    Sincerely,
                                    /RA Jacob Zimmerman For/

                                    Pao-Tsin Kuo, Program Director
                                    License Renewal and Environmental Impacts Program
                                    Division of Regulatory Improvement Programs
                                    Office of Nuclear Reactor Regulation

Docket No.: 52-008

United States Department of the Interior

NATIONAL PARK SERVICE
Colonial National Historical Park
Post Office Box 210
Yorktown, Virginia 23690

IN REPLY REFER TO:

L76

October 25, 2005

Mr. Jack Cushing
Senior Project Manager
U.S. Nuclear Regulatory Commission
Mailstop O-11F1
11555 Rockville Pike
Rockville, MD 20852

Dear Mr. Cushing:

Per your request, we are providing written comments on the National Park Service's concerns with regards to the impacts that the proposed expansion of the Surry Nuclear Power Plant would have on Colonial National Historical Park's resources.

The park includes Jamestown, Yorktown Battlefield, and Colonial Parkway, which are all on the National Register of Historic Places as nationally significant. The view shed from Jamestown and the Colonial Parkway along the James River is a critical component of the integrity of both resources. In addition, there are several bald eagle nests along the Parkway and at Jamestown, which are protected under federal law.

The current design and height of the Surry Nuclear Power Plant structures are barely visible from the Parkway and Jamestown. At the meeting you held with park staff on September 21, you described a structure that would be more intrusive due to its height, design and the visible plume of steam. These additions would have an adverse effect on the viewshed from both Jamestown and the Colonial Parkway, especially the pull-offs along the James River.

Prior to the implementation of this proposal, the impact on the park's and surrounding area's cultural and natural resources would need to be fully assessed through the development of an Environmental Impact Statement. Please keep us advised as to the status of this proposal, as we are requesting to formally review any documents prepared that assess the impacts of the proposed expansion.

Sincerely,

Skip Brooks
Deputy Superintendent

COMMONWEALTH of VIRGINIA

W. Tayloe Murphy, Jr.
Secretary of Natural Resources

Department of Historic Resources
2801 Kensington Avenue, Richmond, Virginia 23221

Kathleen S. Kilpatrick
Director

Tel: (804) 367-2323
Fax: (804) 367-2391
TDD: (804) 367-2386
www.dhr.virginia.gov

November 3, 2005

Mr. Pao-Tsin Kuo, Program Director
Office of Nuclear Reactor Regulation
U.S. Nuclear Regulatory Commission
Washington, D.C. 20555-0001

RE: North Anna Early Site Permit Review (TAC No. MC1128)
       Louisa County, Virginia
       DHR File No. 2000-1210

Dear Mr. Kuo:

We have received your September 27, 2005 letter concerning the action referenced above. According to your letter, the Nuclear Regulatory Commission (NRC) is of the opinion that the consideration given to potential impacts to historical and cultural resources in the draft Environmental Impact Statement (EIS), prepared pursuant the National Environmental Policy Act (NEPA), is sufficient to satisfy NRC's responsibilities under Section 106 of the National Historic Preservation Act. While 36 CFR 800.8 encourages Federal agencies to coordinate their Section 106 compliance with their NEPA responsibilities, it does not support a lower threshold for the identification of historic properties and assessment of effects. These steps of the process can be satisfied during the preparation of an EIS, but must be completed prior to the approval of the undertaking.

It is our opinion that if NRC does not wish to complete the identification and effect determination steps prior to finalizing the EIS, then the only alternative is to execute a Programmatic Agreement, which puts in place a set of procedures for future consultation and would allow this undertaking to proceed according to its stipulations. Such alternate procedures could apply not only to the Early Site Permit, but also to later permitting actions related to construction and operation and could ease and expedite future consultation. The conditional approval of the EIS by NRC without SHPO approval does not afford the Advisory Council on Historic Preservation (ACHP) an opportunity to comment and may be inconsistent with the Federal regulations.

We urge the NRC to reconsider the appropriateness and benefit of a Programmatic Agreement. Pursuant 36 CFR Part 800.2(b)(2), we have requested guidance from the ACHP on this matter. We will forward to you

Administrative Services	Capital Region Office	Tidewater Region Office	Roanoke Region Office	Winchester Region Office
10 Courthouse Avenue	2801 Kensington Ave.	14415 Old Courthouse Way, 2nd Floor	1030 Penmar Ave., SE	107 N. Kent Street, Suite 203
Petersburg, VA 23803	Richmond, VA 23221	Newport News, VA 23608	Roanoke, VA 24013	Winchester, VA 22601
Tel: (804) 863-1624	Tel: (804) 367-2323	Tel: (757) 886-2807	Tel: (540) 857-7585	Tel: (540) 722-3427
Fax: (804) 862-6196	Fax: (804) 367-2391	Fax: (757) 886-2808	Fax: (540) 857-7588	Fax: (540) 722-7535

Page 2
November 3, 2005
Mr. Pao-Tsin Kuo

for consideration any comments received. If you have any questions, please do not hesitate to contact me at (804) 367-2323, ext. 153 or e-mail roger.kirchen@dhr.virginia.gov.

Sincerely,

Roger W. Kirchen, Archaeologist
Office of Review and Compliance

Cc: Mr. Jack Cushing
        Office of Nuclear Reactor Regulation
        U.S. Nuclear Regulatory Commission
        Washington, D.C. 20555-0001

        Mr. David Christian
        Dominion Nuclear North Anna, LLC
        5000 Dominion Blvd.
        Glen Allen, VA 23060

        Mr. Don Klima
        Advisory Council on Historic Preservation
        1100 Pennsylvania Avenue, NW, Suite 803
        Washington, DC 20004

COMMONWEALTH of VIRGINIA

DEPARTMENT OF ENVIRONMENTAL QUALITY
NORTHERN VIRGINIA REGIONAL OFFICE
13901 Crown Court, Woodbridge, Virginia 22193
(703) 583-3800 Fax (703) 583-3801
www.deq.virginia.gov

L. Preston Bryant, Jr.
Secretary of Natural Resources

David K. Paylor
Director

Jeffery A. Steers
Regional Director

June 16, 2006

Mr. Jack Cushing
U. S. Nuclear Regulatory Commission
Washington, D.C. 20555

Re: Dominion Nuclear North Anna Early Site Permit Application, § 401 Certification of
the Federal Clean Water Act

Dear Mr. Cushing:

The Department of Environmental Quality (DEQ) is providing the Commission with justification as to why, at this time, we cannot provide §401 Certification under the Federal Clean Water Act for Dominion Nuclear North Anna's Early Site Permit Application. It is our understanding that a specific scope and schedule for pre-construction activities and determination of specific activities that would result in impacts to state/federal waters or wetlands have not been established. Before the § 401 Certification, through the issuance of a Virginia Water Protection Permit can be authorized, DEQ must know the extent of the surface water impacts and conduct a project review to ensure that all avoidance and minimization to surface waters has occurred.

To address the timing of this certification, DEQ recommends that the ESP should include a condition prohibiting Dominion from conducting any pre-construction activity that would result in a discharge into navigable waters without first submitting to the NRC a Virginia Water Protection Permit (which under Virginia's State Water Control Law at Va. Code § 62.1-44.15:5(A) constitutes the certification required under FWPCA § 401) or a determination by the Virginia DEQ that no certification is required. This condition would make it clear that the ESP does not constitute a license or permit to conduct any activity resulting in a discharge, and therefore, a 401 certification would not be required for issuance of the ESP.

Mr. Jack Cushing
June 16, 2006
Page Two

Our inability to make a permitting determination outside of having received an application should not be construed to mean that this project would not be approvable for permitting under the VWP program. Clearly, should any future application submitted by Dominion meet the appropriate regulatory requirements for permitting, the agency would be prepared to act on said application, while following all administatrative procedures.

Should you need any further information or clarification in this matter, please contact Ms. Joan Crowther of this office at (703) 583-3828.

Sincerely,

Jeffery A. Steers
Regional Director

cc: Nitin Patel, NRC
Tony Banks, Dominion
Jud White, Dominion
Mike Murphy, DEQ/CO
Joan Crowther, DEQ/NVRO

Appendix G

Environmental Impacts of Transportation

Appendix G

Environmental Impacts of Transportation

Changes to this appendix reflect the impacts of the higher power level proposed by Dominion in Revision 6 to its Environmental Report.

In April 2006, Dominion, Nuclear North Anna, LLC (Dominion) submitted Revision 6 to its application for an early site permit (ESP) for proposed Units 3 and 4 at the North Anna Power Station (Dominion 2006a). The Environmental Report (ER) of the revised application addressed, among other things, an increased thermal power rating for the economic simple boiling water reactor (ESBWR) from 4300 MW(t) to 4500 MW(t).

In the Draft EIS (NRC 2004), the NRC staff reported the environmental impacts of transporting unirradiated fuel, spent fuel, and solid radioactive waste to and from the North Anna ESP site and any alternative sites. The higher power level for the surrogate ESBWR would not significantly affect the quantity of unirradiated fuel needed for the initial core and annual refueling requirements previously evaluated in the Draft Environmental Impact Statement (DEIS) (NRC 2004). In addition, the higher power level would not affect the amount of solid waste generated or the number of spent fuel assemblies that would be shipped to a spent fuel repository. A higher power level would normally use more fuel and, therefore, result in more fuel shipments. However, the surrogate ESBWR has a higher unit capacity than was assumed in the Draft EIS (96 percent versus 95 percent) and the fuel would have a higher average burnup. In addition, in it's response to RAIs dated May 24, 2006 (Dominion 2006b), Dominion states that the ESBWR fuel assemblies are about 28 percent lighter than ABWR fuel assemblies. This lower weight is offset by the 30 percent higher number of ESBWR fuel assemblies than ABWR fuel assemblies. On balance, the total number of fuel shipments for the surrogate ESBWR would increase by 1 to 2 percent. The amount of radioactive waste generated more closely aligns with operational practices rather than the power level. The staff concludes that differences in the amount of solid radioactive waste generated by the surrogate ESBWR would be small and within the uncertainty of the estimates. Therefore, the staff concludes that the number of shipments of unirradiated fuel, spent fuel, and radioactive waste did not change the estimates presented in the Draft EIS.

The higher power level would not affect the radiation dose rates from the fuel and waste shipments and other parameters used to define the shipping routes and receptors. Consequently, the higher power level does not affect the per-shipment or annual impacts presented in the Draft EIS. The higher power level would affect the radionuclide characteristics of spent fuel. In ER Revision 6 (Dominion 2006a) and its response to the RAIs dated May 24, 2006 (Dominion 2006b), Dominion presented the estimates for radionuclide concentrations in the spent fuel for the surrogate ESBWR. The most significant radionuclides in spent fuel

1 contributing to dose were compared and, in general, actinide inventories decreased slightly and
2 fission product inventories increased with the power level increase. Overall, the transportation
3 accident risk for the surrogate ESBWR would be about 5 percent higher per year than the
4 ESBWR risk considered in the Draft EIS.
5
6 The higher power level could also change the shipment and impact estimates once normalized
7 to the reference LWR in WASH-1238 and used in Table S-4 of Title 10 of the Code of Federal
8 Regulations (CFR) Section 51.52. As discussed in Section 6.2 the staff determined that the
9 changes are small and well within the range of uncertainty associated with the shipment and
10 impact estimates.
11
12 In summary, the changes in the impact estimates from the higher power level once normalized
13 to the reference 1100 MW(e) LWR are small, and are uniformly lower than the normalized
14 impact estimates calculated in the Draft EIS (see Table G-1). The impacts for the
15 ABWR/ESBWR in the Draft EIS are larger than the surrogate ESBWR except for transportation
16 accidents because the impacts are normalized to net electric output (i.e., impacts per MW(e)).
17 Because the surrogate ESBWR has a higher net electric output than the ABWR/ESBWR
18 previously evaluated, and the un-normalized impacts are essentially the same, the impacts per
19 MW(e) are slightly smaller. Nevertheless, the differences in normalized impacts are small and
20 well within the uncertainty of the estimates.
21
22 The staff concludes that the environmental impacts of transportation of fuel and radioactive
23 wastes to and from advanced LWR designs with the higher power level did not change. The
24 impact level category would still be SMALL and would be consistent with the risks associated
25 with transportation of fuel and wastes to and from current generation reactors presented in
26 Table S–4 of 10 CFR 51.52. The staff concludes that the impacts associated with gas-cooled
27 reactor designs at the higher power level did not change. The impact level category would
28 likely be SMALL, but the issue is not resolved because of the lack of data to validate impacts
29 from gas-cooled designs.
30

Table G-1. Summary of Changes to Transportation Impact Estimates that Result from the Data Provided in Revision 6 of the ER

Impact Category	ABWR/ESBWR in Draft EIS	ESBWR in ER Revision 6	Percent Change [a]
Unirradiated Fuel			
Normalized number of shipments	165	162	-2%
Normalized annual normal condition population doses, person-Sv/yr			
Workers	7.1×10^{-5}	6.9×10^{-5}	-2%
Public – Onlookers	2.7×10^{-4}	2.7×10^{-4}	-2%
Public – Along Route	6.6×10^{-6}	6.5×10^{-6}	-2%
Accident Risks	Section is unaffected by changes in ER Rev. 6		
Spent Fuel			
Normalized annual shipments	41	40	-2%
Normal condition population doses – North Anna, person-Sv/yr[b]			
Workers	4.2×10^{-2}	4.1×10^{-2}	-2%
Public – Onlookers	1.4×10^{-1}	1.4×10^{-1}	-2%
Public – Along Route	3.7×10^{-3}	3.6×10^{-3}	-2%
Accident Risks – North Anna, person-Sy/yr[b]	4.7×10^{-6}	5.0×10^{-6}	+5%
Solid Radioactive Waste			
Normalized waste generation rate, $m^3/1100$ MW(e)	62	60	-2%
Normalized annual shipment per 1100 MW(e)	27	26	-4%

(a) Due to rounding, the percent changes may not exactly match percent changes calculated using the values in the table.

(b) Example results are presented for the North Anna site. The effects on the transportation impacts for the Portsmouth, Savannah River, and Surry sites are consistent with the effects on transportation impacts for the North Anna site.

References

Dominion Nuclear North Anna, LLC (Dominion). 2006a. *North Anna Early Site Permit Application – Part 3 – Environmental Report*. Revision 6, Glen Allen, Virginia.

Dominion Nuclear North Anna, LLC (Dominion). 2006b. Letter from E. Grecheck (DNNA) to NRC dated May 24, 2006, submitting additional information in response to an NRC request dated May 10, 2006, (ML061510131).

U.S. Nuclear Regulatory Commission (NRC). 2004. *Draft Environmental Impact Statement (EIS) for an Early Site Permit (ESP) at the North Anna ESP Site*. NUREG-1811, Washington, D.C.

Appendix H

Supporting Documentation on Radiological Dose Assessment

Appendix H

Supporting Documentation on Radiological Dose Assessment

The detailed material in this appendix supported the analyses in Section 5.9.2 of the Draft EIS (NRC 2004), but was not previously published. The staff elected to incorporate this information into the EIS. It now reflects the radiological analysis at the higher output level of 4500 MW(t).

The staff performed an independent dose assessment on the radiological impacts of normal operation for new nuclear units at the Dominion Nuclear North Anna, LLC (Dominion) early site permit (ESP) site. The results of this assessment are presented in this appendix and are compared to the results from Dominion found in Section 5.9 (Radiological Health Impacts). The appendix is divided into three sections: (1) H.1 – Dose Estimates to the Public from Liquid Effluents, (2) H.2 – Dose Estimates to the Public from Gaseous Effluents, and (3) H.3 – Dose Estimates to the Biota from Liquid and Gaseous Effluents.

To facilitate comparison with Dominion's estimates, all doses and radioactivity levels are reported in millirem (mrem) and curies (Ci), respectively.

H.1 Dose Estimates to the Public from Liquid Effluents

To estimate doses to the maximally exposed individual (MEI) and the population from the liquid effluent pathway, the staff used the LADTAP II code (Strenge et al. 1986) and input parameters supplied by Dominion as part of their ESP Environmental Report (ER) (Dominion 2006).

H.1.1 Scope

Doses to the MEI were calculated for the following:

- *Total Body* – Dose was the total from all pathways (i.e., drinking water, fish consumption, shoreline usage, swimming exposure, and boating) with the highest value for either the adult, teen, child, or infant compared to the 0.03 mSv/yr (3 mrem/yr) per reactor design objective in Title 10 of the Code of Federal Regulations (CFR) Part 50, Appendix I.

- *Organ* – Dose was the total for each organ from all pathways (i.e., drinking water, fish consumption, shoreline usage, swimming exposure, and boating) with the highest value for either the adult, teen, child, or infant compared to the 0.1 mSv/yr (10 mrem/yr) per reactor design objective in 10 CFR Part 50, Appendix I.

1   The values of input parameters used by Dominion were reviewed by the staff and determined to
2   be appropriate to use as inputs into the LADTAP II code for its independent calculation. Default
3   values from Regulatory Guide 1.109 (NRC 1977) were used when input parameters were not
4   available.

5
6   ## H.1.2 Resources Used
7
8   The staff used a version of the LADTAP II code entitled NRCDOSE version 2.3.4 (Bland 2000),
9   obtained through the Oak Ridge Radiation Safety Information Computational Center (RSICC) to
10   calculate doses to the public from liquid effluents.

11
12   ## H.1.3 Input Parameters
13
14   Table H-1 provides a listing of the major parameters used by the staff in calculating dose to the
15   public from liquid effluent releases during normal operation. Table H-2 lists the liquid effluent
16   releases used by the staff in calculating dose to the public. This table is the same as
17   Table 5.4-6 of Dominion (2006).

18
19   **Table H-1**. Parameters Used in Calculating Dose to Public from Liquid Effluent Releases
20

Parameter	Staff Value	Comments
Source term (Ci/yr)[a]	Table 5.4-6 of the ER (Dominion 2006)	Table 5.4-6 of Dominion (2006) represents the bounding liquid effluent source term based on the plant parameter envelope approach.
Discharge flow rate m^3/s (ft^3/s)	0.62 (22)	Site-specific value from Table 5.4-1 of the ER (Dominion 2006).
Source term multiplier	1	Site-specific value from the ER (Dominion 2006).
Site type	Fresh water	Site-specific value from the ER (Dominion 2006).
Reconcentration model	None	Table 5.4-1 of the ER (Dominion 2006).
Effluent discharge rate from impoundment system to receiving water body	N/A	Not applicable because reconcentration was not assumed
Impoundment total volume	N/A	Not applicable because reconcentration was not assumed

Table H-1. (contd)

Parameter	Staff Value	Comments
Shore width factor	0.3	Value from Regulatory Guide 1.109 (NRC 1977).
Dilution factors for aquatic food and boating, shoreline and swimming, and drinking water	10	Site-specific value from the ER (Dominion 2006). Table 5.4-1 of the ER provides an effluent discharge rate of 0.006 m^3/s (100 gpm) with a dilution flow of 0.6 m^3/s (10,000 gpm). This yields a dilution factor of 100. Dominion used a factor of 10, which is more conservative.
Transit time (h)	0	Site-specific value from Table 5.4-1 of the ER (Dominion 2006). The value is conservative.
Consumption and usage factors for adult, teen, children, and infant	Values from Table 5.4-2 of the ER (Dominion 2006)	Values were the default values from Regulatory Guide 1.109 (NRC 1977).
Population supplied by drinking water (population)	22,100	Site-specific value provided by Dominion (Dominion 2004a).
Dilution factor for water intake locations, shoreline exposure location, swimming usage location, and boating usage location (population)	10	Site-specific value from the ER (Dominion 2006). Table 5.4-1 of the ER provides an effluent discharge rate of 0.006 m^3/s (100 gpm) with a dilution flow of 0.6 m^3/s (10,000 gpm). This yields a dilution factor of 100. Dominion used a factor of 10, which is more conservative.
Total shoreline usage time (person-hours/year) (population)	1.31×10^6	Site-specific value provided by Dominion (Dominion 2004a).
Total exposure time for swimming usage location (person-hours/year) (population)	8.76×10^5	Site-specific value provided by Dominion (Dominion 2004a).
Total exposure time for boating activities (person-hours/year) (population)	2.19×10^6	Site-specific value provided by Dominion (Dominion 2004a).

(a) To convert Ci/yr to Bq/yr, multiply the value by 3.7×10^{10}.

Table H-2. Liquid Effluent Release Source Terms from the ER (Dominion 2006)[a,b]

Isotope	Release (Ci/yr)	Isotope	Release (Ci/yr)	Isotope	Release (Ci/yr)	Isotope	Release (Ci/yr)
C-14	4.4×10^{-4}	Rb-88	2.7×10^{-4}	Rh-103m	4.9×10^{-3}	Cs-138	2.1×10^{-4}
Na-24	3.5×10^{-3}	Rb-89	4.8×10^{-5}	Rh-106	7.4×10^{-2}	Ba-137m	1.2×10^{-2}
P-32	6.6×10^{-4}	Sr-89	3.6×10^{-4}	Ag-110m	1.1×10^{-3}	Ba-139	2.5×10^{-5}
Cr-51	2.1×10^{-2}	Sr-90	3.8×10^{-5}	Ag-110	1.4×10^{-4}	Ba-140	5.5×10^{-3}
Mn-54	2.8×10^{-3}	Sr-91	9.8×10^{-4}	Sb-124	6.8×10^{-4}	La-140	7.4×10^{-3}
Mn-56	4.2×10^{-3}	Sr-92	8.8×10^{-4}	Te-129m	1.4×10^{-4}	La-142	2.5×10^{-5}
Fe-55	6.4×10^{-3}	Y-90	3.4×10^{-6}	Te-129	1.5×10^{-4}	Ce-141	1.3×10^{-4}
Fe-59	2.0×10^{-4}	Y-91M	1.0×10^{-5}	Te-131m	1.0×10^{-4}	Ce-143	1.9×10^{-4}
Co-56	5.7×10^{-3}	Y-91	2.4×10^{-4}	Te-131	3.0×10^{-5}	Ce-144	3.2×10^{-3}
Co-57	7.9×10^{-5}	Y-92	6.6×10^{-4}	Te-132	2.4×10^{-4}	Pr-143	1.4×10^{-4}
Co-58	3.4×10^{-3}	Y-93	9.8×10^{-4}	I-131	1.4×10^{-2}	Pr-144	3.2×10^{-3}
Co-60	1.0×10^{-2}	Zr-95	1.0×10^{-3}	I-132	2.8×10^{-3}	W-187	2.1×10^{-4}
Ni-63	1.5×10^{-4}	Nb-95	1.9×10^{-3}	I-133	2.4×10^{-2}	Np-239	1.4×10^{-2}
Cu-64	8.2×10^{-3}	Mo-99	3.9×10^{-3}	I-134	1.9×10^{-3}	Total w/o H-3	3.7×10^{-1}
Zn-65	7.5×10^{-4}	Tc-99m	5.1×10^{-3}	I-135	8.2×10^{-3}	H-3	3.1×10^{3}
Zn-69m	6.0×10^{-4}	Ru-103	4.9×10^{-3}	Cs-134	9.9×10^{-3}		
Br-83	7.5×10^{-5}	Ru-105	1.0×10^{-4}	Cs-136	1.2×10^{-3}		
Br-84	2.0×10^{-5}	Ru-106	7.4×10^{-2}	Cs-137	1.3×10^{-2}		

(a) Table 5.4-6 of Dominion (2006).
(b) To convert from Ci/yr to Bq/yr, multiply the value by 3.7×10^{10}.

H.1.4 Comparison of Results

Table H-3 compares Dominion's results with those calculated by the staff. Doses calculated were similar.

Table H-3. Comparison of Doses to the Public from Liquid Effluent Releases for One ESP Unit

Type of Dose	Applicant's ER (Dominion 2006)[a]	Staff's Calculation[a]	% Difference
Total body (mrem/yr)	1.3 (adult)	1.3 (adult)	0
Organ dose (mrem/yr)	2.5 (child bone)	2.5 (child bone)	0
Thyroid (mrem/yr)	1.3 (infant)	1.4 (infant)	7.7
Population dose from liquid pathway (person-rem)	14	14.7	5

(a) To convert mrem/yr to mSv/yr divide by 100.

H.2 Dose Estimates to the Public from Gaseous Effluents

To estimate doses to the maximally exposed individual and to the population within an 80-km (50-mi) radius of the ESP site from the gaseous effluent pathway, the staff used the GASPAR II code (Strenge et al. 1987) and input parameters supplied by Dominion in the ER (Dominion 2006).

H.2.1 Scope

The staff calculated gamma air dose, beta air dose, total body dose, and skin dose from noble gases at the nearest site boundary located 1.4 km (0.88 mi) east-southeast of the North Anna ESP site. Dose to the MEI was also calculated for the following locations:

- nearest site boundary (plume and inhalation)
- nearest residence (plume and inhalation)
- nearest garden (vegetable)
- nearest meat cow (meat).

MEI doses were not calculated for the nearest dairy cow and goat, within 8 km (5 mi) as specified in NUREG-1555 because as stated in the ER (Dominion 2006), there were no milk cows or goats within 8 km (5 mi) of the proposed ESP units.

The values of input parameters used by Dominion are given in the ER (Dominion 2006) or in a response to a Request for Additional Information (RAI) dated May 17, 2004 (Dominion 2004b). These values were reviewed by the staff and determined to be appropriate to use as input into GASPAR II for its independent calculation. Default values from Regulatory Guide 1.109 (NRC 1977) were used when input parameters were not available.

Population doses were calculated for all types of releases (noble gases, iodine and particulates, and H-3 and C-14) using the GASPAR II code.

H.2.2 Resources Used

The staff used a version of GASPAR II code entitled NRCDOSE, Version 2.3.5 (Bland 2000), obtained through the Oak Ridge RSICC to calculate doses to the public from gaseous effluents.

H.2.3 Input Parameters

Table H-4 provides a listing of the major parameters used in calculating dose to the public from gaseous effluent releases during normal operation. Table H-5 lists the gaseous effluent releases used by the staff in calculating dose to the public. This table is the same as Table 5.4-7 of Dominion (2006).

Table H-4. Parameters Used in Calculating Dose to Public from Gaseous Effluent Releases

Parameter	Staff Value	Comments
Source term for calculating noble gas dose at site boundary and dose to the maximally exposed individual (Ci/yr)[a]	Table 5.4-7 of the ER (Dominion 2006)	These are the bounding plant parameter envelope (PPE) values.
Population distribution	2.784×10^6 – from Table 2.5-8 of the ER (Dominion 2006)	Site-specific data provided by Dominion.
Atmospheric dispersion factors (sec/m^3)	Table 2.7-17 to Table 2.7-19 of the ER (Dominion 2006)	Site-specific data provided by Dominion.
Ground deposition factors ($^{-2}$)	Table 2.7-20 of the ER (Dominion 2006)	Site-specific data provided by Dominion.
Milk production rate within 80 km (50 mi) (L/yr)	7.2×10^8 – Table 5.4-3 of the ER (Dominion 2006)	Site-specific data provided by Dominion.
Meat production rate within 80 km (50 mi) (kg/yr)	1.7×10^9 – Table 5.4-3 of the ER (Dominion 2006)	Site-specific data provided by Dominion.
Vegetable/fruit production rate within 80 km (50 mi) (kg/yr)	5.4×10^8 – Table 5.4-3 of the ER (Dominion 2006)	Site-specific data provided by Dominion.
Pathway receptor locations (direction, distance, and atmospheric dispersion factors): nearest site boundary, vegetable garden, residence, meat animal	Table 5.4-4 and Table 2.7-14 of the ER (Dominion 2006)	Site-specific data provided by Dominion.
Consumption factors for leafy vegetable, meat, milk, and vegetable/fruit	Table 5.4-5 of the ER (Dominion 2006)	Factors taken from Regulatory Guide 1.109 (NRC 1977).
Fraction of year leafy vegetables that are grown	0.5	Site-specific data provided by Dominion.
Fraction of year that milk cows that are on pasture	0.67	Site-specific data provided by Dominion.
Fraction of milk-cow intake that is from pasture while on pasture	1	Default value of GASPAR II code.
Average absolute humidity over the growing season (g/m^3)	8.0	Default value of GASPAR II code.
Average temperature over the growing season (°F)	0	Default value of GASPAR II code.
Fraction of year goats are on pasture	0.75	Site-specific data provided by Dominion.
Fraction of year beef-cattle are on pasture	0.67	Site-specific data provided by Dominion.
Fraction of beef-cattle intake that is from pasture while on pasture	1	Default value of GASPAR II code.

(a) To convert Ci/yr to Bq/yr, multiply the value by 3.7×10^{10}.

Table H-5. Gaseous Effluent Release Source Term from Dominion (2006)[a,b]

Isotope	Release (Ci/yr)	Isotope	Release (Ci/yr)	Isotope	Release (Ci/yr)	Isotope	Release (Ci/yr)
H-3	3.5×10^3	Kr-85	4.1×10^3	Ru-103	3.8×10^{-3}	Xe-135m	7.7×10^2
C-14	1.2×10^1	Kr-87	4.9×10^1	Rh-103m	1.2×10^{-4}	Xe-135	8.2×10^2
Na-24	4.4×10^{-3}	Kr-88	7.4×10^1	Ru-106	7.8×10^{-5}	Xe-137	9.8×10^2
P-32	1.0×10^{-3}	Kr-89	4.7×10^2	Rh-106	2.1×10^{-5}	Xe-138	7.8×10^2
Ar-41	3.0×10^2	Kr-90	4.2×10^{-4}	Ag-110m	2.2×10^{-6}	Xe-139	5.3×10^{-4}
Cr-51	3.8×10^{-2}	Rb-89	4.7×10^{-5}	Sb-124	2.0×10^{-4}	Cs-134	6.8×10^{-3}
Mn-54	5.9×10^{-3}	Sr-89	6.2×10^{-3}	Sb-125	6.1×10^{-5}	Cs-136	6.5×10^{-4}
Mn-56	3.8×10^{-3}	Sr-90	1.2×10^{-3}	Te-129m	2.4×10^{-4}	Cs-137	1.0×10^{-2}
Fe-55	7.1×10^{-3}	Y-90	5.0×10^{-5}	Te-131m	8.3×10^{-5}	Cs-138	1.9×10^{-4}
Co-57	8.2×10^{-6}	Sr-91	1.1×10^{-3}	Te-132	2.1×10^{-5}	Ba-140	3.0×10^{-2}
Co-58	2.3×10^{-2}	Sr-92	8.6×10^{-4}	I-131	5.1×10^{-1}	La-140	2.0×10^{-3}
Co-60	1.4×10^{-2}	Y-91	2.6×10^{-4}	I-132	2.4	Ce-141	1.0×10^{-2}
Fe-59	8.9×10^{-4}	Y-92	6.8×10^{-4}	I-133	1.9	Ce-144	2.1×10^{-5}
Ni-63	7.1×10^{-6}	Y-93	1.2×10^{-3}	I-134	4.1	Pr-144	2.1×10^{-5}
Cu-64	1.1×10^{-2}	Zr-95	1.7×10^{-3}	I-135	2.6	W-187	2.1×10^{-4}
Zn-65	1.2×10^{-2}	Nb-95	9.2×10^{-3}	Xe-131m	1.8×10^3	Np-239	1.3×10^{-2}
Kr-83m	1.3×10^{-3}	Mo-99	6.5×10^{-2}	Xe-133m	8.7×10^1	Total	1.8×10^4
Kr-85m	3.6×10^1	Tc-99m	3.3×10^{-4}	Xe-133	4.6×10^3		

(a) Table 5.4-7 of the ER (Dominion 2006).
(b) To convert from Ci/yr to Bq/yr, multiply the value by 3.7×10^{10}.

H.2.4 Comparison of Doses to the Public from Gaseous Effluent Releases

Table H-6 compares Dominion's results for doses from noble gases at the site boundary with the results calculated by the staff. The calculated doses were similar.

Table H-6. Comparison of Doses to the Public from Noble Gas Releases from One ESP Unit

Type of Dose	Dominion's ER (Dominion 2005)	Staff's Calculation	% Difference
Gamma air dose at site boundary – noble gases only (mrad/yr)[a]	3.2	3.2	0
Beta air dose at site boundary – noble gases only (mrad/yr)[a]	4.8	4.7	-2
Skin dose at site boundary – noble gases only (mrem/yr)[a]	6.2	6.2	0

(a) To convert from mrad/yr or mrem/yr to mGy/yr or mSv/yr, divide by 100.

1   Table H-7 compares doses to MEI calculated by Dominion and the staff. Doses to the MEI were
2   calculated at the nearest site boundary, nearest residence, nearest garden, and nearest meat
3   cow. The calculated doses were similar.
4
5   **H.2.5 Comparison of Results – Population Doses**
6
7   Table H-8 compares Dominion's population dose estimates taken from Table 5.4-12 of the ER
8   (Dominion 2006) with the staff's estimate. The calculated doses were similar.
9
10 # H.3 Dose Estimates to the Biota from Liquid and Gaseous Effluents
11
12 To estimate doses to the biota from the liquid and gaseous effluent pathways, the staff used the
13 LADTAP II code (Strenge et al. 1986) and GASPAR II code (Strenge et al. 1987) and input
14 parameters supplied by Dominion as part of its ER (Dominion 2006).
15
16 **H.3.1 Scope**
17
18 Doses to both terrestrial and aquatic biota were calculated using the LADTAP II code. Aquatic
19 biota include fish, invertebrate, and algae. Terrestrial biota include muskrat, raccoon, heron, and
20 duck. The code calculates an internal dose component and external dose component and sums
21 them for a total body dose. The values of input parameters used by Dominion were reviewed by
22 the staff and determined to be appropriate to use in its independent calculation. Default values
23 from Regulatory Guide 1.109 (NRC 1977) were used when input parameters were not available.
24
25 The LADTAP II code calculates biota dose from the liquid effluent pathway only. Terrestrial biota
26 will also be exposed via the gaseous effluent pathway. These values would be the same as
27 those for the MEI calculated using the GASPAR II code. Dominion used the MEI doses at a
28 location 0.40 km (0.25 mi) east-southeast from the proposed ESP site to estimate these doses.
29 To account for the closer proximity of the main body mass of animals to the ground compared to
30 humans, the MEI calculation for the biota assumed a ground deposition factor twice that used in
31 the MEI calculation for a member of the public.
32
33
34 **H.3.2 Resources Used**
35
36 To calculate doses to the public from liquid releases, the staff used a computer code entitled
37 NRCDOSE, Version 2.3.5 (Bland 2000), which is a version of the LADTAP II code and the
38 GASPAR II code, obtained through the Oak Ridge RSICC.
39
40 **H.3.3 Input Parameters**
41
42 Most of the LADTAP II input parameters are specified in Section H.1.3 to include the source
43 term, discharge flow rate, reconcentration model, effluent discharge rate from the impoundment

1 **Table H-7.** Comparison of Doses to the MEI from Gaseous Effluent Releases for One
2 ESP Unit
3

Location	Pathway	Total Body Dose (mrem/yr)[a,b]	Skin Dose (mrem/yr)[a,b]	Thyroid Dose (mrem/yr)[a,b]
Nearest site boundary (1.4 km [0.88 mi] east-southeast)	Plume	2.1 (2.1)	6.2 (6.2)	[d]
Nearest site boundary (1.4 km [0.88 mi] east-southeast)	Inhalation Adult Teen Child Infant	0.3 (0.3) 0.31 (0.3) 0.27 (0.27) 0.16 (0.16)	[c] [c] [c]	1.6 (1.6) 2.0 (2.0) 2.3 (2.3) 2.0 (2.0)
Nearest garden (1.5 km [0.94 mi] northeast)	Vegetable Adult Teen Child	0.44 (0.43) 0.57 (0.57) 1.1 (1.1)	[c] [c] [c]	4.9 (4.9) 6.6 (6.6) 13.1 (12.6)
Nearest residence (1.5 km [0.96 mi] north-northeast)	Plume	1.4 (1.4)	4.0 (4.0)	[d]
Nearest residence (1.5 km [0.96 mi] north-northeast)	Inhalation Adult Teen Child Infant	0.2 (0.19) 0.2 (0.2) 0.18 (0.17) 0.10 (0.10)	[c] [c] [c] [c]	1.0 (1.0) 1.3 (1.3) 1.5 (1.5) 1.3 (1.3)
Nearest meat cow (2.2 km [1.37 mi] southeast)	Meat Adult Teen Child	0.067 (0.067) 0.049 (0.049) 0.079 (0.079)	[c] [c] [c]	0.15 (0.15) 0.11 (0.11) 0.17 (0.17)

25 (a) Values in parentheses represent the values that the staff calculated. The Dominion values (those not in
26   parentheses) were taken from Table 5.4-9 of the ER (Dominion 2006).
27 (b) To convert from mrem/yr to mSv/yr, divide by 100.
28 (c) Skin dose is not applicable for the inhalation, vegetable, and meat pathways.
29 (d) Thyroid dose is not applicable for the plume pathway.
30
31 **Table H-8.** Comparison of Population Doses from Gaseous Effluent Releases for One ESP Unit
32

Pathway	Applicant's Estimate (person-rem/yr)[a]	Staff's Estimate (person-rem/yr)	% Difference
Liquid	14	14.7 (see Section H.1.4)	5
Noble gases	3.5	2.8 (plume)	-20
Iodine and particulates	1.4	1.2[b]	-14
H-3 and C-14	14	13.7	-2.1
Total	34	32.4	-5

39 (a) Estimated population dose for one ESP unit (see Table 5.4-12 of the ER [Dominion 2006]).
40 (b) Dose represents the summation of doses from iodine and particulates.
41

1   system to the receiving water body, impoundment total volume, and shore width factor.
2   Parameters unique to the biota dose calculation were taken from Table 5.4-14 (terrestrial biota
3   parameters) and Table 5.4-15 (shoreline and swimming exposures) of the ER (Dominion 2006).
4   These parameter values were default values used in the LADTAP II code (Strenge et al. 1986),
5   and are appropriate values to use in calculating biota dose.

6

7   ## H.3.4 Comparison of Results

8

9   Table H-9 compares Dominion's biota dose estimates from liquid effluents taken from
10   Table 5.4-16 of the ER (Dominion 2006) with the staff's estimate. The estimated doses were
11   similar.

12

13   Table H-10 compares Dominion's biota dose estimates for gaseous effluents taken from
14   Table 5.4-16 of the ER (Dominion 2006) with the staff's estimate. The staff calculated the biota
15   dose from gaseous effluents by summing the annual beta air dose, the annual gamma air dose,
16   and two times the ground deposition dose at a location 0.40 km (0.25 mi) east-southeast from
17   the proposed North Anna ESP site. Atmospheric dispersion factors used in the calculation were
18   taken from Table 2.7-15, Table 2.7-18, Table 2.7-19, and Table 2.7-20 of the ER
19   (Dominion 2006). The estimated doses were similar.

20

21   **Table H-9**. Comparison of Dose Estimates to the Biota from Liquid Effluents from One
22   ESP Unit

23

Biota	Dominion's ER (mrad/yr)[a]	Staff's Calculation (mrad/yr)[a]	% Difference
Fish	9.9	9.9	0
Invertebrate	46	47	2.2
Algae	54	55	1.9
Muskrat	44	45	2.3
Raccoon	5.1	5.1	0
Heron	56	56	0
Duck	44	45	2.3
(a) To convert from mrad/yr to mGy/yr, divide by 100.			

33

1   **Table H-10**. Comparison of Dose Estimates to the Biota from Gaseous Effluents from
2   One ESP Unit
3

Biota	Dominion's ER (mrad/yr)[a]	Staff's Calculation (mrad/yr)[a,b]	% Difference
Fish	[c]	[c]	[c]
Invertebrate	[c]	[c]	[c]
Algae	[c]	[c]	[c]
Muskrat	34	38	12
Raccoon	34	38	12
Heron	34	38	12
Duck	34	38	12

12   (a) To convert from mrad/yr to mGy/yr, divide by 100.
13   (b) Dose equals the sum of the annual beta air dose, the annual gamma dose, and two times
14   the ground deposition dose at 0.4 km (0.25 mi) east-southeast of the North Anna ESP
15   site.
16   (c) Fish, invertebrate, and algae would not be exposed to gaseous effluents.
17

H.4 References

20   10 CFR Part 50. Code of Federal Regulations, Title 10, *Energy*, Part 50, "Domestic Licensing of
21   Production and Utilization Facilities."

23   Bland, J.S. 2000. NRC DOSE for Windows. Radiation Safety Information Computational
24   Center, Oak Ridge National Laboratory, Oak Ridge, Tennessee.

26   Dominion Nuclear North Anna, LLC (Dominion). 2004a. Letter dated July 12, 2004 from
27   E.S. Grecheck (DNNA) to the NRC submitting additional information in response to an NRC
28   request RAI E5.4.2-2. dated March 12, 2004. Glen Allen, Virginia.

30   Dominion Nuclear North Anna, LLC (Dominion). 2004b. Letter dated May 17, 2004, from
31   E.S. Grecheck (DNNA) to the NRC submitting additional information in response to an NRC
32   request dated March 12, 2004. Glen Allen, Virginia.

34   Dominion Nuclear North Anna, LLC (Dominion). 2006. *North Anna Early Site Permit
35   Application – Part 3 – Environmental Report*. Revision 6, Glen Allen, Virginia.

37   Strenge D.L., R.A. Peloquin, and G. Whelan. 1986. *LADTAP II - Technical Reference and User
38   Guide*. NUREG/CR-4013, Pacific Northwest Laboratory, Richland, Washington.

1   Strenge D.L., T.J. Bander, and J.K. Soldat. 1987. *GASPAR II - Technical Reference and User*
2   *Guide.* NUREG/CR-4653, Pacific Northwest Laboratory, Richland, Washington.
3
4   U.S. Nuclear Regulatory Commission (NRC). 1977. *Regulatory Guide 1.109 – Calculation of*
5   *Annual Doses to Man from Routine Releases of Reactor Effluents for the Purpose of Evaluating*
6   *Compliance with 10 CFR Part 50, Appendix I.* Washington, D.C.
7
8   U.S. Nuclear Regulatory Commission (NRC). 2000. *Standard Review Plans for Environmental*
9   *Review for Nuclear Power Plants.* NUREG-1555, Washington, D.C.

Appendix I

ESP Site Characteristics and Plant Parameter Envelope

Appendix I

ESP Site Characteristics and Plant Parameter Envelope

This appendix has changed to reflect the changes to the plant parameter envelope caused by the changes presented in ER Revision 6.

The site specific plant parameter envelope (PPE) values and the Early Site Permit (ESP) site characteristics are from Environmental Report (ER) Table 3.1-9 unless otherwise specified. The staff used time dependent atmospheric dispersion factors from this Environmental Impact Statement (EIS) instead of the non-time dependent dispersion factors.

In its ER, Dominion Nuclear North Anna, LLC (Dominion) listed its proposed ESP site characteristics and plant parameter. These characteristics and parameters were used by the Nuclear Regulatory Commission (NRC) staff in its independent evaluation of the environmental impacts of the ESP Units 3 and 4. The ESP site characteristics specifically used in the staff's evaluation are presented in Table I-1. PPE parameters that are relevant to the environmental review are presented in Table I-2. The staff used the values in both Tables I-1 and I-2 in its evaluation.

Table I-1. ESP Site Characteristics

Item	Single Unit Value [Second Unit Value]	Description and References
Atmospheric Dispersion (χ/Q) (Accident)		• Time-dependent values as listed in Table 5-13 of this EIS
• Exclusion Area Boundary (EAB)	3.34×10^{-5} sec/m³ [Same for 2nd unit]	0 to 2 hr interval
• Low Population Zone (LPZ)	2.17×10^{-6} sec/m³ [Same for 2nd unit]	0 to 8 hr interval
	1.5×10^{-6} sec/m³ [Same for 2nd unit]	8 to 24 hr interval
	1.2×10^{-6} sec/m³ [Same for 2nd unit]	1 to 4 day interval
	9.0×10^{-7} sec/m³ [Same for 2nd unit]	4 to 30 day interval

Appendix I

Table I-1. (contd)

Item	Single Unit Value [Second Unit Value]	Description and References
Gaseous Effluents Dispersion, Deposition (Annual Average)		
• Atmospheric Dispersion (χ/Q)	χ/Q values presented in ER Table 2.7-14 [Same for 2nd unit]	• The atmospheric dispersion coefficients used to estimate dose consequences of normal airborne releases.
Residence	2.4×10^{-6} sec/m^3	No decay
	2.4×10^{-6} sec/m^3	2.26-day decay
	2.1×10^{-6} sec/m^3	8-day decay
EAB	3.7×10^{-6} sec/m^3	No decay
	3.7×10^{-6} sec/m^3	2.26-day decay
	3.3×10^{-6} sec/m^3	8-day decay
Meat animal	1.4×10^{-6} sec/m^3	No decay
	1.4×10^{-6} sec/m^3	2.26-day decay
	1.2×10^{-6} sec/m^3	8-day decay
Vegetable garden	2.0×10^{-6} sec/m^3	No decay
	2.0×10^{-6} sec/m^3	2.26-day decay
	1.8×10^{-6} sec/m^3	8-day decay
• Ground Deposition (D/Q)	D/Q values presented in ER Table 2.7-14 [Same for 2nd unit]	• The ground deposition coefficients used to estimate dose consequences of normal airborne releases
Residence	7.2×10^{-9}/m^2	
EAB	1.2×10^{-8}/m^2	
Meat animal	3.1×10^{-9}/m^2	
Vegetable garden	6.0×10^{-8}/m^2	

Table I-1. (contd)

Item	Single Unit Value [Second Unit Value]	Description and References
1 Dose Consequences		
2 • Normal	10 CFR 20, 10 CFR 50 Appendix I, and 40 CFR 190 dose limits	• Radiological dose consequences due to gaseous releases from normal operation of the plant
3 Liquid effluent	2.6 mrem/yr 2.7 mrem/yr 5.0 mrem/yr	Total body (Value for two units, see ER Table 5.4-11) Thyroid (Value for two units, see ER Table 5.4-11) Other organ/bone (Value for two units, see ER Table 5.4-11)
4 Gaseous effluent	4.8 mrem/yr 25 mrem/yr 6.5 mrem/yr 6.2 mrem/yr	Total body (Value for two units, see ER Table 5.4-11) Thyroid (Value for two units, see ER Table 5.4-11) Other organ/bone (Value for two units, see ER Table 5.4-11) Skin (Value for one unit, see ER Table 5.4-10)
5 Total	7.5 mrem/yr 28 mrem/yr 11 mrem/yr 6.2 mrem/yr	Total body (Value for two units, see ER Table 5.4-11) Thyroid (Value for two units, see ER Table 5.4-11) Other organ/bone (Value for two units, see ER Table 5.4-11) Skin (Value for one unit, see ER Table 5.4-10)
6 • Post-Accident	10 CFR 50.34(a)(1) and 10 CFR 100 dose limits [Same for 2nd unit]	• Radiological dose consequences due to gaseous releases from postulated plant accidents. • Design basis accidents (DBA) as listed in Tables 5-15 and 5-16 of this EIS • Severe accidents as listed in Tables 5-17 and 5-18 of this EIS
7 • Minimum Distance 8 to Site Boundary	2854.9 ft [Same for 2nd unit]	• Minimum lateral distance from the ESP PPE boundaries to the EAB
9 **Liquid Radwaste** 10 **System**		
11 • Normal Dose 12 Consequences	10 CFR 50, Appendix I, 10 CFR 20, and 40 CFR 190 dose limits	
	2.6 mrem/yr 2.7 mrem/yr 5.0 mrem/yr	Total body (Value for two units, see ER Table 5.4-11) Thyroid (Value for two units, see ER Table 5.4-11) Other organ/bone (Value for two units, see ER Table 5.4-11)

Table I-1. (contd)

	Item	Single Unit Value [Second Unit Value]	Description and References
1	**Population Density**		
2 3 4 5 6	• Population density at the time of initial site approval and within about 5 years thereafter	Population density meets the guidance of RS-002, Section 2.1.3 for RG 4.7, Regulatory Position C.4 [Both units]	• At the time of initial site approval and within about 5 years hereafter, the population densities, including weighted transient population, averaged over any radial distance out to 20 miles (cumulative population at a distance divided by the circular area at that distance), would not exceed 500 persons per square mile.
7 8 9	• Population density at the time of initial operation	Population density meets the guidance of RS-002, Section 2.1.3 [Both units]	• The population densities, including weighted transient population, averaged over any radial distance out to 30 miles (cumulative population at a distance divided by the area at that distance), would not exceed 500 persons per square mile at the time of initial operation.
10 11 12 13	• Population density over the lifetime of the new units until 2065	Population density meets the guidance of RS-002, Section 2.1.3 [Both units]	• The population densities, including weighted transient population, averaged over any radial distance out to 30 miles (cumulative population at a distance divided by the area at that distance), would not exceed 1000 persons per square mile over the lifetime of new units.
14 15	**Population Center Distance**	10 CFR 100.21(b) Meets requirement [Both units]	• The distance from the ESP PPE to the nearest boundary of a densely populated center containing more than about 25,000 residents is not less than one and one-third times the distance from the ESP PPE to the outer boundary of the LPZ.
16	**EAB**	10 CFR 100.21(a) Meets requirement [Both units]	• The exclusion area boundary is the perimeter of a 5000-ft-radius circle from the center of the abandoned Unit 3 containment.
17	**LPZ**	10 CFR 100.21(a) Meets requirement [Both units]	• The LPZ is a 6-mile-radius circle centered at the Unit 1 containment building.
18			

Table I-2. Plant Parameter Envelope (PPE)

Item	Single Unit Value [Second Unit Value]	Description and References
Normal Plant Heat Sink		
• Maximum Inlet Temperature Condenser/ Heat Exchanger	100°F [Same for the 2nd unit]	• Maximum intake temperature at condenser and heat exchanger inlet
• Evaporation Rate	8707 gpm, average (96% plant capacity factor with wet cooling tower) 11,532 gpm maximum (MWC mode) 16,695 gpm maximum (EC mode)	• Expected rates at which water is lost by evaporation resulting from operation of the plant cooling towers.
Structure Height	≤234 ft [Same for 2nd unit]	• The height from finished grade to the top of the tallest power block structure, excluding cooling towers
Structure Foundation Embedment	≤140 ft [Same for 2nd unit]	• The depth from finished grade to the bottom of the basemat for the most deeply embedded power block structure
Normal Plant Heat Sink		
• Condenser/Heat Exchanger Duty	≤1.03 x 10^{10} Btu/hr [Additional ≤1.03 x 10^{10} Btu/hr for 2nd unit]	• Waste heat rejected from the main condenser and the auxiliary heat exchangers during normal plant operation at full station load
• Unit 3 Closed-Cycle, Dry and Wet Tower		
Height	≤180 ft	• The height above finished grade of the cooling towers
Make-Up Flow Rate	15,384 gpm, maximum (MWC mode) 22,268 gpm, maximum (EC mode)	• The expected rate of removal of water from Lake Anna to replace water losses from the closed-cycle cooling water system

Appendix I

Table I-2. (contd)

Item	Single Unit Value [Second Unit Value]	Description and References
Evaporation Rate	8707 gpm, average (96% plant capacity factor with wet tower cooling? 11,532 gpm, maximum (MWC mode) 16,695 gpm, maximum (EC mode)	• Expected rates at which water is lost by evaporation resulting from operation of the plant cooling towers.
Drift Rate	8 gpm, maximum (MWC mode) 8 gpm, maximum (EC mode)	• Expected rates at which water is lost by drift resulting from operation of the plant cooling towers based on 0.001% of cooling water flow
Blowdown Flow Rate	3844 gpm, maximum (MWC mode) 5565 gpm, maximum (EC mode)	• Flow rate of the blowdown stream from the closed-cycle cooling water system to the WHTF
Blowdown Temperature	100°F, maximum	• The maximum expected temperature of the cooling tower blowdown stream to the WHTF
Blowdown Constituents and Concentrations		• The maximum expected concentrations for anticipated constituents in the cooling water system blowdown to the WHTF
• Free Available Chlorine	<0.3 ppm	
• Copper	<1 ppm	
• Iron	<1 ppm	
• Sulfate	<300 ppm	
• Total Dissolved Solids	<3000 ppm	
Noise	<65 dbA EAB	• Maximum expected sound level produced by operation of the cooling towers
• Unit 4 Dry Cooling Towers		
Evaporation Rate	None or negligible (on the order of 1 gpm, average)	• The expected rate at which water is lost by evaporation from the cooling water system
Height	≤150 ft	• The vertical height above finished grade of the cooling towers

Table I-2. (contd)

Item	Single Unit Value [Second Unit Value]	Description and References
1 Makeup Flow Rate	None or negligible (on the order of 1 gpm, average)	• The expected rate of removal of water from Lake Anna to replace evaporative water losses from the cooling water system
2 Noise	<60 dbA at EAB	• Maximum expected sound level produced by operation of the cooling towers
3 Heat Rejection Rate	$\leq 1.03 \times 10^{10}$ Btu/hr	• Waste heat rejected to the atmosphere from the cooling water system, during normal plant operation at full station load

4   Ultimate Heat Sink (UHS)
5   Mechanical Draft Cooling
6   Towers

Item	Single Unit Value [Second Unit Value]	Description and References
7 • Blowdown Constituents 8 and Concentrations 9 10 11	Envelope values [Same for 2nd unit] <0.3 ppm	• The maximum expected concentrations for anticipated constituents in the UHS blowdown to the WHTF
12 • Free Available 13 Chlorine 14 • Copper 15 • Iron 16 • Sulfate 17 • Total Dissolved 18 Solids 19	<1 ppm <1 ppm <300 ppm <3000 ppm	
20 • Blowdown Flow Rate	144 gpm expected, 850 gpm maximum [Same for 2nd unit]	• The normal expected and maximum flow rate of the blowdown stream from the UHS system to the WHTF
21 • Evaporation Rate	411 gpm normal, 850 gpm shutdown [Same for 2nd unit]	• The expected (and maximum) rate at which water is lost by evaporation from the UHS system
22 • Height	$\leq$60 ft [Same for 2nd unit]	• The vertical height above finished grade of mechanical draft cooling towers associated with the UHS system

Table I-2. (contd)

	Item	Single Unit Value [Second Unit Value]	Description and References
1 2	• Maximum Consumption of Raw Water	850 gpm, nominal [Same for 2nd unit]	• The expected maximum short-term consumptive use of water from Lake Anna by the UHS system (evaporation and drift losses)
3 4 5	• Monthly Average Consumption of Raw Water	411 gpm [Same for 2nd unit]	• The expected normal operating consumption of water from Lake Anna by the UHS system (evaporation and drift losses)
6	**Release Point**		
7	• Elevation	Ground Level	• The elevation above finished grade of the release point for routine operational and accident sequence releases
8	**Source Term**		
9	• Gaseous (Normal)	Maximum values presented in Table H-5 of this EIS and ER Table 5.4-7 [Same for 2nd unit]	• The annual activity, by isotope, contained in routine plant airborne effluent streams
10 11	• Atmospheric (Design Basis Accidents)	Ci as indicated in ER Table 7.1-6c RAI Table 1-1 ER Table 7.1-20 RAI Table 1-2 ER Table 7.1-24a RAI Table 7.1-29 ER Table 7.1-31 RAI Table 15.4-5a	ESBWR Feedwater System Pipe Break ESBWR Failure of Small Lines Carrying Primary Coolant Outside Containment ESBWR Main Steam Line Break ESBWR Loss-of-Coolant (0 to 8 hr) ESBWR Loss-of-Coolant (8 to 720 hr) ESBWR Fuel Handling Accident ESBWR Cleanup Water Line Break ABWR Cleanup Water Line Break
12	• Tritium	3530 Ci/y [Same for 2nd unit] (maximum values)	• The annual activity of tritium contained in routine plant airborne effluent streams

Table I-2. (contd)

Item	Single Unit Value [Second Unit Value]	Description and References
1 **Liquid Radwaste System**		
2 • Release Point Dilution **3** Factor	10 (minimum) [Same for 2nd unit]	• The ratio of liquid potentially radioactive effluent streams to liquid non-radioactive effluent streams from plant systems to the WHTF through the discharge canal used for NAPS Units 1 and 2
4 • Liquid	Values presented in Table H-2 of the EIS and ER Table 5.4-6 (maximum values) [Same for 2nd unit]	• The annual activity, by isotope, contained in routine plant liquid effluent streams
5 • Tritium	≤3100 Ci/yr [Same for 2nd unit]	• The annual activity of tritium contained in routine plant liquid effluent streams
6 **Solid Radwaste System**		
7 • Activity	≤2700 Ci/yr [Same for 2nd unit]	• The annual activity contained in solid radioactive wastes generated during routine plant operations
8 • Volume	≤9041 cu ft/yr [Same for 2nd unit]	• The expected volume of solid radioactive wastes generated during routine plant operations
9 **Plant Characteristics**		
10 • Acreage	Approximately 128.5 acres [Both units]	• Approximate area on the NAPS site that would be affected on a long-term basis as a result of additional permanent facilities
11 • Megawatts Thermal	≤4500 MWt [Same for 2nd unit]	• The thermal power generated by one unit (may be the total of several modules)
12 • Plant Population – **13** Operation	Approximately 720 permanent employees [Both units]	• Anticipated number of new employees that would be required for operation of the new units
14 • Plant Population – **15** Refueling / Major **16** Maintenance	Approximately 700 to 1000 temporary workers during planned outages [Same for 2nd unit]	• Anticipated number of additional workers onsite during planned outages of the new units

Table I-2. (contd)

	Item	Single Unit Value [Second Unit Value]	Description and References
1 2	• Plant Population – Construction	5000 people maximum [simultaneous construction]	• Peak workforce of 5000 for construction of both new units
3 4 5	• Maximum Fuel Enrichment for Light-Water-Cooled Reactors	5% [Same for 2nd unit]	• Concentration of U-235 in fuel
6 7 8	• Maximum Fuel Burn-up for Light-Water-Cooled Reactors	62,000 MWd/MTU [Same for 2nd unit]	• The value derived by calculating the reactor thermal power multiplied by the time of irradiation divided by fuel mass (expressed as megawatt-days per metric ton of irradiated fuel)
9 10	• Maximum Fuel Enrichment for Gas-Cooled Reactors	19.8% [Same for 2nd unit]	• Concentration of U-235 in fuel
11 12	• Maximum Fuel Burn-up for Gas-Cooled Reactors	133,000 MWd/MTU [Same for 2nd unit]	• The value derived by calculating the reactor thermal power multiplied by the time of irradiation divided by fuel mass (expressed as megawatt-days per metric ton of irradiated fuel)

13

Appendix J

Dominion Nuclear North Anna, LLC Commitments and Assumptions Relevant to the Analysis of Impacts

A tabulation to be provided in the Final EIS

Appendix K

Staff's Independent Review of
Water Budget Impacts

Appendix K

Staff's Independent Review of
Water Budget Impacts

K.1 Summary

This appendix discusses the methods used for the U.S. Nuclear Regulatory Commission (NRC) staff's independent review of Dominion Nuclear North Anna, LLC's (Dominion) assessment of the impacts of the proposed North Anna early site permit (ESP) Unit 3's closed-cycle, combination dry and wet cooling system and the staff's findings. The NRC staff computed impacts of plant operations on the Lake Anna reservoir lake level elevation and discharge to the North Anna River downstream of North Anna Dam. Dominion has proposed dry cooling for Unit 4, and the staff concluded that any resulting impacts to the water resources from Unit 4 would be undetectable and were not analyzed further. Therefore, no further mention of Unit 4 is included in this appendix.

The Lake Anna reservoir (or "the reservoir") was formed by impounding the North Anna River above the North Anna Dam. Construction of the dam was permitted by the Virginia State Corporation Commission in 1969 (Virginia State Corporation Commission 1969). The Lake Anna reservoir is divided into two distinct bodies of water, Lake Anna and the Waste Heat Treatment Facility (WHTF). The WHTF is composed of three lagoons and is designated by the Commonwealth of Virginia as a waste heat treatment facility in Dominion's Virginia Pollutant Discharge Elimination System (VPDES) permit (VDEQ 2001) for the North Anna Power Station (NAPS) (Figure K-1). The lagoons have a total surface area of approximately.400 ac and are separated from the rest of Lake Anna by a series of dikes. The main body of the lake is approximately 17 mi long with 272 mi of irregular shoreline and approximately 9600 ac of water surface.

The scope of the staff's evaluation was limited to an assessment based on the relevant values stated in Dominion's plant parameter envelope (PPE). The staff evaluated whether or not PPE values were reasonable. If the ESP is granted and an applicant for a construction permit (CP) or combined license (COL) references the ESP, the applicant would be required to demonstrate that the design of the facility falls within the parameters specified in the ESP. The staff's evaluation also relied on a variety of environmental data that were obtained independently of Dominion. For instance, streamflow data were obtained from the U.S. Geological Survey, meteorological data were obtained from the National Weather Service, and lake geometry data were independently digitized from maps of the lake.

Heat rejected from proposed Unit 3 would be rejected to the atmosphere via the wet and dry cooling system, but blowdown associated with cooling system operation would be discharged into the WHTF. The cooling system blowdown would be discharged into the existing discharge

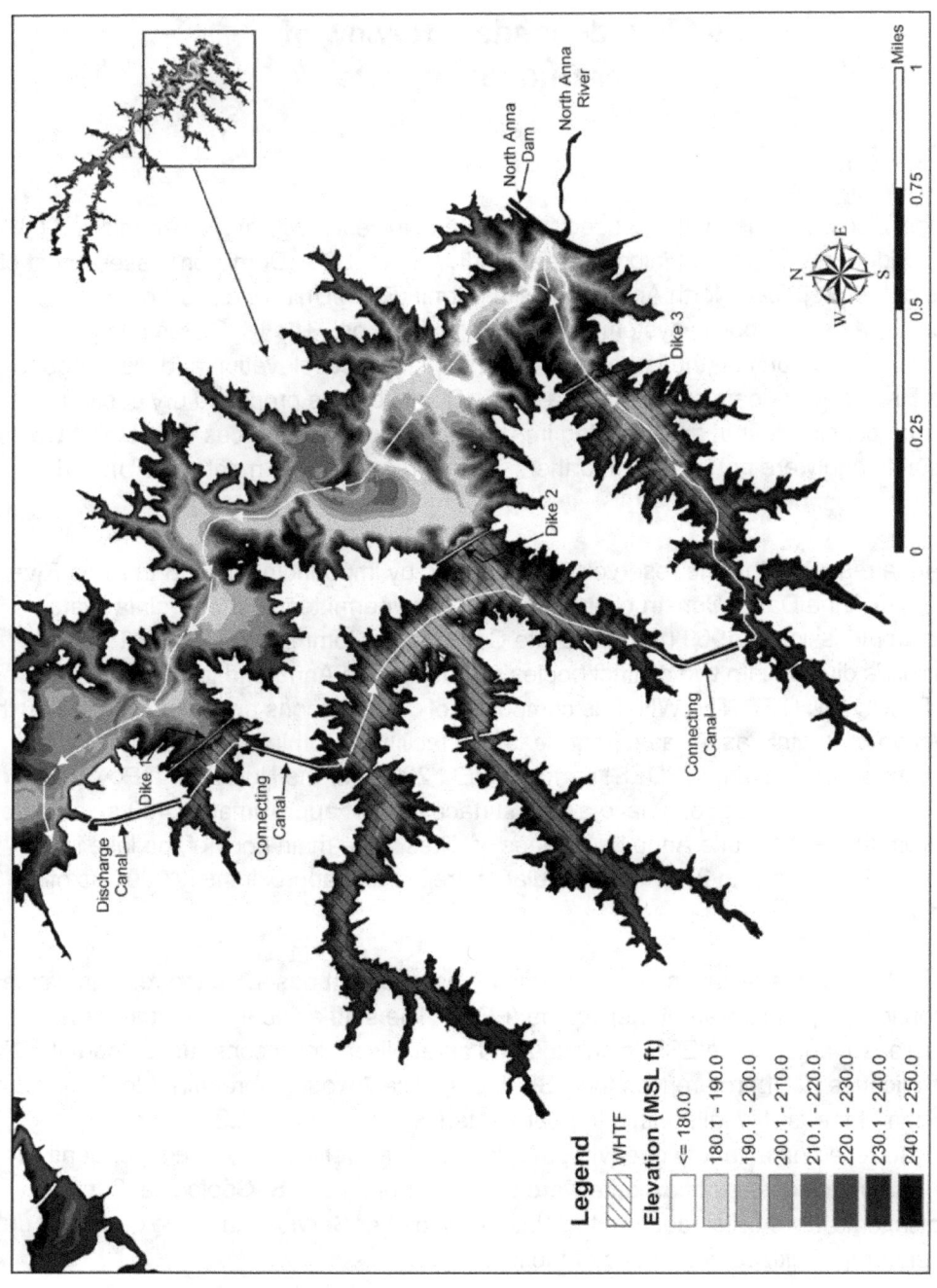

Figure K-1. Main Features of Lake Anna Reservoir

Legend

WHTF

Elevation (MSL ft)

	<= 180.0
	180.1 - 190.0
	190.1 - 200.0
	200.1 - 210.0
	210.1 - 220.0
	220.1 - 230.0
	230.1 - 240.0
	240.1 - 250.0

1
2
3

1 canal at a maximum PPE discharge rate of 12.4 cfs and a maximum PPE temperature of 100°F.
2 Existing Units 1 and 2 employ once-through cooling water systems. With Units 1 and 2, and the
3 proposed Unit 3 operating simultaneously, the blowdown discharge from Unit 3 would mix in the
4 discharge canal with the circulating once-through water discharged from Units 1 and 2.
5 The combined discharge of Units 1 and 2 is approximately 4300 cfs, which is almost 350 times
6 greater than the proposed Unit 3 system maximum blowdown rate of 12.4 cfs. Therefore, the
7 waste heat associated with Unit 3 blowdown is not expected to alter water temperature in either
8 the WHTF or Lake Anna.
9
10 The quantity of water consumed by the Unit 3 wet cooling tower system would reduce the net
11 discharge from North Anna Dam. In addition, during periods of drought when the lake is below
12 elevation 250 ft above mean sea level (MSL), the consumptive use of water from the operation
13 of the Unit 3 wet cooling tower system would reduce the water volume in the lake. This
14 reduction of volume would result in a warming of the reservoir, assuming that the waste heat
15 load from Units 1 and 2 to the reservoir remains constant. Warming is expected to be minimal,
16 as shown by staff's independent assessment, because the difference in overall reservoir volume
17 is slight. For example, the difference in lake level elevation with and without operation of the
18 Unit 3 cooling system was computed to be to be less than 3 in. for 69 percent of the simulated
19 period (a 23 year period) and less than 1.0 ft for 94 percent of the simulation period.
20
21 The water budget assessment examined hydrological impacts from both the existing NAPS
22 Units 1 and 2 and the proposed Unit 3 to bound the minimum lake level elevations. A period of
23 record of more than 23 years was examined to determine a critical historical period for
24 comparison between the existing conditions with Units 1 and 2 and the proposed conditions with
25 Units 1, 2, and 3. The critical period selected was the 34-month period between June 2000 and
26 April 2003, specifically targeting the minimum lake level elevations occurring during October of
27 2002. The staff estimated the following minimum lake level elevations for the critical period:
28
29 • Units 1 and 2 (existing/observed conditions): 245.2 ft
30 • Units 1 and 2 plus Unit 3 (proposed conditions): 243.5 ft
31

K.2 Plant Parameter Envelope

33
34 An ESP is a Commission approval of a location for siting one or more nuclear power facilities.
35 An ESP application may refer to the characteristics of a specific reactor design, or a PPE, which
36 is a set of postulated design parameters representing the characteristics of a reactor or reactors
37 that might be built on a selected site.
38
39 The PPE values are a surrogate for actual reactor design information. Analysis of
40 environmental impacts based on a PPE approach permits an ESP applicant to defer the
41 selection of a reactor design until the construction permit (CP) or COL stage.
42

1   In this evaluation, the staff relied on the following PPE values from Dominion as listed in its
2   Environmental Report (ER) as ESP site characteristics and design parameters (Table 3.1-9)
3   (Dominion 2005) and summarized in Appendix I of this Supplement to the Draft Environmental
4   Impact Statement (SDEIS):
5
6   • Unit 3 Evaporation Rate – Dominion defined this parameter as "...expected rates at
7   which water is lost by evaporation resulting from operation of the plant cooling towers."
8   Dominion stated that the maximum flow rate varies from 16,695 gpm in Energy
9   Conservation (EC) model to 11,532 gpm in Maximum Water Conservation (MWC) mode.
10   Dominion stated that the average evaporation rate is 8303 gpm with an associated
11   96 percent plant capacity factor with wet tower cooling.
12
13   • Unit 3 Blowdown Flow Rate – Dominion defined this parameter as, "...flow rate of the
14   blowdown stream from the closed-cycle cooling water system to the WHTF." Dominion
15   stated that the maximum flow rate in EC mode is 5565 gpm and the maximum flow rate
16   in MWC mode is 3844 gpm.
17
18   • Unit 3 Blowdown Discharge Temperature – Dominion defined this parameter as, "...the
19   maximum expected temperature of the cooling tower blowdown stream to the WHTF."
20   Dominion provided a value of 100°F.
21

K.3 Plant and North Anna Dam Operation Assumptions

23
24   The existing two NAPS units are able to operate at a lake level elevation as low as 242 ft MSL.
25   Dominion is proposing that Unit 3 also be allowed to operate to a lake level elevation as low as
26   242 ft MSL.
27
28   Normal plant cooling for Unit 3 would be accomplished by a closed-cycle, combination dry and
29   wet cooling tower system. The cooling system would operate in EC and MWC modes. In EC
30   mode, all of the rejected heat would be dissipated through use of the wet tower system. When
31   the reservoir water surface elevation is at or above elevation 250 ft MSL, EC mode would be
32   used. In MWC mode, a minimum of one-third of the rejected heat from Unit 3 would be
33   removed by the dry tower system. During periods of favorable atmospheric conditions, more
34   than one-third (and possibly as much as 100 percent) of the rejected heat may be dissipated
35   through the dry towers. MWC mode would be used when the lake level elevation falls below
36   250 ft MSL for seven consecutive days, and would continue to be used until the lake level
37   elevation is restored to 250 ft MSL.
38
39   Operating rules for the North Anna Dam were assumed to be unchanged if the proposed Unit 3
40   is constructed. North Anna Dam is operated in accordance with the Lake Level Contingency
41   Plan (a condition of the NAPS Virginia pollution discharge elimination system [VPDES] permit

1 issued to Virginia Electric and Power Company [VEPCo] by the Virginia Department of
2 Environmental Quality [VDEQ]). Releases from the dam are designed to maintain the lake level
3 elevation as close to elevation 250 ft MSL as possible. When the lake level elevation drops
4 below elevation 250 ft MSL because of inadequate inflows to offset natural and induced
5 evaporative losses, the releases from North Anna Dam are reduced to 40 cfs. If the lake level
6 elevation continues to declined below elevation 248 ft MSL, releases are decreased to 20 cfs.
7 Discharges are increased to 40 cfs when the lake level elevation rises again to elevation 248 ft
8 MSL, and are increased further when the lake level rises above elevation 250 ft MSL.
9

10 ## K.4 WHTF and Lake Anna Bathymetry
11

12 The staff obtained digital 1:24,000 scale digital raster graphic quadrangles of Lake Anna from
13 the Department of Geography at Radford University (http://www.runet.edu/~geoserve/
14 Virginia.html). These images served as the source data set for bathymetry. A mosaic of the
15 raw images was used to generate a geo-referenced base map that was then digitized using the
16 ESRI™ software package ArcMap™ 9.0. The resulting 10 ft interval contours from elevation
17 180 to 250 ft MSL are shown in Figure K-1.
18

19 A continuous surface was created from these contours. This surface was broken into three
20 zones based on observed water temperatures in the reservoir (see Figure K-1): (1) the WHTF,
21 (2) Lake Anna from North Anna Dam upstream to the Highway 208 Bridge, and (3) Lake Anna
22 arms upstream of the Highway 208 Bridge. Impounded surface areas and volumes were then
23 calculated for each section as a function of water surface elevation, the results of which are
24 presented in Table K-1.
25

26 The Lake Anna reservoir, which was formed when North Anna Dam began to impound water, is
27 comprised of numerous fingers and arms. The reservoir is approximately 17 mi long, and
28 several dikes have been constructed to increase travel time of water exiting from the NAPS
29 discharge canal exit and flowing through the WHTF and the lake to the intake for existing
30 Units 1 and 2. Connecting canals, which are trapezoidal in cross section, have been
31 constructed to convey flow from each of the three ponds formed by these dikes. The collection
32 of ponds and connecting canals are collectively labeled as the WHTF.
33

34 Water leaving the discharge canal may only exit the WHTF through Dike 3. This dike contains a
35 submerged discharge structure with adjustable stop logs to constrict the exiting discharge. This
36 structure creates a positively buoyant high velocity (typically >6 ft/s) jet, which was designed to
37 quickly entrain cooler Lake Anna water.

1   **Table K-1.**   Computed Areas and Volumes as a Function of Lake Level Elevation for the
2   Various Zones of Lake Anna Reservoir (See Figure K-1)

Lake Anna Reservoir			WHTF		
Elevation (ft)	Area (ac)	Volume (ac-ft)	Elevation (ft)	Area (ac)	Volume (ac-ft)
250	13,068	31,2171	250	3,194	64,082
240	9,219	20,0737	240	2,120	37,515
230	6,553	12,1877	230	1,374	20,045
220	4,418	67,021	220	830	9,026
210	2,715	31,354	210	418	2,787
200	1,281	11,377	200	139	
190	523	3,257			
180	129				

Lake Anna			Lake Anna Arms		
Elevation (ft)	Area (ac)	Volume (ac-ft)	Elevation (ft)	Area (ac)	Volume (ac-ft)
250	5,540	17,4374	250	4,334	73,715
240	4,528	12,4032	240	2,571	39,190
230	3,614	83,323	230	1,565	18,509
220	2,803	51,240	220	786	6,755
210	2,034	27,055	210	263	1,512
200	1,101	11,377	200	40	
190	523	3,257			
180	129				

K.5 Dominion's Assessment

Dominion developed a water balance model that simulated releases from the North Anna Dam
and lake level elevations in the reservoir on a weekly basis for the period between October
1979 and April 2003. During this period, Units 1 and 2 were both operating (Existing Units
Scenario). The model was separately used to predict releases from the dam and water surface
elevations in the reservoir had Unit 3 been operating during the same period (Existing Units plus
Unit 3 Scenario). Variations in dam outflow frequencies, periods below various lake level
elevations, and the relative difference in lake level elevations between the two scenarios were
then examined and presented by Dominion.

The minimum lake level elevation of the reservoir was 245.1 ft MSL for the Existing Units
Scenario and 244.2 ft for the Existing Units plus Unit 3 Scenario, a difference of 0.9 ft. The
percent of time North Anna Dam discharge was 20 cfs was 5.2 percent of the period for the
Existing Units Scenario and 7.3 percent for the Existing Units plus Unit 3 Scenario, a difference
of 2.1 percent.

K.6 Boundary Condition for Staff's Assessment

The staff performed an independent water balance calculation to predict impacts of the proposed Unit 3 on the reservoir and releases from North Anna Dam (Cook et al. 2005). This was accomplished by first simulating the more than 23 year period between October 1979 and April 2003, when only Units 1 and 2 were operating. The assessment was then used to predict how the reservoir and downstream releases would be altered had Unit 3 been operating.

The model required input of time-series boundary condition data. Inflows to the lake were input for each time step. Outflows were computed based on the previous time-step lake level elevation and the relationship between lake level and discharge for North Anna Dam. Meteorological data were used to estimate the volume of precipitation falling directly on the lake and to compute volume lost from the reservoir through evaporation. Lake inflows and meteorological data were held constant for all scenarios; however, outflows varied between scenarios according to the lake level elevation.

K.6.1 Inflows

The principal tributaries of Lake Anna are the North Anna River, Pamunkey Creek, and Contrary Creek. Unfortunately, no stream flow gauges were installed on these tributaries. Estimates of inflows to lake Anna were derived from measurements of streamflow in an adjacent basin. Daily average stream flows for the Little River near Doswell, Virginia, were obtained from U.S. Geological Survey (http://waterdata.usgs.gov/nwis) gauge 01671100. The Little River is a tributary to the North Anna River downstream of North Anna Dam. The size of the Little River watershed at this gauging station is 107 mi^2, which is approximately one-third the size of the North Anna watershed where it enters Lake Anna. Inflows to Lake Anna were therefore computed during the simulation period by multiplying the watershed scale ratio to the daily average Little River discharges.

K.6.2 Meteorology

Meteorological information about the atmosphere above the lake is necessary to compute evaporation for this assessment. Dew point temperature and wind speed were obtained from the Richmond airport (EarthInfo 2003), which was the nearest location that collected data during the critical drought period. Hourly observed data were used as model inputs for the simulated drought period. Precipitation falling onto Lake Anna was considered an inflow boundary condition for the water budget assessment. Total accumulated precipitation on each day was obtained from National Climate Data Center (NCDC), and was originally collected at the Richmond airport (NCDC 2004).

1   Based on precipitation data measured at the Richmond airport from January 1, 1921, to
2   May 31, 2004, Figure K-2 shows the long-term mean monthly precipitation and monthly
3   precipitation for the three driest water years in the Richmond record (water years 1924, 2002,
4   and 1954). The total precipitation during the 2002 water year was 26.4 in., which is
5   60.6 percent of mean annual precipitation. The precipitation for the 2001 water year totaled
6   33.1 in., which is 75.9 percent of mean annual precipitation. Combined precipitation during
7   water years 2001 and 2002 was the driest 2-year period in the precipitation record. Table K-2
8   shows the monthly precipitation during water years 2001 and 2002 as a percentage of the long
9   term corresponding monthly mean.

10

11   ## K.7 Staff's Assessment Approach

12

13   The staff's assessment was completed in two steps. In the first step, the natural evaporation
14   rate from the lake, the induced evaporation from Units 1 and 2, and a small monthly-averaged
15   inflow adjustment that force computed water surface elevations to match observed values were
16   computed. Once these variables were computed, they were then held constant during the
17   second step, which evaluated the impact of the proposed Unit 3 on the reservoir and releases
18   from North Anna Dam.

19

20   Evaporation rate at the water's surface represents the volume per surface area per unit time
21   of liquid water that is vaporized into the atmosphere. Numerous formulations to compute
22   evaporation rate exist in the technical literature. The formulation used in this analysis is that
23   recommended by TVA (1972), which is also reported in Bras (1990) and is credited to
24   Marciano-Harbeck (1954). Additional details regarding the formulation as applied in the North
25   Anna analysis can be found in Cook et al. (2005).

26

27   Water temperatures during the historical period were based on results from the Lake Anna
28   Cooling Pond Model developed by the Massachusetts Institute of Technology (Ho and Adams
29   1984). This calibrated and validated model predicts water temperatures at various locations
30   around the WHTF and lake with the two existing reactor operating. These water temperatures
31   were used only to compute natural evaporation from the reservoir (i.e., the background
32   evaporation rate in the case with no reactors operating) and the induced evaporation resulting
33   from operation of NAPS Units 1 and 2.

34

35   The volumetric water balance for the first step was computed using the appropriate watershed
36   inflows, precipitation falling onto the lake, natural evaporation, induced evaporation from Units 1
37   and 2, and observed and estimated outflows from North Anna Dam. The resulting volume was
38   then converted to a water surface elevation in the reservoir and compared to observed data.
39   Differences in computed elevations were removed using a monthly-averaged inflow adjustment.
40   The inflow adjustment was small and averaged 4.6 cfs over the 1978 to 2003 simulation period.

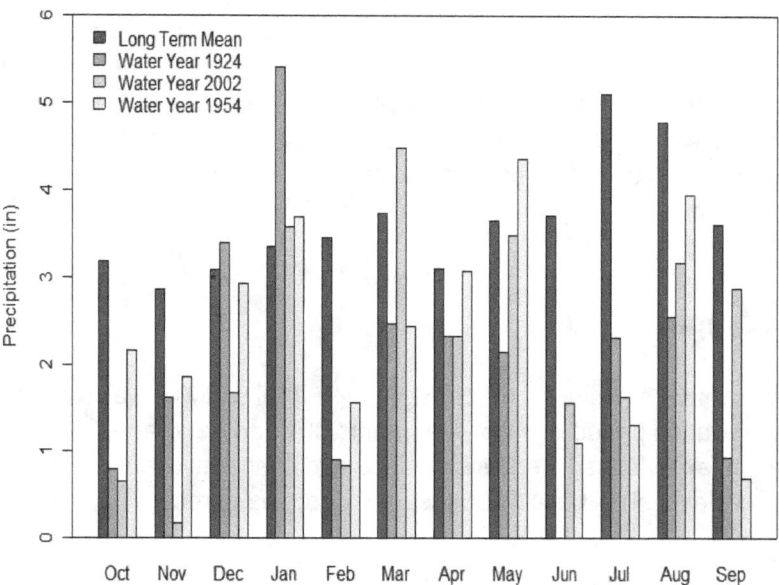

Figure K-2. Monthly Precipitation (in.) at the Richmond Airport

Table K-2. Monthly Precipitation as a Percentage of Long-Term Monthly Means During Water Years 2001 and 2002

| | Percentage of Long-Term Monthly Mean | |
Month	Water Year 2001	Water Year 2002
October	0.3	20.4
November	59.9	5.9
December	76.8	53.9
January	61.4	106.8
February	73.9	23.8
March	100.9	119.9
April	68.9	75.1
May	55.5	95.4
June	176.2	42.1
July	53.5	32.0
August	106.4	66.6
September	59.2	79.6
Total Annual	75.9	60.6

1 The second step of the assessment evaluated the relative impacts of Unit 3 operations on the
2 reservoir and releases from the North Anna Dam. The time-series of natural evaporation rate,
3 induced evaporation from NAPS Units 1 and 2, and inflow adjustment were applied from step
4 one. Constant evaporation rates for the proposed Unit 3, based on PPE values, were applied
5 and the volumetric water balance was computed. In these calculations, changes in surface
6 area and volume as a result of reservoir drawdown were explicitly considered and influenced
7 both the volume of natural evaporation leaving the lake and precipitation volume falling on
8 the lake.
9

10 ## K.8 Assessment Results
11

12 While the entire period of October 15, 1978, through April 9, 2003, was simulated, the critical
13 water surface elevation period was between April 2001 and February 2003. During this critical
14 period, the region experienced a severe drought, and concerns over water use conflicts arose
15 as the lake level elevation in Lake Anna reservoir dropped to record lows in October 2002.
16

17 Figure K-3 displays the computed time-series of lake level elevation throughout the entire
18 simulation period. The Existing Units scenario represents the historical variation in lake level
19 elevation during the more than 23 years of simulation with both Units 1 and 2 operating. The
20 Existing Units plus Unit 3 scenario includes a constant loss rate of 8707 gpm from the lake,
21 which represents the long-term average PPE evaporative loss rate from the proposed use of a
22 wet cooling tower system for Unit 3. Figure K-4 displays the computed time-series of results
23 during the critical drought period when minimum lake level elevation values were reached. As
24 shown in the figure, the decline in lake level elevation is gradual, declining from elevation 250 ft
25 above MSL in July 2001 to the minimum level in October 2002, a 15-month period. The return
26 of lake level to elevation 250 ft MSL was rapid in comparison, and occurred over a 4-month
27 period between October 2002 and February 2003.
28

29 Table K-3 presents the percentage of time the lake level elevation of the reservoir was near
30 several threshold levels, which correspond to prescribed outflow discharge rates from North
31 Anna Dam. Simulation results indicate that the percent of time the reservoir was at or below
32 elevation 248 ft MSL and North Anna Dam was discharging 20 cfs would have increased from
33 6 percent with only the existing Units 1 and 2 operating to 11 percent if the proposed Unit 3 was
34 also operating. The percent of time the reservoir elevation was at or below 246 ft was predicted
35 to increase by 0.9 percent, from 1 to 2 percent, during the simulation period. The minimum lake
36 level elevation reached in October 2002 during the critical drought period fell by 1.7 ft, from
37 245.2 ft MSL to 243.5 ft MSL. At no time during the simulation did the lake level elevation reach
38 the minimum operational plant intake elevation of 242.0 ft MSL, when the plant would shut
39 down.

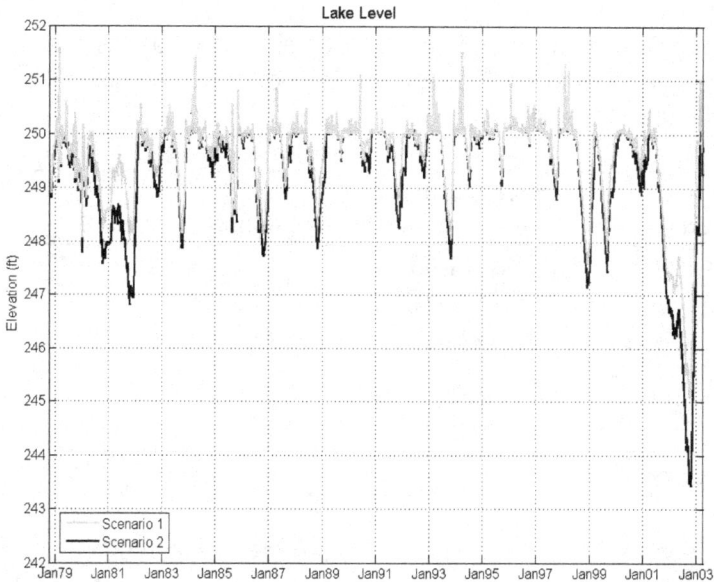

1   **Figure K-3**. Time Series of Lake Anna Lake Level Elevations for the Entire Simulation Period

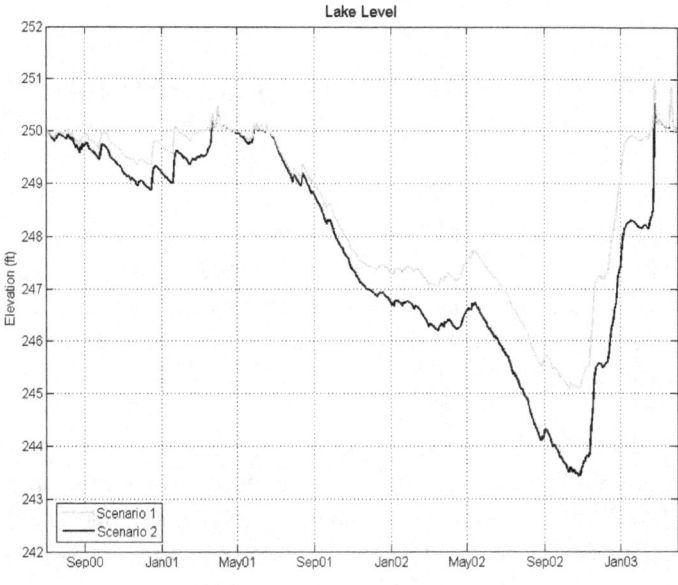

2   **Figure K-4**. Time Series of Lake Anna Lake Level Elevations During the Critical Drought Period

1   **Table K-3**. Lake Level Elevation Exceedance Table and the Minimum Water Surface Elevation
2   for the 1978 to 2003 Simulation Period (values expressed as percent of the total
3   simulation period)

4

Lake Level Elevation (ft)	North Anna Dam Discharge (cfs)	Existing Units 1 & 2	Existing Units 1 & 2 plus Unit 3
At or above 250 ft	Follows rating curve (>40 cfs)	37.3%	33.6%
Between 250 and 248 ft	40 cfs	57.0%	55.4%
At or below 248 ft	20 cfs	5.7%	11.0%
At or below 246 ft	20 cfs	1.1%	2.0%
Minimum elevation		245.2 ft	243.5 ft

13 Table K-4 presents differences in water surface elevation computed by subtracting the
14 time-series of elevations computed for the Existing Units Scenario from the Existing Units plus
15 Unit 3 scenarios (see Figure K-3). As a percent of the total simulation period, differences were
16 less than 3 in. for over 69 percent of the simulation and less than 1 ft for over 94 percent of the
17 simulation. The time-averaged difference in lake level elevation between the two scenarios
18 was 2.8 in.

20 Figure K-5 presents the cumulative distribution frequency of Lake Anna lake level elevation for
21 the simulation period for scenarios:

23 1. Existing Units
24 2. Existing Units plus Unit 3 at 8707 gpm average evaporation rate
25 3. Existing Units plus Unit 3 at 16,695 gpm (EC mode) when above 250 ft MSL dropping to the
26   8707 gpm average evaporation rate when below 250 ft MSL
27 4. Existing Units plus Unit 3 at 16,695 gpm (EC mode) when above 250 ft MSL decreasing to
28   11,532 gpm when below 250 ft MSL.

30 Figure K-5 shows that, because the lake does not store this water for later use during times of
31 water scarcity, any additional Unit 3 evaporative losses that occur when the lake is above
32 elevation 250 ft MSL do not impact the frequency of lake elevations below 250 ft MSL.
33 However, if the Unit 3 evaporative loss is increased during periods when the lake is below
34 250 ft MSL, the duration of lake levels less than elevation 250 ft MSL would increase. For
35 example, the frequency of lake elevation at or below 246 ft MSL increased from 1.1 percent for
36 the existing units only scenario to 2.0 percent for the scenario where Unit 3 evaporative losses
37 are 8707 gpm and 2.6 percent for the scenario when Unit 3 evaporation losses are 11,532 gpm.

1 **Table K-4**. Differences in Lake Level Elevation between the Existing Units 1 and 2 Scenario
2   and the Existing Units 1 and 2 Plus Unit 3 Scenario for the 1978 to 2003 Simulation
3   Period
4

Elevation Difference	Percent of Time of the Total Simulation
Less than 3 in.	69.0%
Less than 6 in.	85.0%
Less than 12 in.	94.2%
Average difference	2.8 in.
Maximum difference	1.7 ft

11

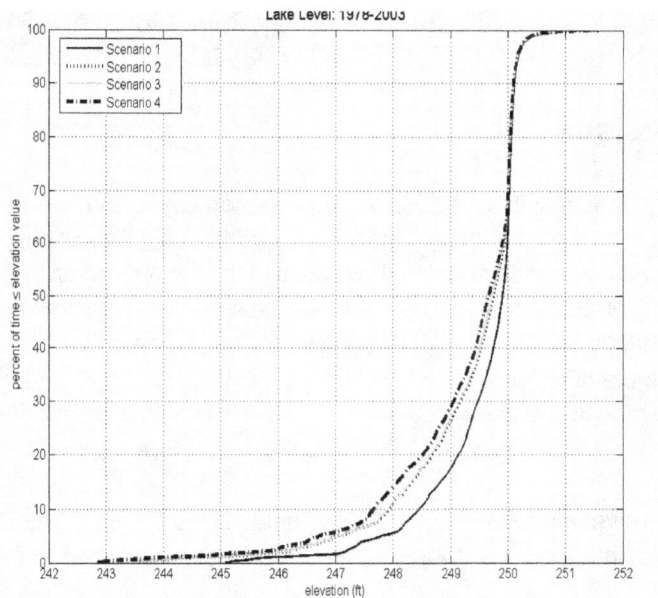

12 **Figure K-5**. Cumulative Distribution Function of Lake Anna Lake Level Elevation
13
14 Additional cumulative distribution frequencies, like those shown in Figure K-5, were developed
15 for NAPS Units 1 and 2 plus proposed Unit 3 with modified lake level. The staff determined the
16 increase in normal pool elevation necessary to maintain the current frequency of occurrence of
17 20 cfs discharge from North Anna Dam (i.e., the occurrence of 20 cfs releases). Inherent in this
18 analysis is the assumption that the 23-yr period of record simulated would be representative of
19 future conditions (e.g., inflows, precipitation, etc.) at the site.
20
21 As shown in Table K-3, the frequency of occurrence of 20 cfs discharges occurred with the
22 NAPS Units 1 and 2 for 5.7 percent of the simulated period. With a normal pool elevation of

1   250 ft, the predicted occurrence of 20 cfs discharges with NAPS Units 1 and 2 and proposed
2   Unit 3 operating increased to 11.0 percent of the simulation period. However, if the normal pool
3   elevation is raised by 10 in. to 250.8 ft, the frequency of occurrence fell to 5.9 percent.
4   Therefore, assuming the return period of drought and wet periods were approximately the same
5   as the simulated period, the frequency of occurrence of 20 cfs would remain the same if Unit 3
6   were constructed and the normal pool elevation was raised to elevation 250.8 ft MSL.
7
8   Alternatively, the frequency of occurrence of 20 cfs releases from North Anna Dam could be
9   reduced, if the threshold for decreasing releases from 40 cfs was lowered below an elevation of
10 248 ft. The staff evaluated lowering the threshold elevation for reducing releases from the dam
11 to 20 cfs for the NAPS Units 1 and 2 plus proposed Unit 3 8707 gpm scenario. Results from
12 this analysis indicate that if Unit 3 were operated and the threshold elevation reduced by
13 approximately 8 in. to elevation 247.3 ft MSL, the occurrence frequency of 20 cfs releases
14 would not change from the Units 1 and 2 only scenario.
15

K.9 Conclusions

17
18 The only operational activity with respect to proposed Unit 3 that would result in a detectable
19 hydrological alteration of the environment is the additional consumptive use of water to cool the
20 unit. Although some blowdown from the closed-cycle, combination dry and wet cooling system
21 would occur, the quantity of discharge is much less than the discharge from the existing Units 1
22 and 2. The additional withdrawal of cooling water for the new Unit 3 would increase the
23 duration of time the lake level elevation is below 250 ft MSL, and hence periods of reduced
24 releases from North Anna Dam would occur. Unit 3 evaporative cooling withdrawals would also
25 reduce the minimum lake level elevation of Lake Anna during periods of drought.
26
27 Calculated lake level elevations during the critical period between June 2000 through April 2003
28 predicted minimum elevations during the second week of October 2002. The minimum
29 elevation predicted for continuous operation of Units 1 and 2 is 245.2 ft, whereas, the addition of
30 Unit 3 would result in further declines to 243.5 ft.
31
32 The periods of minimum releases from the dam would also increase with the operation of the
33 proposed Unit 3. For example, the percentage of time of minimum release (20 cfs) would
34 increase from 5.7 percent to 11 percent, and the percentage of time release greater than 40 cfs
35 would decrease from 37 percent to 34 percent.
36

K.10 References

Bras R.L. 1990. *Hydrology: An Introduction to Hydrologic Science.* Addison Wesley Publishing, Reading, Massachusetts.

Cook C.B., L.W. Vail, and D.L. Ward. 2005. *Report on the North Anna Early Site Permit Water Budget Model (LakeWBT) for Lake Anna.* Pacific Northwest National Laboratory, Richland, Washington.

Dominion Nuclear North Anna, LLC (Dominion). 2005. *North Anna Early Site Permit Application – Part 3 – Environmental Report.* Revision 5, Glen Allen, Virginia.

EarthInfo Inc. 2003. National Climatic Data Center Surface Airways TD-3280, East:2, CD-ROM Database. Boulder, Colorado.

Ho E. and E.E. Adams. 1984. *Final Calibration of the Cooling Lake Model, North Anna Power Station.* Report No. 295, Ralph M. Parsons Laboratory, Massachusetts Institute of Technology, Cambridge, Massachusetts.

National Climatic Data Center (NCDC). 2004. TD-3200, Summary of the Day, Richmond International Airport and Richmond Byrd Field, WBAN ID 13740. Purchased online via http://nndc.noaa.gov/onlinestore.html. Accessed August 3, 2005.

Tennessee Valley Authority (TVA). 1972. *Heat and Mass Transfer between a Water Surface and the Atmosphere, Water Resources Research*, Laboratory Report No 14 prepared for the Tennessee Valley Authority, Division of Water Control Planning, Engineering Laboratory, Report No. 0-6803, Norris, Tennessee.

Marciano T.T. and G.E. Harbeck, Jr. 1954. "Mass Transfer Studies." In *Water Loss Investigations*, Lake Hefner Studies, Technical Report, U.S. Geological Survey, Professional Paper 269, Washington, D.C.

Virginia Department of Environmental Quality (VDEQ). 2001. *Authorization to Discharge Under the Virginia Pollutant Discharge Elimination System and the Virginia State Water Control Law, Virginia Electric and Power company, North Anna Nuclear Power Station.* Permit No. VA0052451, Commonwealth of Virginia, Department of Environmental Quality, Richmond, Virginia.

Virginia State Corporation Commission. 1969. Order of June 12, 1969, O.B. 58, p. 353.

Appendix L

Authorizations and Consultations

Appendix L

Authorizations and Consultations

This appendix was previously presented as Appendix H in the Draft Environmental Impact Statement. Although the location is changed, the appendix tabulation was not affected by the changes presented in ER Revision 6.

1   Table L-1 contains a list of the environmental-related authorization, permits, certifications, and
2   consultations, potentially required by Federal, State, regional, local, and affected Native
3   American tribal agencies for activities related to site preparation, construction, and operation of
4   potential new nuclear units at the North Anna ESP site.
5

Table L-1. Federal, State, and Local Authorizations and Consultations

Agency	Authority	Requirement	Activity Covered
Federal Aviation Administration	49 USC 1501 14 CFR 77.13	Construction Notice	Notice of erection of structures (>200 feet) potentially impacting air navigation
NRC	10 CFR 52, Subpart C	Combined License	NRC requirements and procedures applicable to issuance of combined licenses for nuclear power facilities
NRC	10 CFR 52, Subpart A	Early Site Permit	NRC requirements and procedures applicable to issuance of Early Site Permits for approval of a site for one or more nuclear power facilities
NRC	10 CFR 30	Byproduct License	NRC license to possess special nuclear materials
NRC	10 CFR 70	License	NRC license to possess nuclear fuel
ACE	CWA 33 USC 1251	Section 404 Permit	Disturbing or crossing wetland areas or navigable waters
ACE	Rivers and Harbors Act 33 USC 403	Section 10 Permit	Impacts to navigable waters of the United States
FWS and NOAA Fisheries Service	Endangered Species Act 16 USC 1531	Consultation regarding potential to adversely impact protected species	Consultation concerning potential impacts to threatened and endangered species
FWS	Migratory Bird Treaty Act 16 USC 703	Consultation	Consultation concerning potential impacts to migratory birds
Virginia State Corporation Commission	Code of Virginia 56-580D	Permit	Approval for construction of new generating facility
VDEQ	9 VAC 5-20-160	Registration	Annual re-certification of air emission sources
VDEQ	Clean Air Act Title V 9 VAC 5-80-50	Operating Permit	Operation of air emission sources
VDEQ	9 VAC 5-80-120	Minor Source - General Permit	Construction and operation of minor air emission sources
VDEQ	CWA 9 VAC 25-10	Virginia Pollutant Discharge Elimination System Permit (VPDES)	Regulate limits of pollutants in liquid discharge to surface water

Table L-1. (contd)

Agency	Authority	Requirement	Activity Covered
VDEQ	9 VAC 25-150	General Permit Registration Statement for storm water discharges from industrial activity (VAR5)	General permit to discharge storm water during operations
VDEQ	9 VAC 25-210	Virginia Water Protection Permit (Individual or General)	Permit to dredge, fill, discharge pollutants into or adjacent to surface water. Joint application with ACE Section 404 permit.
VDEQ	CWA 33 USC 1341	Section 401 Certification	Compliance with water quality standards
VDEQ	9 VAC 25-220	Surface Water Withdrawal Permit	Permit to withdraw water from Lake Anna (unless otherwise regulated by State Water Control Board)
VDEQ	Coastal Zone Management Act 16 USC 1456	Consistency determination	Compliance with Virginia Coastal Program
VDEQ	9 VAC 25-180	General Permit Registration Statement for storm water discharges from construction activities (VAR10)	General permit to discharge storm water from site during construction
VDEQ	9 VAC 25-180	General Permit Notice of Termination (NOT) for storm water discharges from construction activities (VAR4)	Termination of coverage under the general permit for storm water discharge from construction site activities
VDEQ	9 VAC 25-180	General Permit NOT for storm water discharges from industrial activity (VAR5)	Termination of coverage under the general permit for storm water discharge associated with operational site activities
Virginia Department of Historical Resources	National Historic Preservation Act 36 CFR 800	Cultural Resources Survey/Review	Confirm ESP site does not contain protected historic/cultural resources
Virginia Marine Resources Commission	9 VAC 25-210	Permit	Permit to fill submerged land. Joint application with ACE Section 404 permit.

NRC FORM 335 (9-2004) NRCMD 3.7	U.S. NUCLEAR REGULATORY COMMISSION	1. REPORT NUMBER (Assigned by NRC, Add Vol., Supp., Rev., and Addendum Numbers, if any.)
BIBLIOGRAPHIC DATA SHEET *(See instructions on the reverse)*		NUREG 1811 Supplement 1

2. TITLE AND SUBTITLE	3. DATE REPORT PUBLISHED	
Draft Environmental Impact Statement for an Early Site Permit (ESP) at the the North Anna ESP Site Draft Report for Comment	MONTH	YEAR
	July	2006
	4. FIN OR GRANT NUMBER	

5. AUTHOR(S)	6. TYPE OF REPORT
See Appendix B of Report	Technical
	7. PERIOD COVERED *(Inclusive Dates)*

8. PERFORMING ORGANIZATION - NAME AND ADDRESS *(If NRC, provide Division, Office or Region, U.S. Nuclear Regulatory Commission, and mailing address; if contractor, provide name and mailing address.)*

Division of New Reactor Licensing

Office of Nuclear Reactor Regulation

U. S. Nuclear Regulatory Commission

Washington, D.C. 20555-0001

9. SPONSORING ORGANIZATION - NAME AND ADDRESS *(If NRC, type Same as above ; if contractor, provide NRC Division, Office or Region, U.S. Nuclear Regulatory Commission, and mailing address.)*

Same as above.

10. SUPPLEMENTARY NOTES

Docket No. 52-008

11. ABSTRACT *(200 words or less)*

Dominion proposed (1) changing its approach for cooling proposed Unit 3 from once-through cooling to a closed cycle cooling system and (2) proposing an increase in the maximum power output from 4300 MW(t) to 4500 MW(t) for proposed Units 3 and 4. The NRC determined the changes to the proposed action were substantial and therefore a Supplement to its Draft Environmental Impact Statement (SDEIS) was necessary pursuant to 10 CFR 51.72. The scope of this SDEIS is limited to the environmental impacts associated with the change in to the cooling system for Unit 3 and in the maximum power level for both units.

The staff's preliminary recommendation to the Commission related to its environmental review of the proposed action is that the ESP should be issued. In making its recommendation, the staff has concluded that there are no sites that are obviously superior to the proposed site. The staff has concluded that the site preparation and preliminary construction activities allowed by 10 CFR 50.10(e)(1) will not result in any significant adverse environmental impact that cannot be redressed.

12. KEY WORDS/DESCRIPTORS *(List words or phrases that will assist researchers in locating the report.)*	13. AVAILABILITY STATEMENT
North Anna Early Site Permit ESP National Environmental Policy Act NEPA Dominion Supplement	unlimited
	14. SECURITY CLASSIFICATION
	(This Page) unclassified
	(This Report) unclassified
	15. NUMBER OF PAGES
	16. PRICE